POLYMER SCIENCE AND TECHNOLOGY
Volume 9A

ADHESION SCIENCE AND TECHNOLOGY

Edited by
Lieng-Huang Lee
Xerox Corporation
Rochester, New York

PLENUM PRESS · NEW YORK AND LONDON

Library of Congress Cataloging in Publication Data

American Chemical Society Macromolecular Symposium on Science and Technology of Adhesion, Philadelphia, 1975.
Adhesion science and technology.

(Polymer science and technology; v. 9)
Includes bibliographical references and indexes.
1. Adhesives—Congresses. 2. Adhesion—Congresses. I. Lee, Lieng-Huang, 1924- II. American Chemical Society. III. Title. IV. Series.
TP967.A55 1975   668'.3   75-35744
ISBN 978-1-4615-8203-8
ISBN 978-1-4615-8201-4 (eBook)
DOI 10.1007/978-1-4615-8201-4

First half of the Proceedings of the American Chemical Society
Macromolecular Symposium on Science and Technology of Adhesion
held in Philadelphia, Pennsylvania, April, 1975

© 1975 Plenum Press, New York
A Division of Plenum Publishing Corporation
227 West 17th Street, New York, N.Y. 10011
Softcover reprint of the hardcover 1st edition 1975

United Kingdom edition published by Plenum Press, London
A Division of Plenum Publishing Company, Ltd.
Davis House (4th Floor), 8 Scrubs Lane, Harlesden, London, NW10 6SE, England

All rights reserved

No part of this book may be reproduced, stored in a retrieval system, or transmitted, in any form or by any means, electronic, mechanical, photocopying, microfilming, recording, or otherwise, without written permission from the Publisher

POLYMER SCIENCE AND TECHNOLOGY
Volume 9A

ADHESION SCIENCE AND TECHNOLOGY

POLYMER SCIENCE AND TECHNOLOGY

Editorial Board:

William J. Bailey
University of Maryland
College Park, Maryland

J. P. Berry
Rubber and Plastics Research Association
of Great Britain
Shawbury
Shrewsbury, England

A. T. DiBenedetto
The University of Connecticut
Storrs, Connecticut

C. A. J. Hoeve
Texas A&M University
College Station, Texas

Yōichi Ishida
Osaka University
Toyonaka, Osaka, Japan

Frank E. Karasz
University of Massachusetts
Amherst, Massachusetts

Osias Solomon
Polytechnical Institute of Bucharest
Bucharest, Romania

Volume 1 • STRUCTURE AND PROPERTIES OF POLYMER FILMS
Edited by Robert W. Lenz and Richard S. Stein • 1972

Volume 2 • WATER-SOLUBLE POLYMERS
Edited by N. M. Bikales • 1973

Volume 3 • POLYMERS AND ECOLOGICAL PROBLEMS
Edited by James Guillet • 1973

Volume 4 • RECENT ADVANCES IN POLYMER BLENDS, GRAFTS, AND BLOCKS
Edited by L. H. Sperling • 1974

Volume 5 • ADVANCES IN POLYMER FRICTION AND WEAR (Parts A and B)
Edited by Lieng-Huang Lee • 1974

Volume 6 • PERMEABILITY OF PLASTIC FILMS AND COATINGS
TO GASES, VAPORS, AND LIQUIDS
Edited by Harold B. Hopfenberg • 1974

Volume 7 • BIOMEDICAL APPLICATIONS OF POLYMERS
Edited by Harry P. Gregor • 1975

Volume 8 • POLYMERS IN MEDICINE AND SURGERY
Edited by Richard L. Kronenthal, Zale Oser, and E. Martin • 1975

Volume 9 • ADHESION SCIENCE AND TECHNOLOGY (Parts A and B)
Edited by Lieng-Huang Lee • 1975

A Continuation Order Plan is available for this series. A continuation order will bring delivery of each new volume immediately upon publication. Volumes are billed only upon actual shipment. For further information please contact the publisher.

Preface

The first ACS Adhesion Symposium was held in Washington, D.C., September 1971. During the four years since that meeting, much interest in adhesion has been generated among six divisions of the American Chemical Society. Then, in 1974, the Macromolecular Secretariat appointed me to work closely with the six Session chairmen in organizing this Symposium on Science and Technology of Adhesion. Needless to say, the success of the Symposium which took place between April 7 and 10, 1975 in Philadelphia, Pa., is due to their excellent cooperation and the enthusiastic response of contributors. As originally planned, each division was responsible for one session, and most of the papers, including several late contributions, are published in these two volumes of proceedings.

During the Symposium, we held a banquet in honor of Professor Herman Mark in celebration of his eightieth birthday. His Plenary Lecture and the Symposium Address by Professor Murray Goodman are published in full at the beginning of the first volume. I thank Professors Mark and Goodman for their excellent presentations on this memorable occasion.

This book consists of the following eight parts including a discussion section at the end of each part:

1. Interfacial Phenomena and Adhesion
2. Synthetic Polymers and Adhesives
3. Rubber Adhesives and Sealants
4. Natural Products and Structural Adhesives
5. Growth and Change in Adhesives
6. Performance of Adhesive Joints
7. Trends in Adhesion Research
8. Surface Energetics of Printing Processes

Three papers of the last part were recently presented to the Symposium on Polymers for Lithography, August 1975, in Chicago, Illinois. On the other hand, four papers originally presented to the Symposium were published elsewhere. Thus, the order of papers in the Proceedings differ somewhat from the earlier program. The first four parts constitute Volume 6A and the last four, Volume 6B, of the Polymer Science and Technology series.

I sincerely appreciate all those who contributed and the following session chairmen:

Professor R.J. Good, Division of Colloid and Surface Chemistry,
Dr. F.D. Petke, Division of Polymer Chemistry,
Dr. M.E. Gross, Rubber Division,
Dr. H.G. Arlt, Cellulose, Paper and Textile Division,
Dr. I. Skeist, Division of Chemical Marketing and Economics,
Dr. R.L. Patrick, Division of Organic Coatings and Plastics Chemistry.

We acknowledge the partial assistance to two speakers by the Program Development Fund of the American Chemical Society. Lastly, I would like to express my sincere appreciation to Robert M.S. Lee, now at Princeton University, for his editorial assistance in preparing both Volume 5 and 6 for this Polymer Science and Technology Series.

Lieng-Huang Lee

September, 1975
Webster, New York
U.S.A.

Contents of Volume 9A

Significant Advances and Developments in
Adhesion and Adhesives   1
   L. H. Lee

Polymer Progress - Worldwide (Plenary Lecture) 19
   H. Mark

A Tribute to Herman Mark, Mr. Polymer Science
(Symposium Address) 31
   M. Goodman

PART ONE: Interfacial Phenomena and Adhesion

Introductory Remarks: How to Solve the
Problem of Adhesion - A Prescription 37
   R. J. Good

The Conformation of Adsorbed Polystyrene 43
   W. H. Grant, B. W. Morrissey,
   and R. R. Stromberg

Recent Advances in Wetting and Adhesion 55
   W. A. Zisman

Polymer Reactions on Solid Surfaces 93
   K. Hamann, R. Laible, and J. Horn

The Role of Interfacial Processes in Adhesion 107
   R. J. Good

Surface Chemical Criteria for Maximum Adhesion
and Their Verification against the Experimentally
Measured Adhesive Strength Values 129
   K. L. Mittal

Discussion 169

PART TWO: Synthetic Polymers and Adhesives

Introductory Remarks 175
    F. D. Petke

Structure - Property - Performance Relationships
in Synthetic Polymeric Adhesives 177
    F. D. Petke

Solvent and Structure Studies of Novel
Polyimide Adhesives 187
    T. L. St. Clair and D. J. Progar

Block Copolymers as Adhesives 199
    D. H. Kaelble

Effects of Adhesive Structure on Impact Resistance
and Optical Properties of Acrylic/Polycarbonate
Laminates 217
    J. L. Illinger, R. W. Lewis, and
    D. B. Barr

Physical Property - Performance Correlations
in Contact Adhesive Systems 233
    R. G. Azrak, D. L. Joesten,
    and W. F. Hale

Discussion 249

PART THREE: Rubber Adhesives and Sealants

Introductory Remarks 257
    M. E. Gross

Mechanisms of Adhesion in Elastomer-to-Textile
Bonding 259
    F. H. Sexsmith and E. L. Polaski

On Bonding Rubber to Glass 281
    A. Ahagon, A. N. Gent, and E. C. Hsu

The Microscopy of Polymeric Adhesive
Systems 289
    R. W. Smith

High Speed Testing of Rubber-to-Metal Bonds 315
    D. A. Given and R. E. Downey

CONTENTS OF VOLUME 9A

The Nature of Rubber Reinforcement by
Silane-Treated Mineral Fillers 329
 E. P. Plueddemann and W. T. Collins

Catalysis of Silicone Elastomer Adhesion 339
 T. W. Greenlee

Discussion 355

PART FOUR: Natural Products and Structural Adhesives

Introductory Remarks 361
 H. G. Arlt, Jr.

The Use of SEM, ESCA, and Specular Reflectance
IR in the Analysis of Fracture Surfaces in
Several Polyimide/Titanium 6-4 Systems 365
 T. A. Bush, M. E. Counts, and
 J. P. Wightman

The Chemistry of Tackifying Terpene Resins 395
 E. R. Ruckel, H. G. Arlt, Jr., and
 R. T. Wojcik

Some Factors for Achieving Environmental
Resistance in 120°C Structural Adhesives 413
 J. S. Noland

A Fluoro-Anhydride Curing Agent for
Heavily Fluorinated Epoxy Resins 429
 J. R. Griffith, J. G. O'Rear,
 and J. P. Reardon

Discussion 437

Author Index xv

Subject Index xxvii

Contents of Volume 9B

PART FIVE: Growth and Change in Adhesives

Introductory Remarks: Adhesive Economics - Growth and Change 441
    I. Skeist

Growth and Change in Adhesives 443
    J. Miron and I. Skeist

Growth and Change in Textile Adhesives 449
    M. D. Hurwitz

Growth and Trends in Specialty Adhesives 467
    J. R. Elliott and R. J. Dauksys

Growth and Change in Epoxy and Urethane Adhesives 473
    M. E. Edwards

Growth and Change in Hot Melts 485
    P. O. Powers

Discussion 495

PART SIX: Performance of Adhesive Joints

Introductory Remarks 499
    R. L. Patrick

Mixed-Mode Fracture of Structural Adhesives 501
    W. D. Bascom, R. L. Jones, and
    C. O. Timmons

Flaw Tolerance of a Number of Commercial
and Experimental Adhesives 513
    S. Mostovoy and E. J. Ripling

The Attachment Site Theory of Adhesive
Joint Strength 563
    A. F. Lewis and R. T. Natarajan

A Technology for Achieving Durable
Joining of Ferrites 577
    S. G. Seger, Jr. and L. H. Sharpe

Durability of Adhesive Joints 597
    A. J. Kinloch, W. A. Dukes, and
    R. A. Gledhill

Discussion 615

PART SEVEN: Trends in Adhesion Research

Capillary Adhesion 621
    W. J. O'Brien and P. L. Fan

Adhesion in Deformable Isolated
Capillaries 635
    P. L. Fan and W. J. O'Brien

Wettability of Functional Polysiloxanes 647
    L. H. Lee

Phthalocyanine Resins: A New Class of
Thermally Stable Resins for Adhesives,
Coatings, and Plastics 665
    T. R. Walton, J. R. Griffith,
    and J. G. O'Rear

The Limits of Crack Detection by Ultrasonic
Transmissibility 677
    D. J. Beyer, K. C. Ludema,
    T. R. Bates, Jr., and J. R. Frederick

Adhesion of Brittle Films on a
Polymeric Substrate 687
    T. S. Chow

Discussion 707

CONTENTS OF VOLUME 9B

PART EIGHT: Surface Energetics of Printing Processes

Polymers for Lithography: State of the Art 711
   L. H. Lee

Mechanism of Lithography 725
   M. C. Wilkinson, M. P. Aronson,
   J. W. Vanderhoff, and A. C. Zettlemoyer

Surface Energetics Analysis of Lithography 735
   D. H. Kaelble, P. J. Dynes, and D. Pav

Toner Particle - Photoceptor Adhesion 763
   N. Goel and P. R. Spencer

Thermal Fixing of Electrophotographic
Images 831
   L. H. Lee

Concluding Remarks 853
   J. R. Elliott

About the Contributors 855

Author Index 861

Subject Index 873

Significant Advances and Developments in Adhesion and Adhesives

Lieng-Huang Lee

Wilson Center for Technology, Xerox Corporation

Webster, New York   14580

This paper is a review of significant advances in adhesion studies and developments in adhesives between the 1971 Adhesion Symposium held in Washington, D. C. and the 1975 Adhesion Symposium held in Philadelphia, Pa. The survey is not exhaustive; the adhesive systems selected for review are those considered by the author to include some interesting concepts.

Several adhesive systems which have future value because of ecological and energy requirements are briefly described. Aerospace adhesives, electronic adhesives, high solid adhesives, hot melt adhesives and photocured adhesives are among those of high growth potentials.

INTRODUCTION

The growth of adhesive materials[1a] will, at the least, double in the next ten years (Table 1). Various forms of new adhesives are likely to appear as a result of environmental requirements. For example, volatile fluid forms of adhesives are now 87% of the total (Table 2), but we expect that by 1980 most of those fluids will be converted into nonfluid systems because of environmental reasons or energy limitations (Table 3). We anticipate more hot melt adhesives, powder adhesives, and film or webs will be used in the future. In this paper, we shall discuss some of those promising systems with high growth potential.

We anticipate that wood products will be the leading end uses of the various kinds of adhesives[1b]. Other major applications will be in paper/packaging, construction, auto, apparel, etc. (Fig. 1). In addition to these major applications, there will be increasing demands for adhesives in the aerospace, textile and electronic industries. In the medical field, we are likely to find more surgical and dental adhesives being developed parallel to the progress of bioengineering.

In the following sections, we shall briefly review recent understandings of adhesion phenomena and some developments in adhesive materials and processes. In the last section we shall examine several new systems which may help meet ecological and energy requirements. This review does not include most of papers recently presented to the Symposium on Science and Technology of Adhesion held in Philadelphia, Pa. in April, 1975.

Table 1. Shipments of Fasteners (U.S.A.)
(Ref. 1a)

	1960	1970	1980	Annual Growth
Adhesives	424	655	1475	6.8%
All other Fasteners	2033	3695	7225	
Total	2457	4350	8700	
% Adhesive	17.3	15.1	17.0	
% of GNP	0.47	0.45	0.44	

Table 2. Alternate Ways to Deposit an Adhesive Film or Sealant in 70's
(Ref. 1a)

	Distribution
Volatile Fluid Systems	
Organic	18%
Aqueous	69%
Non-Volatile Fluid Systems	
Hot Melt	10%
Reactive Fluids	1%
Powder	nil
Plastisol	1%
Non-Volatile Solid Systems	
Films and Web	2%

Table 3. Energy Consumption to Apply and/or Dry
(Ref. 1a)

	kcal/100g
Aqueous (@50% solids)	200
Organic Solvent (@20% solids)	60
Hot Melt	18
Powder	6
Films and Webs	6
Reactive Fluids	nil

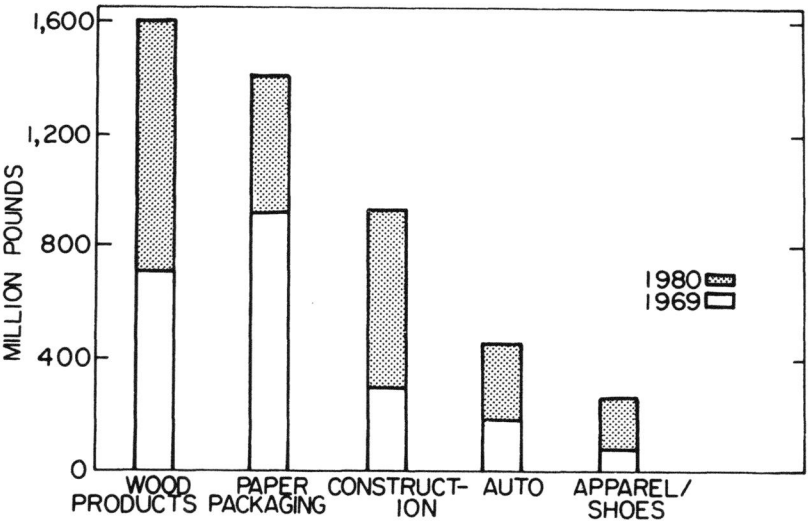

Fig. 1. Adhesives consumption and expected growth - U.S.A.

ADHESION PHENOMENA

The science of adhesion, or "adhesiology"[2], started to take shape in the last decade. With the cooperation of surface scientists, polymer chemists, and rheologists, we gradually started to understand the mechanisms of adhesion. Interestingly, adhesion is a universal phenomenon which can take place even in the absence of an adhesive. New terminologies have been introduced to describe various aspects of adhesion, e.g. bio-adhesion, electro-adhesion, particle adhesion, and photo-adhesion.

In the presence of an adhesive, a joint (or adhint) is formed.

The formation and destruction of an adhesive joint have been shown to be nonequilibrium and non-thermodynamic processes. In the past several years, the principles of fracture mechanics[3,4] have been employed to study the breaking of an adhesive joint; several papers on this subject are published in this volume on <u>Adhesion Science and Technology</u>[5].

The development of new tools in surface characterization[6] (Table 4 and Fig. 2)[7] has enabled us to examine interfaces at various stages. One of the important tools is ESCA (Electron Spectroscopy for Chemical Applications). ESCA has been increasingly used to investigate both inorganic and organic surfaces[8]. Another invaluable tool, SEM, is being extensively used to study the morphology of interfaces[9].

Table 4. Methods Used for the Study of Solid Surfaces

Macrostructure
    Microscopy
    Surface Topography
    Friction Measurements
    Ellipsometry

Molecular Composition and Structure
    Multiple Interference Reflection Spectroscopy
    Raman Spectroscopy
    Ellipsometry
    Contact Angle Measurements
    Surface Potential Measurements

Crystal Structure
    X-ray Diffraction
    Low Energy Electron Diffraction, LEED
    Reflected High Energy Electron Diffraction, RHEED

Elemental Composition
    Photoemission Spectroscopy
        ESCA
        Auger
    Ion Scattering Spectroscopy
        Low Energy
        High Energy
    Ion-probe Mass Spectrometry
    Microprobe
    Appearance Potential Spectroscopy

	Incident Beam					
Reflected beam	Infrared	Visible	Ultraviolet	X-Ray	Electrons	Ions
Infrared	Reflection spectroscopy (MIRS)					
Visible		Raman Spectroscopy				
		Ellipsometry "light" microscopy Reflection Spectroscopy				
Ultraviolet			Ellipsometry Reflection Spectroscopy			
X-ray				X-ray diffraction	Microprobe Appearance Potential Spectroscopy	
				X-ray microscopy		
Electron		Photoemission spectroscopy	ESCA	Photoelectron spectroscopy ESCA Auger Spectroscopy	LEED, RHEED electron microscopy Auger spectroscopy	
Ions						Ion scattering spectroscopy Ion probe mass spectroscopy

Fig. 2. Methods of surface analysis based on the examination of reflected radiation and/or particles. Abbreviations: MIRS, multiple interference reflection spectroscopy; LEED, low energy electron diffraction; RHEED, reflected high energy electron diffraction; ESCA, electron spectroscopy for chemical analysis (Ref. 7).

In Figure 2, various surface analyses[7] are listed according to the types of reflected beam versus incident beam. In the infrared range, the reflection spectroscopy (MIRS) has found interesting applications in identifying organic contaminants as well as polymers on the surface. In the visible and UV ranges, ellipsometry has been used to study adsorption of polymers on metal surfaces. In the x-ray range, besides ESCA and Auger spectroscopy, x-ray diffraction is commonly used to probe the metal surface and the crystalline structure of polymer. In the electron range, microprobe analysis has been employed to determine trace metallic elements. Though not all of these surface analyses have been used to study metal-polymer adhesion, the trend is to find more new methods employed in solving adhesion problems.

Bullett and Prosser[10] have reviewed, critically, various methods for measuring paint film adhesion. An empirical method was developed to determine metal-ceramic and metal-glass bonds[11]. The adhesion of gold-underlayer film combinations to glass[12] has also been investigated qualitatively. Those two studies actually involved no adhesives.

Several papers by Kendall[13-16] treated the subject of adhesion rather explicitly with simple models of mechanics. Andrews and Kinloch[17,18] attempted to separate "adhesion" and "adhesive joint strength" and established that the intrinsic failure energy is close to the work of adhesion when pure interfacial failure occurs. Kloubek[19] attributed the interaction of polar forces to the contribution of the work of adhesion. The effect of surface energetics and wetting on adhesion has been summarized by Kaelble[20], Mittal[5], and Zisman[5,21].

Papers on other mechanisms of adhesion have also appeared recently. The diffusion of rubbers has been related to "free-volume" by Campion[22]. Mechanical interlocking has been claimed to be the main cause of polyethylene adhesion to porous aluminum[23,24].

Electro-adhesion was established by Krupp and Schnabel[25] as the cause of light-modulated adhesion. Further evidence for electro-adhesion is given in a new book by Derjaguin[26]. The importance of electrostatic adhesion in electrophotography or xerography will be further discussed at the end of this book. For example, the attachment of toner particles on the surface of a carrier is determined predominantly by the electrostatic attraction between the two materials. This attraction is also called triboelectricity on the surfaces. Besides electrostatic attraction, Van der Waal forces can influence the toner particle adhesion depending upon the size of the toner.

The role of adhesion and surface energetics on polymer friction has been critically examined by Lee[27]. The application of molecular kinetic theory to adhesion and friction of elastomers has been carried out by Lavrentev[28], and molecular interaction for solids at contact have been reviewed[29].

ADHESIVE JOINT STRENGTH

McCarvill and Bell[30] have described the use of torsional test for the determination of adhesive joint strength, and dynamic loss[31] has been used to measure tire cord adhesion to rubber. Holographic method[10,32] has been suggested as one non-destructive method for examining adhesive bonding. Standard methods for adhesive testing[33] have also been compared.

Surfactants[34,35] have been reported to affect adhesive joint strength, and the crosslink density of elastomers has been studied with respect to adhesive strength[36]. The effect of moisture on joint strength was investigated by Schonhorn and Frisch[37].

RECENT DEVELOPMENTS IN ADHESIVES

Modified acrylics[38] have been reported to be good metal adhesives. Radial block copolymers[39], in addition to S-B block copolymers, have been offered as new contact adhesives. Cyanoacrylates[40] have found wide applications beyond their use as surgical adhesives in the past several years. In-situ toughened epoxies[41] have been found convenient for adhering honeycomb structures.

Microvoid epoxies[42] have been recently reported to be promising as toughened adhesives. The curing agents leading to microvoids are shown in Table 5, while those which do not are shown in Table 6. These microvoid adhesives have good moisture and chemical resistance.

Table 5. Representative Microvoid Curing Agents

DETA	2-Ethyl-4-methyl imidazole
TETA	
AEA	Versamid 115, 125, 140
*AEP	Advacure-10
TEPA	Shell "U", "T"
DICY	*Ajicure B-001
*DMP-30	Thiokol LP-3

* Does not give microvoids when cured at high temperature, about 250°F.

Table 6. Curing Agents Which do not Give Microvoids

HHPA	Jeffamine D-230
MPDA	Piperidine
BF_3MEA	Ajicure B-002
MDA	Nadic methyl anhydride Advacure-20
	Isophorondiamine

A new class of sealants - <u>polymercaptane</u> (PM Polymer) - has been introduced for construction applications[43]. Their formulations and performance properties have been described.

The curing system developed for butyl rubbers has been extended to <u>ethylene-propylene terpolymer</u> (EPDM)[44]. Mechanisms of curing by p-dinitrosobenzene have been discussed. The cured cements are promising as outdoor sealants in terms of oxidative stability.

<u>Silane-modified</u> urethane <u>adhesives</u>[45] offer moisture-resistance. The gain of peel strength at high R.H. is remarkable. A mercapto-silane is used to react with the urethane prepolymer.

<u>High-speed-cured</u> <u>wood</u> <u>adhesive</u> has been developed[46], and a "honeymoon" gluing scheme (Fig. 3) has been proposed. With the use of m-aminophenol, the boil-dry test results are comparable to those of phenol-resorcinol system.

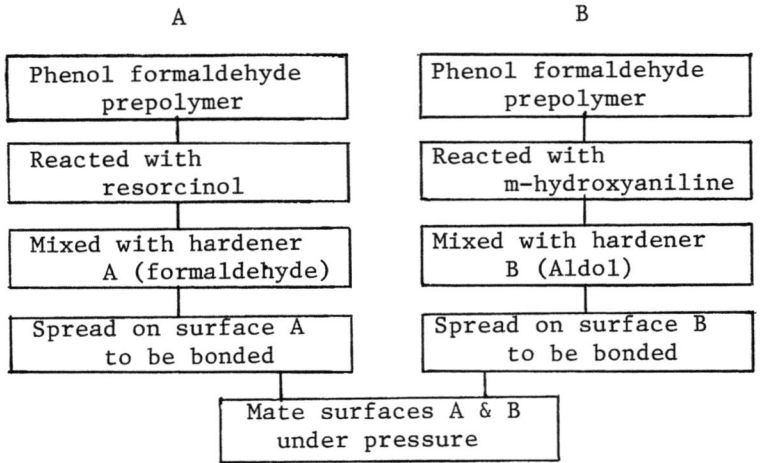

Fig. 3. "Honeymoon" Gluing Scheme

ADHESIVES WITH FUTURE VALUE

Aerospace Adhesives

New military and civil aircrafts[47] use higher amounts of adhesives. In space systems, adhesives offer the advantages of a light weight structure and the joining of dissimilar materials. For example, approximately 80% of the F-111 is adhesive bonded sandwich. In Boeing 747, there are 3,600 ft^2 of bonded honeycomb, 9,960 ft^2 of bonded metal-to-metal and 12,000 ft^2 of bonded fiberglass reinforced thermoplastics. The complexity of bonding systems are multiplied in space-craft, e.g. Mariner venus/mercury system[48].

Aerospace adhesives require moisture resistance, fatigue resistance and high temperature resistance. Several papers in this volume have described structural adhesives developed for extreme environments. There are many publications dealing with high temperature resistant polymers of exotic structures. However, there are only three representative polymers[49]: polyimide (PI), polyphenylquinoxaline (PPQ), and polyimidazoquinazoline (PIQ) (Fig. 4). Among these, polyimides have gained wide acceptance.

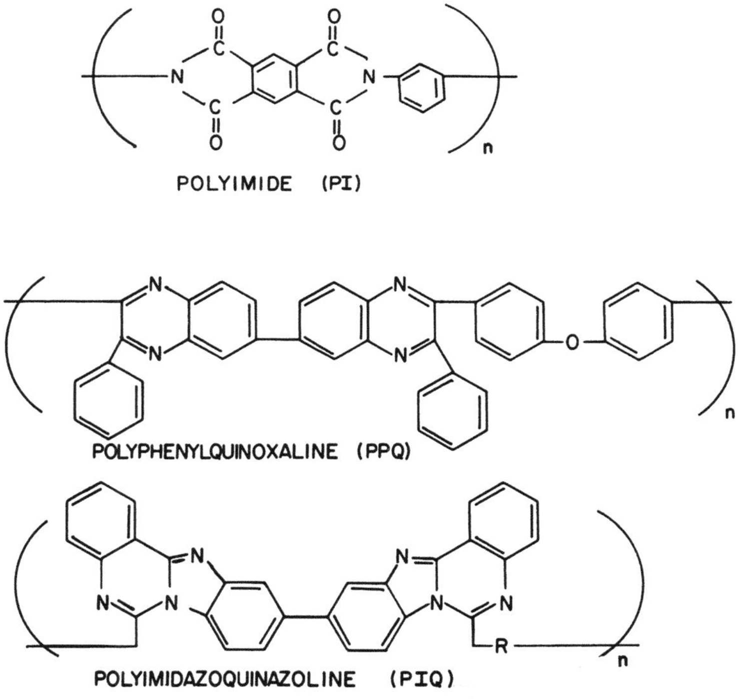

POLYIMIDE (PI)

POLYPHENYLQUINOXALINE (PPQ)

POLYIMIDAZOQUINAZOLINE (PIQ)

Fig. 4. High Temperature Polymers for Adhesives

For use in spacecraft[48], there is another important requirement for adhesives, the volatile condensable material (VCM). Most polymeric materials tend to emit volatile components when exposed to thermal-vacuum space environment. In this regard, thermosetting resin is superior to thermoplastic or elastomeric adhesives. Generally, an adhesive having a total weight loss of $\leq 1\%$ and a VCM of $\leq 0.1\%$ is acceptable for spacecraft usage. One of the aerospace adhesives chosen for the 1973's Mariner Venus/Mercury was American Cyanamid FM-960 film adhesive which was suitable for a 350°F service temperature, possessed minimum weight and outgassing characteristics and provided adequate strength for metal-to-metal and sandwich constructions.

Electronic Adhesives

Adhesives have been used to bond electronic components and to prepare flexible circuits[50]. The advantages of bonded flexible circuits are as follows:

(1) Superior dielectric films not suitable for melt bonding can be used. Advantages of the films include:

    (a) Better dimensional stability
    (b) Higher melting points
    (c) Greater strength
    (d) Higher dielectric strength
    (e) Coefficient of expansion better matched to metal foil to reduce residual stress from lamination or operating temperature extremes.

(2) Adhesive bonding is done without melting the dielectric film giving less residual stress for later relief in etching or elevated temperature exposure.

(3) Lower bonding temperatures are employed than in melt-bonding further reducing residual stress.

(4) Dielectric film and thermoset adhesives used in base laminates do not re-melt when covercoat is laminated eliminating the problem of conductor movement and distortion of the circuit pattern.

The steps in manufacturing flexible circuits are illustrated in Fig. 5. A list of adhesives for this application is given in Table 7.

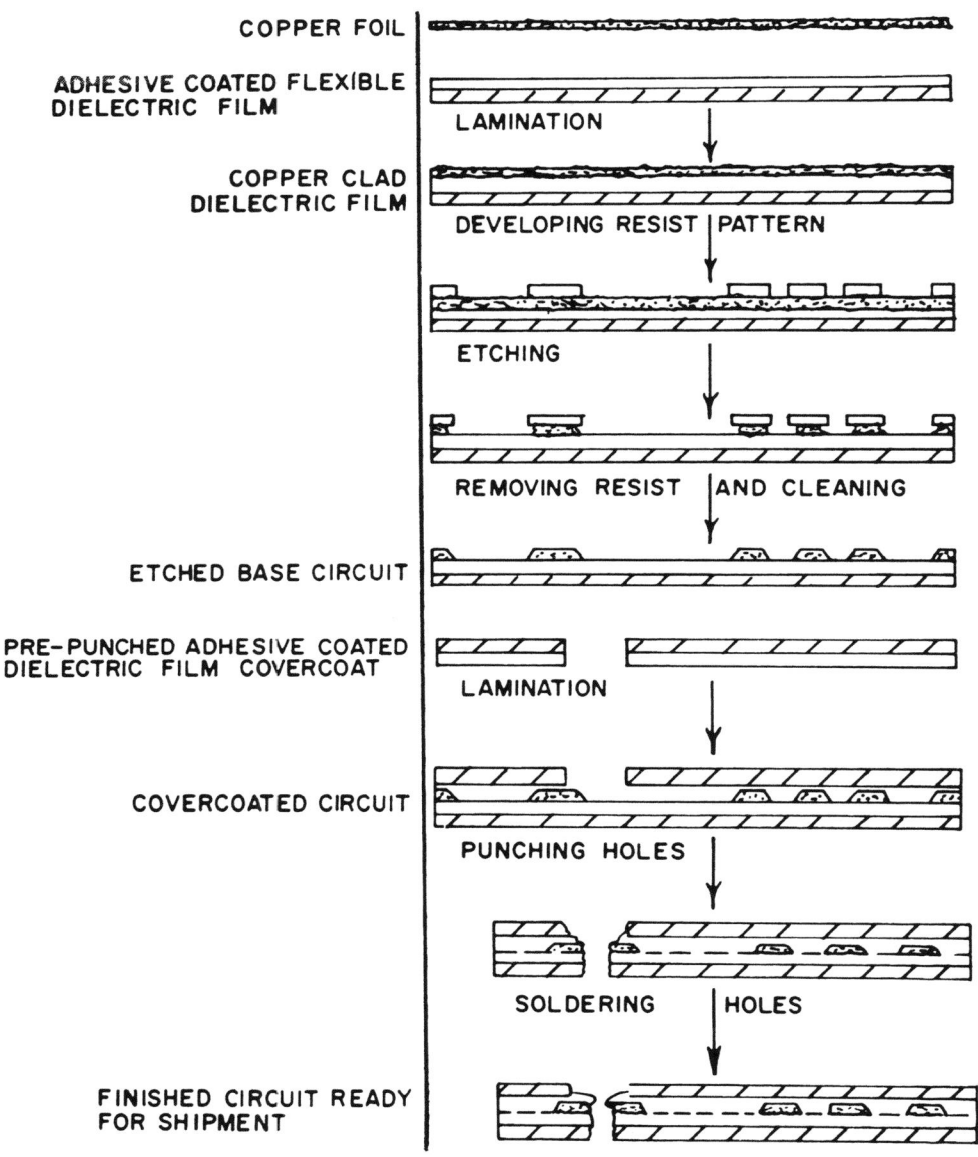

(Ref. 50)
Fig. 5. Steps in the manufacturing of a simple flexible circuit.

Table 7. Adhesives and Adhesive-Bonded Materials
for Flexible Circuit Manufacture (Ref. 50)

(1) Liquid adhesives
   polyester-isocyanate
   phenolic butyral
   phenolic nitrile
   modified epoxy
(2) Dry film adhesives
   phenolic-butyral
   epoxy (modified)
   epoxy-nylon
   modified polyethylene
   FEP Teflon
(3) Adhesive-coated dielectric films
   Polyimide film/B stage adhesive (epoxy, polyester or phenolic)
   Polyimide/FEP Teflon heat-sealable film
   Polyester film/polyester adhesive
   Polyester film/polyethylene heat-sealable film
   Bonded FEP Teflon/B stage adhesive
(4) Copper-clad dielectric film (Copper one or both sides bonded with thermoplastic or B stage adhesive)
   Polyimide film
   Polyester film
   Bondable FEP Teflon film
(5) Copper-clad flexible fiber-reinforced dielectric
   (Binder usually functions as adhesive)
   Epoxy-nonwoven polyester
   Flexible epoxy fiberglass

High Solid Adhesives

Future energy and environmental requirements will encourage the development of high solid adhesives, including powder adhesives. A recent high solid adhesive[51] is based on Neoprene AH. An aliphatic solvent is added to water to form high solid dispersion. For example, Neoprene AH is dissolved in heptane with a small amount of thiuram E and an accelerator 552; the solution is then added to water to form a dispersion (40% solid).

Hot Melt Adhesives

Hot melt adhesives are also high solid systems which can meet environmental requirements. Thus, the growth of hot-melt[52] has been projected to double by 1980 (Fig. 6). It is likely that new hot-melt systems will be introduced to the market. To demonstrate the promising features of hot-melt, we choose the ethylene-vinyl-acetate system (EVA) as an example. EVA can be formulated with a tackifier and a wax. The formulation can be facilitated with the use of computer to plot the viscosity-composition profile.

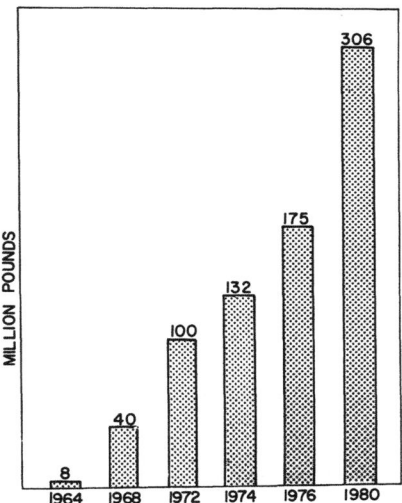

Fig. 6. Consumption of hot melt adhesives, in millions of pounds with estimates to 1980 (Ref. 52).

Photo-cured Adhesives

To meet environmental requirements, we shall see more systems developed involving photo-curing processes. Two interesting systems which have already used photo-curing principles are the USM process[53] and the Japanese Photobond process[54].

The USM process involves the use of a sensitizer; e.g. benzophenone to create polar sites for crosslinking and/or reaction on nonpolar polymers. The beneficial effects of using isocyanate were found for EPDM and PVF, polyvinyl fluoride.

The Japanese process[54] also involves a sensitizer but not much details have been disclosed. Their process is somewhat like the photoresist system (Fig. 7). The photosensitive layer after irradiation is washed off and the photocured adhesive is then washed with water.

In brief, photo-curing systems may offer interesting adhesive properties which have not been obtained with conventional adhesives.

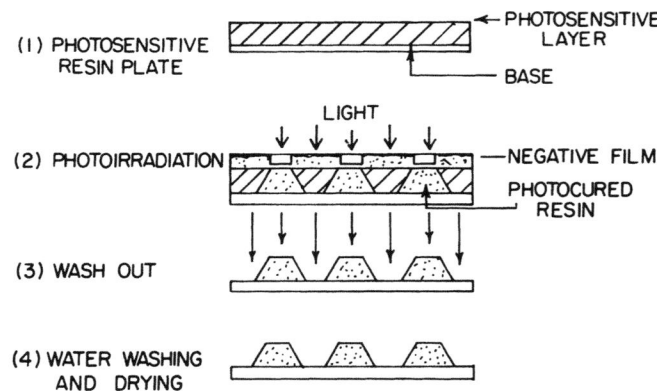

Fig. 7. Formation of photosensitive resin relief (Ref. 54).

CONCLUSIONS

This survey briefly reviewed work published during the five years prior to this Symposium on Science and Technology of Adhesion. We first discussed several interesting papers on adhesion and on adhesive joint strength, then cited several systems illustrating recent developments of adhesive technology from concepts to products. Our choices for discussion should not be considered inclusive or representative.

Several systems may be important in the future as aerospace adhesives, electronic adhesives, high solid adhesives, hot melt adhesives, and photo-cured adhesives. We believe that future adhesives will be less likely to pollute the environments and will somehow save energy.

REFERENCES

1. (a) D. Kuespert, "Adhesives: Past, Present and Future", Adhesives Age, 18 26, Jan. 1975. (b) T. K. Slack, Adhesives Age 17, 7, March 1974.
2. L. H. Lee, "Advances in Adhesiology" in *Recent Advances in Adhesion*, L. H. Lee, Ed., Gordon and Breach, New York (1973).

3. M. L. Williams, "The Relation of Continum Mechanics to Adhesive Fracture" in *Recent Advances in Adhesion*, L. H. Lee, Ed., Gordon and Breach, New York (1973).
4. G. G. Trantina, "Fracture Mechanics Approach to Adhesive Joints", J. Composite Mat. $\underline{6}$, 192, April, 1972.
5. L. H. Lee, Ed., *Adhesion Science and Technology*, Plenum Press (New York) 1975.
6. P. F. Kane and G. B. Larrabee, Ed., *Characterization of Solid Surface*, Plenum Press, New York (1974).
7. W. D. Bascom and R. L. Patrick, "The Surface Chemistry of Bonding Metals with Polymer Adhesives", Adhesives Age, $\underline{17}$, 25, Oct. 1974.
8. D. T. Clark, "The Application of ESCA to Studies of Structure and Bonding Polymers", in *Advances in Polymer Friction and Wear*, L. H. Lee, Ed., Plenum Press, New York (1974), p. 243.
9. R. L. Patrick, "The Use of Scanning Electron Microscopy", in *Treatise on Adhesion and Adhesives*, Dekker, New York (1973), Vol. 3, p. 163.
10. T. R. Bullett and J. L. Prosser, "The Measurement of Adhesion", Prog. in Org. Coatings, $\underline{1}$, 45, (1972).
11. J. T. Klomp, "Binding of Metals to Ceramics and Glasses", Cera. Bull. $\underline{51}$, No. 9, 683 (1972).
12. K. E, Haq, K. H. Behrndt and I. Kobin, "Adhesion Mechanism of Gold-Underlayer Film Combinations to Oxide Substrates", J. Vac. Sci. and Tech., $\underline{6}$, No. 1, 148 (1969).
13. K. Kendall, "The Adhesion and Surface Energy of Elastic Solids", J. Phys. D: Appl. Phys., $\underline{4}$, 1186 (1971).
14. K. L. Johnson, K. Kendall and A. D. Roberts, "Surface Energy and the Contact of Elastic Solids", Proc. Roy. Soc. Lond. A. $\underline{324}$, 301 (1971).
15. K. L. Kendall, "The Shape of Peeling Solid Films", J. Adhesion, $\underline{5}$, 105 (1973).
16. K. L. Kendall, "Kinetics of Contact Between Smooth Solids", J. Adhesion, $\underline{7}$, 55 (1975).
17. E. H. Andrews and A. J. Kinloch, "Mechanics of Adhesive Failure I", Proc. Roy. Soc. Lond. A $\underline{332}$, 385 (1973).
18. E. H. Andrews and A. J. Kinloch, "Mechanics of Adhesive Failure II', Proc. Roy. Soc. Lond. A $\underline{332}$, 401 (1973).
19. J. Kloubek, "Interaction of Polar Forces and Their Contribution to the Work of Adhesion", J. Adhesion, $\underline{6}$, 293 (1974).
20. D. H. Kaelble, *Physical Chemistry of Adhesion*, Wiley-Interscience New York (1971).
21. W. A. Zisman, "Surface Energetics of Wetting, Spreading and Adhesion", J. Paint Tech. $\underline{44}$, No. 564, 41 (1972).
22. R. P. Campion, "The Influence of Structure on Autohesion (Self-Tack) and other Forms of Diffusion into Polymers", J. Adhesion, $\underline{7}$, 1 (1975).
23. D. E. Packham, K. Bright and B. W. Malpass, "Mechanical Factors in the Adhesion of Polyethylene to Aluminum", J. Appl. Poly. Sci. $\underline{18}$, 3237 (1974).

24. B. W. Malpass, D. E. Packham and K. Bright, "A Study of the Adhesion of Polyethylene to Porous Aluminum Films Using the Scanning Electron Microscope", J. Appl. Poly. Sci., 18, 3249 (1974).
25. H. Krupp and W. Schnabel, "Light Modulated Electrostatic Double Layer Adhesion", J. Adhesion, 5, 269 (1973).
26. B. V. Derjaguin, *Adhesion of Solids*, (Russ.) Eng. Ed. to be published by Plenum Press, New York (1975).
27. L. H. Lee, "Effect of Surface Energetics on Polymer Friction and Wear", in *Advances in Polymer Friction and Wear*, L. H. Lee, Ed., Plenum Press, New York (1974).
28. V. V. Lavrentev in *Polymer Friction and Wear*, (Russ.) G. M. Bartenev and V. V. Lavrentev, tran. into Eng. by D. B. Payne and edited by L. H. Lee, to be published (1975).
29. M. Van Den Temple, "Interaction Forces Between Condensed Bodies in Contact", Advan. Colloid Interface Sci., 3, 137 (1972).
30. W. T. McCarvill and J. P. Pell, "Torsional Test Method for Adhesive Joints", J. Adhesion, 6, 185 (1974).
31. T. Murayama and E. L. Lawton, "Dynamic Loss Energy Measurement of Tire Cord Adhesion to Rubber", J. Appl. Poly. Sci., 17, 669 (1973).
32. R. K. Erf, R. M. Gagosz and J. P. Walters, "Holography: A New NDT Tool Comes of Age", SAMPE Quart. 26, April, 1975.
33. N. C. McDonald, "Standard Test Methods for Adhesives", Adhesives Age, 15, 21, Sept. 1972.
34. L. A. Akopyan, E. V. Gronskaya and G. M. Bartenev, "Adhesion Properties of Rubbers Modified by Surface Active Substances", J. Adhesion, 6, 177 (1974).
35. F. Van Voorsf Vader and H. Dekker, "The Influence of Surfactants on the Adhesion Between Solids", J. Adhesion, 7, 73 (1975).
36. E. H. Andrews and A. J. Kinloch, "Elastomeric Adhesives: Effect of Cross Link Density on Joint Strength", J. Poly. Sci. Poly. Phys., 11, 269 (1973).
37. H. Schonhorn and H. L. Frisch, "Environmental Aspects of Adhesion and Adhesive Joint Strength", J. Poly. Sci., Poly. Phys., 11, 1005 (1973).
38. L. E. Toy, "Plastic/Metals: Can They be United", Adhesive Age, 17, 32, Sept. 1974.
39. B. D. Simpson and P. R. Fowler, "Contact Adhesives Based on Radial Block Copolymers of Butadiene and Styrene", Adhesives Age, 17, 32, Sept. 1974.
40. T. M. Brumit, "Cyanoacrylate Adhesives - When You Should Use Them", Adhesives Age, 18, 17, Feb. 1975.
41. C. D. Weber and M. E. Gross, "In Situ Toughened High-Flow Structural Adhesives", Adhesives Age, 17, 18, Feb. 1974.
42. R. A. Peters and T. J. Logan, "Microvoid Epoxy Adhesives For High Peel and Shear Strength", Adhesives Age, 18, 17, April, 1975.
43. R. C. Doss and O. L. Marrs, "New Polymercaptan Polymer For Elastomeric Sealants", Adhesives Age, 17, 25, Nov. 1974.

44. S. E. Cantor, "RTV Adhesive System Based on Ethylene - Propylene - Diene Terpolymer", Adhesives Age, 17, 17, June 1974.
45. F. D. Swanson and S. J. Price, "Chemistry of Urethane Adhesives With Silane Coupling Agents", Adhesives Age, 16, 23, June 1973.
46. R. E. Kreibich, "High-Speed Adhesives for the Wood-Gluing Industry", Adhesives Age, 17 26, Jan. 1974.
47. R. J. Dauksys, "Research and Development of Aerospace Adhesive Bonded Systems and Concepts", SAMPE Quart. 1, Oct. 1973.
48. B. Pascuzzi, "Adhesives' Role in the Development of Mariner Venus/Mercury 1973 Spacecraft", Adhesives Age, 16, 20, Sept. 1973.
49. P. M. Hergenrother, "High-Temperature Organic Adhesives", SAMPE Quart. 3, 1, Oct. 1971.
50. D. L. Holland, "Adhesives for Flexible Printed Circuits", Adhesives Age, 16, 17, April, 1973.
51. J. L. Nyce and R. W. Keown, "High-Solid Adhesives", Adhesives Age, 18, 25, March 1975.
52. R. J. Litz, "Development in Ethylene-Based Hot Melt Adhesives", Adhesives Age, 17, 35, Aug. 1974.
53. R. A. Bragole, "Adhesive Bonding of Polyolefins", Adhesives Age, 17, 24, April 1974.
54. Y. Nakano, "Photosensitive Adhesives and Sealants", Adhesives Age, 16, 28, Dec. 1973.

Polymer Progress--Worldwide

H.F. Mark

Polytechnic Institute of New York, 33 Jay Street

Brooklyn, New York   11201

This report contains a few observations made on recent polymer progress in various countries.

An interesting type of ion exchange system has been pioneered and developed at the CSIRO in Australia in which the removal of the captured mobile counterion (Na^+, Cl^-, etc.) is effected by heat and does not require the application of chemicals. In Israel, PVC membranes containing tricresylphosphate are being successfully used by Professor Vofsi of the Weizmann Institute to collect uranium from dilute solutions.

Textile fibers, in which a crystalline fibrillae are axially distributed in an amorphous crosslinked matrix are studied in Japan and the USSR. Fine long semicrystalline axially oriented PVC fibrils are contained in a crosslinked matrix of polyvinylalcohol in the Cordelan fiber and fibrils of PAN are similarly distributed in a matrix of a wet spun natural protein in the Chinon fiber. In the USSR Rogowin is studying similar composite fiber systems on the basis of cellulose and its derivatives, together with vinylic and acrylic polymers.

Elastomers based on polyphosphazenes with fluoro side groups are investigated because of their favorable behavior in the cold, their resistance to ignition and their stability against oils at elevated temperatures.

Progress is also made in the synthesis and processibility of polymides. Thin films and fine fibers can be directly prepared from precursors but larger pieces require delicate and severe conditions of forming and finishing. Equally important for thermally and mechanically lightweight composites are successful efforts in the chemistry and technology of high temperature thermosets particuarly of bis-maleiimides and aromatic polyesters. New and promising applications of carbon fibers and fabrics are being studied in their use in brakelining, cutting wheels and fabrics having unusual capacity to absorb hot corrosive and toxic gases.

INTRODUCTION

Worldwide progress in polymer science and technology is essentially influenced by two motivations.

One originates in the creative imagination of scientists and engineers who ask themselves: what else could we do, what new concepts, materials and processes could we imagine and create? Let us try to advance the frontiers of our knowledge and techniques. The other stems from their innate desire to comply with societal needs and let them ask such questions as: how can we make existing products better and less expensive? How can we improve health and safety and the cleanliness and purity of our environment? How can we with better efficiency use the dwindling natural resources--minerals, coal, oil and gas?

The author of this report found these two driving forces to be very alive in countries from Australia[1] to the Soviet Union[2] and from Argentina to Canada and takes great pleasure to present to you a report on his findings and impressions.

FUNDAMENTAL STUDIES

One basic question is that for the ultimate rigidity, strength and toughness of polymeric materials in the form of fibers, films, adhesives and of molded and forged objects. How high are the theoretical values, what percentage of them have been practically reached and how could we get closer to them? Early, crude estimates put the force to break an isolated covalent bond C-C, C-N, or C-D at about 4×10^{-4} dynes[1]. Better subsequent calculations confirmed the order of magnitude and led to values between 2 and 5×10^{-4} [3] dynes per bond. As a result, an isolated infinitely long molecule of polyethylene, if placed in an Instron Tester, would require this force to be broken, whereby its weakest bond--

weakest as a result of statistical fluctuations or of accidental imperfections—would break somewhat below this value. In a completely oriented, ideally crystallized rod of polyethylene with chains going from one end of the rod to the other there are—according to x-ray data—4×10^{14} chains per sq cm, so that its breaking strength would be

$$4 \times 10^{14} \times 5 \times 10^{-4} = 20 \times 10^{10} \text{ dynes per sq cm}$$

$$\text{or} \quad 2 \times 10^{5} \text{ kg/cm} \sim 3{,}000{,}000 \text{ psi}$$

In textile science and industry, one always has fine fibers in hand and a square inch is a very large and unrealistic cross section. The tensile properties of textiles are, therefore, usually expressed in grams per denier (more recently and more adequately in pound per decitex); for materials in the specific gravity range from 1.0 to 1.3, one gram per denier corresponds to about 20,000 psi. The theoretical strength of an idealized polyethylene filament having infinitely long chains would, therefore, be about 150 g/den; highly oriented filaments of linear polyethylene have a "technical" strength up to 10 g per denier. There is obviously a large gap between what we get in practice and what we should expect from the theoretical strength of covalent bonds. This is not surprising because the theoretical value is based on oversimplifying assumptions, namely infinitely long chains (no chain ends) and ideal axial orientation of all chains. In reality, both assumptions are not fulfilled: technical polyethylenes have average degrees of polymerizations (DP) of about 1000 and a chain length of about 2500 Å; they are not infinitely long but very short (2.5×10^{-5} cm). Technical fibers are produced by drawing the polymer which consists of folded chain lamellae and of randomly coiled chains between them. Standard drawing never orients each molecule with all its segments exactly parallel to the fiber axis. Evidently this discrepancy stimulated experiments with higher DP polymers, having better axial alignments. Such tests were actually carried out recently in the USA, Europe, the Soviet Union and Japan with DP's up to 50,000 using careful two stage drawing processes.[4] On the average, values of 25 g per denier were reached; the highest reported figure was 30 g per den. This proves that the approach outlined here was a step in the right direction and that even higher values could be eventually materialized.

Commercial highly oriented polyamide and polyester fibers have tensile strength up to 11 g/den with elongation to break around 20%. Theoretical considerations indicated that considerable strength increase could be expected if one would build up rigid polyamide chains instead of the present flexible ones. From the synthetic point of view, this calls for the replacement of the aliphatic base monomers by corresponding aromatic units. Initial primary successes in the USA were soon followed by similar efforts in all fiber producing countries (Europe, Japan, USSR) and led to a new generation of fibers with tensile strength between 20 and 30 g per denier and elongation to break between 6 and 10%. Frequently, these fibers are used to reinforce thermoplastic and thermosetting resins; their mechanical properties in context with this need are best presented as a modulus of rigidity up to 15 million psi and a tensile strength around 500,000 psi. Again, the limit has been approached but not yet reached.

Since about 15 years, rigid and strong fibers are being produced by the carbonization and graphitization of organic filamentous precursors; they have reached moduli around 100 million psi and tensile strength above 500,000 psi which is not far from the theoretical limit. Presently, in all leading countries teams of experimentalists are trying to reach still higher values and to improve the elongation to break by the use of other precursors and by optimization of the variables--time, temperature, tension--during carbonization and graphitization. There have already been reports of 120 million psi and 650,000 psi for modulus and tensile strength respectively.[5]

What would the consequences be if superfibers with extreme mechanical and thermal properties could be made commercially?

Many metal parts in planes, ships, cars, buildings and bridges could be made from light organic polymers (density between 1.0 and 1.5) instead of from much heavier metals (densities between 3.0 and 7.5); cables for elevators, cranes and even for bridges would be lighter and still stronger. Already now the disk brakes in the SST Concorde in Soviet S,S jets and even in the 747's and DC 10's contain carbon fiber reinforced systems. Altogether the use of these materials instead of metals would amount to a very significant saving in weight and in energy.

Organic polymers are intrinsically electrical <u>insulators</u> but by appropriate design of the chains, such as branching, cyclization, stereo-regulation and by the introduction of specific groups electronically <u>conducting</u> polymers have been made. Their synthesis and application is still in its infancy but there exist already valuable experimental guidelines and promising theoretical hypotheses. No basic law rules out the existence of long chains with electron donating and accepting groups, densely packed and eventually crosslinked which could form strong wires with the specific gravity around 1.3 and the electrical conductivity of copper--or even higher. Imagine what it would mean for the distribution of electric power if we would have wires which are ten times as conducting as copper and, at the same time, four times lighter!

<u>Semiconducting</u> fibers and films have been prepared in the laboratory; they cannot yet rival with the classical inorganic species but high level, systematic efforts are expended in the USA

and USSR to improve their performance so that they may be applied in transistors, rectifiers, solar cells and other advanced electronic systems. There is a flurry of hopes that even superconductive organic polymers with a relatively high transition point, say 30-40°K could be prepared. In fact, some of them become superconductive in the low temperature range (around 4°K) like many metals and alloys. All these inorganic systems lose their superconductivity in the neighborhood of 22°K. If it were possible to synthesize polymeric superconductors which could operate in the liquid hydrogen or even liquid nitrogen range, enormous progress could be made in the design of generators, motors, transmission cables and strong stable magnetic fields for spectrometers and high energy particle accelerators. The challenge is so great that experiments are justified although theoretical predictions are inconclusive or, even, negative.

Other intriguing electric properties of organic polymers are piezoelectricity, pyroelectricity, photoelectricity and photochromism. Films of polyvinylidene fluoride, properly cast, oriented and annealed are piezo- and pyroelectric; they are presently pioneered in the design of loudspeakers and other mechano-electrical transfer systems. If wound around a hot pipe--in any chemical plant--with a coolant on the other side, they would permanently produce electricity on the basis of the temperature gradient which prevails across the film. Polyvinylcarbazole has long been known to be photoelectric but brittle and, hence, difficultly processable. Progress has been made in the USA, Europe, USSR and Japan by the synthesis of linear segmented copolymers which contain long (DP of 100) stereo-regulated blocks of polyvinylcarbazole and short (DP of 10) blocks of a soft polyacrylate; they are very useful for xerography, holography and dry, grainless photography but the present state-of-the-art is still primitive and much fundamental knowledge has to be collected.

The same thing is true for photochromic polymers. Several aromatic photochromic units can be incorporated into macromolecules by appropriate copolymerization or such active molecules may even only be dissolved in a transparent polymeric matrix without chemical bonding. The formation of color under the influence of light is rapid, but, for the time being, the rate of discoloration in the dark is still slow and the addition or incorporation of active color quenchers appears to be necessary to reach systems with practical applicability. If this could be done, many interesting uses are immediately apparent such as in contact lenses, protective glasses, windshields and in windows for shops, offices, schools, etc. In fact, several research teams in Israel, Europe, Japan and the USA are making progress in the desired direction.

ADVANCES IN TECHNOLOGY

The field which attracts the most widespread and intensive international attention is that of man-made fibers. From the point of view of the consumer, fibers--natural and synthetic--may be classified in comfort fibers, safety fibers and industrial fibers.

Comfort fibers are those which we wear on our body: underwear, socks, suits, pants, dresses, coats, etc.

Safety fibers are those with which we live at home, at work, in school and on the move; they are in carpets, curtains, wall and seat covers, draperies, etc.

Industrial or structural fibers are those used as reinforcing elements in composite systems such as in hoses, pipes, tires and in thermoplastic and thermosetting resins to fabricate construction elements of improved stiffness, strength and toughness.

In Japan, novel types of <u>comfort fibers</u> are made by dispersing the emulsion polymer of one material in a viscous solution of another polymer. If one suspends very small polyvinylchloride (PVC) particles, containing some grafted polyvinylalcohol (PVA) in a viscous, aqueous solution of polyvinylalcohol, extrudes this system, coagulates it in an ionic bath and stretches the resulting gel, one obtains a fiber (Cordelan of Kohjin)[6] which consists of very fine semicrystalline, highly oriented PVC fibrils embedded in a crosslinked matrix of PVA. This composite character imparts a series of valuable properties such as adequate strength and extensibility, softness, good moisture regain and dyeing behavior and a high degree of flame retardancy. Another new comfort fiber is produced in Japan (Chinon of Toyobo) from a graft of polyacrylonitrile on a natural protein; it contains fine, highly oriented fibrils of polyacrylonitrile embedded in a crosslinked proteinic matrix and represents a particularly soft, comfortable and dyeable textile material. Systematic work on the grafting of various vinylic and acrylic monomers on cellulosics is now being carried on in the USSR (Rogovin, Textile Research - Moscow) and on proteinic matrices in Australia (CSIRO). In both cases it is intended to reduce the excessive hydrophilicity of the carrier (cellulose or protein) through the introduction of a hydrophobic polymer to a desired degree while, at the same time, maintaining the moisture regain and dye acceptance of the system at a favorable level. Additional dividends resulting from the special choice of the two (or three) components, of the degree of orientation and crosslinking are improved resistance against chemicals, light and ignition. Related efforts to arrive at new and superior comfort fibers are based on the manufacturing of so-called bicomponent fibers in which two (or three) independent polymers are co-spun in such a manner that each filament contains <u>both</u> components in a "side by side" or "core and skin" manner visibly separated but strongly bonded at the interface. Such fibers were pioneered in the USA (Sayelle Orlon, Cantrece Nylon by Dupont) but are now produced in many variations by all major fiber producing companies all over the world.

Interesting work along similar lines is now underway in Australia where systematic studies of wool (CSIRO) Textile Institute) have led to a detailed knowledge of its composite structure. As a consequence, it was decided to try making bicomponent fibers consisting of two different synthetic polypeptides or of a synthetic polypeptide and a natural protein. Benefits coming from the composite structure of all these systems are permanent crimp, bulkiness, abrasion resistance and an adjustable degree of moisture and dye acceptance.

Safety fibers must be strong, tough, durable, abrasion resistant and, most of all, offer maximal safety in the case of fire. To do that, they must be difficult to ignite, have a slow flame spread, a low heat development during burning and, if they finally are in full conflagration, they should develop a minimum of smoke and toxic gases. Elaborate experience in all industrialized countries has shown that all existing comfort fibers—natural and man-made—can be rendered flame retardant by the addition (5-15% by weight) of a variety of substances, usually containing such elements as B, N, Si, P, Cl, Br and Sb. In fact, these flame retardants or flameproofing agents do delay ignition, flame spread and heat development but unfortunately on burning and even on smoldering, they develop copious quantities of smoke and obnoxious (some highly toxic) gases. The objective formulated as a result of this experience was to design a fiber former which, on the basis of its molecular structure, would have flame retardant properties without the addition of a flameproofing agent. Early studies in the USA (Dupont, Monsanto) had shown that aromatic polyamides are resistant against ignition and on burning, form a large proportion of carbonaceous char, thus having slow flame spread and low heat evolution. This lead was taken up by many companies in Europe, Japan and the USSR and there exist today fibers made from several aromatic polyamides, polyimides, polybenzimidazoles and polyoxadiazoles which represent a respectable collection of safety fibers, far superior in every respect to the "flame retarded" comfort fibers of earlier years. One remaining disadvantage of all nitrogen containing safety fibers is their tendency to evolve highly toxic gases, once they are in full conflagration. Recently aromatic polyesters have been developed which approach the corresponding polyamides in flame retardant characteristics but, containing no nitrogen, evolve less toxic gas mixtures, when they burn.

In structural fibers modulus, strength and thermal stability have to be very high with toughness and durability at the same time at adequate level. They are essentially used as reinforcing elements in composite systems such as elastic and rigid pipes, tires, plates, rods and formed parts to be used in the construction of planes, cars, boats and buildings. Presently existing species, such as highly drawn conventional polyamides and polyesters having moduli in the range of 50 to 80 g per denier and around 10 g per denier tensile strength are now gradually replaced by aromatic polyamides and other linear aromatic polymers with moduli up to 1200 g per denier and up to 20 g per denier tensile strength. These materials will permit to design lighter and safer tires and lead to increased use of polymers in vehicles and buildings of all kinds, opening large new markets for synthetics.

One pressing problem of our days is to provide for an adequate supply of water for municipal and industrial use in all countries of the world. This calls for efficient technology to convert sea water into sweet water and to reclaim the waste water of cities and plants. Sedimentation and filtration are the classical methods for the purification of fluids but they remove only relatively coarse dispersed matter. As a consequence, there exist since several years, sustained and intensive efforts in all countries--particularly also in Israel and Australia--to add the more efficient techniques of ultrafiltration and reverse osmosis to the classical procedures. The core of these new techniques are polymeric <u>membranes</u> with selective permeability, high strength, toughness and durability. Originally, they were made from cellulose derivatives, but recently it has been found that rigid chain polymers such as aromatic polyamides, polysulfides and polysulfones allow to cast better membranes and to design more efficient aggregates which have already succeeded to alleviate the traditional severe water shortages in many subtropical and tropical countries, particularly in Asia, Africa and Australia. At the same time, progressing purification of city and plant waste has already removed several serious water pollution cases. In fact, with today's available membranes and filtering and washing devices, the essential elimination of air and water pollution is not any more a problem of technology, but rather of the manner in which the costs should be distributed amongst those interested in the clean-up.

Much progress has been made worldwide in the large field of <u>elastic</u> and <u>foamed</u> polymers with a noticeable comeback of natural rubber from South Asia. The successful use of elastomers in practice depends on the development of characteristics of the most diverse types, all of which, evidently depend on the structure of the individual molecules and on the specific topology of the network. Overriding are modulus, tensile strength and rapid, complete recovery from deformation together with a minimum hysteresis loss. But important are also slow crack growth, adequate adhesive and frictional properties, and for any longer range use, resistance against abrasion and against chemical degradation by reagents, oxygen, light and ionizing radiation. No wonder that a satisfactory compromise of all these properties cannot be accomplished by any single material and that, therefore, a system's approach has to be taken attempting to optimize the preponderantly significant properties for every specific application. Everywhere in the world you find strong participation in these efforts. Polyblends are studied instead of single elastomers including recently discovered hydrocarbon rubbers (polypentenamer) together with chlorinated polyethers, fluoropolymers, silicones, polyurethanes, polyphosphazenes and polysulfazenes. On top of that <u>segmented</u> macromolecules of many kinds are developed consisting of <u>elastomeric</u> blocks having a very low (down to -70°C) glass transition temperature--cis-polydienes, polyalkylacrylates, polyethers and silicones--and of <u>rigid</u> blocks having high (up to 250°C) transition or melting points--polystyrene and its chlorinated derivatives, polyesters, polycarbonates and polyamides. Mixed filler compositions consisting of carbon black, silica or alumina with short inorganic fibers such as filex, fiberfrax, saffil and others appear to offer additional advantages in modulus, strength and abrasion resistance. Capping of elastomeric segments, in the DP range from 50-500, with reactive units (OH, NH_2, COOH, isocyanate or epoxide) should permit the buildup of a more controlled three dimensional network with the elimination of stress accumulating irregularities. The consequences are higher tensile strength, more rapid and complete recovery and reduced hysteresis loss. A new discipline of systematic oligomer chemistry using practically all available reactive endgroups and all conceivable chain-species between them occupies presently a preferred position in the polymer program of the Soviet Academy of Sciences.

Cellular elastic and plastic systems--soft and rigid foams--are studied and developed with the use of the above mentioned principles and are aiming at a wide range of applications in damping layers (carpet backings) and in the insulation against the dissipation of heat and sound avoiding energy losses and improving comfort in vehicles, homes and factories. It appears that <u>any</u> <u>polymer</u> can be foamed with the aid of volatile solvents or of blowing agents as long as it is above its glass point; cell size and cell size distribution are important factors for the behavior of the foam, its modulus, resistance and toughness. Polystyrene--rubber--and polyurethane foams are most widely used; if resistance against ignition and flame spread is important fluoro- or chloro-elastomers are employed. The most advanced technology is the foaming of polyphosphazenes which is being pioneered in the USA and Japan.

Many worldwide polymer efforts are related to the formulation and use of coatings and adhesives. Recent fundamental work on fast polymerization reactions has opened new vistas on the possibility of <u>solventless</u> coatings and adhesives. Adequate monomer blends--usually acrylics or isocyanate capped oligomers--are mixed with thickeners, tackifiers in the case of adhesives, pigments in the case of coatings, stabilizers and other auxiliary products and are polymerized in situ within fractions of a second by heat, radiation or appropriate catalytic systems. Nothing comes off during the entire process, no solvent has to vaporize and no thermal curing is needed. Besides its simplicity, this new technology saves energy and contributes nothing to air and water pollution.

As our fundamental knowledge on polymer structure and properties deepened and, simultaneously, their practical uses widened, it seemed timely to attempt using synthetics in medical science and practice. This is now a new discipline strongly cultivated in all countries of the world and includes the use of certain synthetic polymers for hard and soft implants which replace bones, membranes, arteries, muscles, skin and other parts of our body. The systematic application of synthetics all over the world has actually led to the emergence of a new discipline known as implantology which studies and develops the special characteristics of synthetic polymers which are necessary to make them useful. Preponderant is biocompatibility but, depending on the special use, other properties such as rigidity, strength, abrasion resistance and, most of all, durability are necessary to give a good material for any implant.

During all these efforts it was not ignored that in nature organic polymers do not only provide for the firm and durable receptacles of life but also for the vehicles and carriers of the life functions themselves. Plants are constructed of cellulose, lignin and resins, but their growth and reproduction is controlled by proteins, hemicelluloses, starch and other polysaccharides. Animals are built of bones and of fibrous proteins, but the life functions inside of their bodies are carried on by water soluble or dispersible proteins, nucleic acids, and polysaccharides. Since polymer science has been so successful in producing synthetic analogs for the group of the natural life protecting materials, why should it not be possible to arrive at similar success for the life carrying organic polymers? In any event, it would be worthwhile to try.
In doing so, there would, of course, be other properties and property combinations which would be necessary and important for these new synthetics. Solubility in water, compatibility with other solutes in the liquids of a growing plant or of a living human body, capacity to form reversible or irreversible complexes with certain substrates, degradability and excretion under biological conditions. This approach considers synthetic polymers not as fibers, films, plastics, rubbers, coatings or adhesives but as chemical reagents, catalysts, activators or stabilizers and intends their application together and, eventually, instead of their natural originals.

There exist certain areas in which this concept was already rewarded with notable success in all countries concerned with polymer interests, namely the synthesis and use of polymeric acids and bases in the form of ion exchange beads, membranes and fabrics and the corresponding activities in the field of chelate forming polymers and of polymeric oxidation-reduction systems. In these applications, it was necessary to combine certain physical properties --hardness, abrasion resistance, high softening range--with the essential chemical character--diffusibility, ionic strength and ionic density or chelating capacity--and it has been, indeed, possible to materialize reasonably successful combinations of these properties with a good prospect to arrive at even better results in the not too distant future.

But there are now all over the world, sustained efforts which aim at the use of synthetic organic polymers to improve our understanding of the structure of natural proteins to produce water soluble analogs of them and to provide biochemical, biological and medical research and development with new materials which have already led to improved methods of diagnosis and cure, and will, hopefully, bring other even more important innovations in the not too distant future.

REFERENCES

1. Lectures given at the Australian Polymer Symposium in Canberra; May 1974.
2. Lectures given at the International Textile Conference in Kalinin, U.S.S.R.; May 1974.
3. Compare, f.i., *Polymer Science and Materials*, A.V. Tobolsky and H.F. Mark, Wiley, pp. 231-237, 1971.
4. Compare lectures of A. Peterlin and V. Perepelkin at the International Conference in Kalinin; May 1974.
5. Compare, e.g., H.M. Ezekiel in Applied Polymer Symposia, Vol. 21, Wiley, pp. 153-166, 1973.
6. Compare, e.g., Lecture given by S.M. Atlas at the International Conference in Bad Neuheim, 1974.

A Tribute to Herman Mark, Mr. Polymer Science

Murray Goodman

Department of Chemistry, University of California

San Diego

La Jolla, California 92037

Herman Francis Mark--pioneer of polymer science--is 80 years young. Born in Vienna on May 3, 1895, he received his education there and was awarded his Ph.D. summa cum laude from the University of Vienna in 1921. Dr. Mark became an instructor at the University of Berlin and then a research associate of the Kaiser Wilhelm Institute near Berlin. In 1927, he joined the Central Laboratory of the I. G. Farben Co. in Ludwigshafen.

These were most exciting times for polymer science. Theories were proposed and experiments were undertaken to establish whether covalently bonded long molecular chains could exist. It was in this period that Dr. Mark received his initial and lasting exposure to the complexity of macromolecules. He began by applying the technique of x-ray diffraction to substances such as silk, wool and cellulose. Working with such great figures as K. H. Meyer, H. Hopff and others, Dr. Mark and his associates were able to make the first scientifically valid comments on the structure of natural and synthetic polymers, a group of molecules whose existence was also deduced by H. Staudinger on organochemical grounds.

With the growth of Naziism, Dr. Mark left I. G. Farben in 1932 and accepted the position as Director of the First Chemical Institute of the University of Vienna. It was here in his native Vienna that Herman Mark built an unusual Institute devoted to the study of polymers. He trained many illustrious Austrian and foreign scientists there and developed with E. Guth one of the first important theories on rubber elasticity. Of course, the laboratories were also continuously concerned with the structural analysis of macromolecules.

Once again, Naziism threatened in 1937. After Hitler occupied Austria, Dr. Mark was dismissed as Professor. He, his wife Mimi, and his sons Hans and Peter, left for Canada via a complicated route. He worked at the Canadian International Paper Company in Hawkesbury. Although the work and living conditions in Canada were fine, Herman Mark felt isolated and began looking for greater horizons. In May 1940, he joined the Shellac Bureau and became an Adjunct Professor at the Polytechnic Institute of Brooklyn. Raymond Kirk, head of the Chemistry Department soon realized the fantastic resource he had acquired by this addition to the Shellac Bureau. Dr. Mark was promoted to Full Professor at the Polytechnic Institute in 1942. Throughout the period of World War II, Dr. Mark developed the academic base for the study of polymers and also worked on government projects to help in his adopted country's war effort. In 1945, he became a United States citizen.

After the War, Dr. Mark organized the Polymer Research Institute with the aid of C. G. Overberger and founded the Journal of Polymer Science with the encouragement and support of Eric Proskauer. The Institute and the Journal remain living testimonies to the genius and farsightedness of Herman Mark. Dr. Mark brought many new people and ideas to the study of polymers. As a show of affection, one of his perceptive new colleagues, Isadore Fankuchen, dubbed him the Geheimrat. Never has such a formidable title been altered to fit the man. Usually Geheimrats were formal and distant people. Our Geheimrat is warm, open, and above all, reachable.

The Geheimrat was once asked how he was able to attract so many outstanding polymer scientists to the Polymer Research Institute. His response was typical of his basic approach to the academic profession. "I have always encouraged my associates and colleagues to develop their careers as they saw fit. My role has been to support, stimulate and most of all leave them alone. I continually tried to provide my associates with maximum scope." With a twinkle in his eyes he added, "To those of us who have known the Polytechnic Institute, I certainly could not attract people by promises of sumptuous surroundings." It is difficult to think of polymer scientists the world over who have not had the benefit of interaction with the Polymer Research Institute and Herman Mark - Mr. Polymer Science.

Throughout the years, Dr. Mark has maintained a phenomenal level of productivity. He exhibits "zero energy of activation" necessary to accomplish his tasks and duties. Once he has accepted a lecture or a writing obligation the Geheimrat settles down to digest vast volumes of background material which encompasses the definitive original articles, reviews, books and pertinent patents. With such broad information as a base, he writes his papers, books and reports in a most lucid manner using the shortest imaginable pencils. Dr. Mark has produced approximately 500 original papers and reviews; he has authored, coauthored and edited more than 15 books. His works cover a fantastically wide range of subject matter related to polymer science. In addition to the Journal of Polymer Science, he commenced and edited a series of monographs entitled High Polymers. He was instrumental in creating the journal Biopolymers and many review series including Macromolecular Reviews and Polymer Reviews and continues to serve as chief editor of the Encyclopedia of Polymer Science and Technology.

The medals, awards and honorary degrees accorded to Dr. Mark are so numerous that I am not able to record them in this brief testimony. They come from many countries, organizations and societies. They span more than four decades. In 1928, Dr. Mark was awarded the Heinrich Hertz Medal. This year our Geheimrat will receive the prestigious Willard Gibbs Medal of the Chicago Section of the American Chemical Society.

Herman Mark remains a wonderfully effective teacher and communicator of scientific information. Throughout the years, all sorts of people have come to him for guidance and advice. I am reminded of my first year at the Polytechnic. My laboratory and office were in the Polymer Research Institute on Willoughby Street, approximately a five-street walk from the original Polytechnic buildings on Livingston Street. Professor Mark's office was located there in the South Building. One morning I received a telephone call from the Geheimrat asking me to show two visitors through the laboratories of the Polymer Institute. I fetched them immediately. On our walk from Livingston to Willoughby Street, I discovered to my horror that these gentlemen were not chemists but rather French architects commissioned to plan the Institute Plastique in France. They wanted to pattern their design on the "world famous" Polymer Research Institute at the Polytechnic Institute of Brooklyn. Those of you who remember our laboratories on Willoughby Street recognize the sheer ludicrousness of this situation. When I next saw Dr. Mark, I plaintively asked why he had not alerted me to the fact that the French gentlemen were architects and not chemists. He answered, "Think how spontaneous your answers were to their questions about the Polymer Institute. Had I warned you, even _you_, my dear Murray, would have been struck dumb."

At the age of 80, Herman Mark remains at the forefront of polymer science. He is concerned with applications of polymer chemistry to the needs of modern society. He continually discusses such problems as flammability tests, the medical applications of polymers, and the design of new polymers. It is clear that Dr. Mark is looking ahead. He has demonstrated that polymer science has made many valuable contributions to benefit human beings. We owe our Geheimrat a great debt of gratitude for his work and inspiration.

PART ONE

Interfacial Phenomena and Adhesion

Introductory Remarks: How to Solve the Problem of Adhesion--a Prescription

Robert J. Good

Department of Chemical Engineering

State University of New York at Buffalo

Buffalo, New York   14214

    The symposium in which we are participating is a multidisciplinary one. The symposium was set up in this mode of presentation, in response to the needs of the problem. I would like to offer, at this time, some multidisciplinary remarks as to how we can go about solving the general problem of adhesion. There are, I believe, three components or steps which must be taken, and one of these steps must be taken simultaneously with the other two.

    As a preliminary, we must recognize some basic facts about the problem. First, it is a field of applied science. Second, it is a field where specific disciplines such as polymer science, surface science, mechanics and physical chemistry meet, and where they may or may not overlap. It is safe to say that the solution of the general problem of adhesion will never be found in any <u>one</u> of the existing disciplines of science or engineering.

    The first component of the solution is, of course, to mobilize investigators from all the relevant disciplines. (Normally, this is done separately. That is to say, you don't go to a synthetic polymer chemist and talk to him about the stress tensor, nor do you open the conversation with a theoretical mechanologist by talking about radiation-grafting of polymers.) The practitioners of the various disciplines learn of the existence of a problem in various ways, which I will not go into here. Suffice it that they learn about the intellectual fascination of the problem, and about its technological and economic importance. A scientist or engineer who has been educated in some discipline, say, mechanics, or surface chemistry, responds by applying the methods of his own discipline to the problem.

The strategy that has been collectively adopted, at this point, may be described as abstraction, or isolation of variables. To call it a "strategy" is really too strong, for that implies a strategist who is planning the whole attack. With a multidisciplinary problem, there is almost never a master strategist; and there certainly is no such person at the time a multidisciplinary attack on an arbitrary problem <u>begins</u>. The effect is, however, as if a rather capricious master-strategist had divided up the problem. It is as if he had said to some polymer chemists: Solve the whole problem, by polymer-chemical methods; and to some surface chemists: Solve the whole problem by surface-chemical methods; and to stress analysts: Analyze the stresses; and to the rheologists: Solve the problem by rheological methods; and so forth.

After some time it gradually dawns on each of the investigators that his own discipline may not be adequate for the <u>complete</u> solution. A surface chemist becomes aware that he needs the results that a stress analyst can give him, as well as the techniques which a polymer chemist can give him. He may learn this fact with no aid but his own intellectual brilliance; but it is far, far more common that he gets hints of such facts by talking to other scientists and engineers who are working in other fields--whether or not they are working on the adhesion problem.

This brings me to my point regarding the first of the three components or steps towards solving the adhesion problem. What we are doing in this interdisciplinary symposium is just exactly what I have prescribed: The reporting of results that have been obtained by the methods of the various disciplines, the comparison of those results, and the stimulation of the members of each separate discipline to look at the others, and to see how they complement his own work.

Thus, Session I, which is on Interfacial Phenomena in Adhesion, offers a set of papers by workers who have one (or two) out of the large set of disciplines just referred to, as their principal fields of work: Surface chemistry, or polymer chemistry, or both. These scientists, in their work on adhesion, represent the melding to the disciplines, and this is a first step towards melding all of the needed disciplines for the purpose of solving this problem.

The second step, which must eventually be taken, might be referred to as a "holistic" approach: The study of the problem as a whole, rather than as the sum of a lot of sub-problems, each one in a separate discipline. This means, eventually, creation of a separate discipline: Adhesion Science. There has been talk, over the past 10 or 20 years, about how this step can be taken; and so it might be asked why has this not been accomplished already? Surely the intellectual needs of the problem are great enough!

There must be a number of answers to this question; but first and foremost, we can conclude that, despite the evidence given by the attendance at this Symposium and the one at the Washington, D.C., ACS meeting 3 years ago, and the existence of two journals, the total needs of the problem are not "great enough". If they were, Adhesion would now be competing with problems such as Cancer or Energy, for the best applied scientists in the world.

Now the level of expertise required for advanced research in adhesion is so high, that the practitioners must be at least close to the front ranks of their own disciplines. And scientists who have that kind of talent are seldom at a loss for problems which have challenged them to on-going efforts which are near the limits of their powers, entirely within their own disciplines. They are not likely to be driven by boredom to do research on adhesion.

A second hindrance to the creation of the discipline of Adhesion Science is, that we do not as yet know for sure what the entire problem is. We don't know whether the list of disciplines which we must enroll is complete. And we don't know whether disciplines such as continuum mechanics and synthetic polymer chemistry can be melded together at all.

This brings me to the third step towards the solution of our general problem, a step which can be carried out concurrently with either of the first two. In fact, it is an essential component of the process of going from the first to the second of the two steps discussed above. It involves activating a mechanism by which enlistment in the task force studying adhesion can be greatly improved in quantity and quality. This mechanism is needed because there is an important class of brilliant scientists and engineers who are currently showing a very noteworthy apathy towards the problems of adhesion. These are found among the research scientists in academia.

We may recall that chemistry, as a discipline, grew out of the medieval and renaissance technologies, such as the recovery of metals from ores, and the making of glass, of dyes, and of medicines, and out of the economic motivation of wanting to turn base metal into gold. It is noteworthy that there were real financial rewards for the research workers and engineers, who devoted themselves to these studies. The chemical industry in the US and Europe, from the 1920's to the 50's, endowed many chairs of chemistry, and gave fellowships for chemistry students. Biology, as a discipline, grew out of the needs of medicine, surgery and the pharmaceutical industry. Adhesion Science can become an independent academic discipline only if it is supported, directly and massively, by the industry which it feeds.

When I said, above, that Adhesion is an _applied_ science, I meant that in distinction from the so-called pure sciences. The research workers in pure science have a lot of intellectual freedom, and can choose from a vast number of problems to work on. They can choose how to attack the problems, and can set their own pace. They can carry the problems to completion; this last is very important. They can publish without restraint. (An academic scientist finds this totally essential to his career; he gets no Brownie points in his Dean's books for unpublished reports to a corporation.) Pure research wins Nobel prizes. So there is every motivation for an academic scientist or engineer to do his research in "pure science". And since adhesion is not regarded as part of pure science, there is a lot of _subtle_ pressure on academic scientists and engineers, _not_ to make their careers in the field of adhesion. So, what can be done to counter these subtle pressures? There is a not-so-subtle counter-pressure that can be employed. Let me illustrate.

Recently, a new PhD chemical engineer, interviewing for a teaching job, was asked, "What do you plan to do for your research?" His answer was, "That depends on which project I can get financial support for." If he had been offered a major research grant in adhesion, he would have considered it very seriously.

I could cite a number of other brilliant young academic scientists and engineers, whom I know of personally (and I am sure there are many thousands that I haven't met) who are looking for a field or a problem to make their life work. I was in that situation once myself. It is not venality on their part, not seduction by industry, nor corruption, nor even charity, to offer them money for summer salary, and to support graduate students and buy equipment. A young scientist or engineer, if he is to achieve tenure and promotion in a major university, must start producing research without delay--and he is hard put to do so, without support. And for that matter, many senior academics, of reputation so high that it would surprise you, are finding their former sources of funds drying up --and are looking for new fields. Or if their research (e.g. on adhesion) goes unsupported for very much longer, they will have no choice but to turn to other problems.

The Office of Coal Research has great gobs of money this year, to support research on coal. Congress, last year, appropriated a large chunk of "add-on" funds for the NSF, to be earmarked for energy research. The NIH has been told to devote a lot of new money, which Congress happily provided, for the "war on cancer".

Of course, the devoting of money to the solution of a problem is not a _sufficient_ condition for solving the problem. But it comes very, very close to being a _necessary_ condition---particularly in

<u>applied</u> science. If the industries that are concerned with adhesion, and the Government, do not start supporting research on adhesion, we can be 99.44% sure that the real problems of adhesion will not soon be solved. And if the industrial corporations that are concerned with adhesion and adhesives wait for the Government to support such research, in the massive way that is needed, then we can be sure that the <u>serious</u> attack on the problem of adhesion will not come until after the problems of energy, pollution, cancer, arthritis, and the common cold have all been solved.

The Conformation of Adsorbed Polystyrene

W. H. Grant, B. W. Morrissey, and R. R. Stromberg

Institute for Materials Research

National Bureau of Standards

Washington, D.C.   20234

    The properties of synthetic polymers at the solid-solution interface are relevant to the fields of adhesion, composites, coatings, and flocculation. The description of both the microscopic and macroscopic systems requires the determination of both the conformation and changes in the conformation of the polymer molecule upon interaction with a surface. A number of theoretical and experimental techniques have been employed to characterize adsorbed polystyrene and to test the theoretical models describing the system. These include investigations of: 1) the extension of the adsorbed molecule normal to the surface by ellipsometry and viscosity studies, 2) the distribution of segments of the adsorbed molecule normal to the surface by attenuated total reflection, 3) the fraction of segments of the polymer chain attached to the surface using infrared difference spectroscopy, and 4) the rates of polymer adsorption and desorption using radiotracer techniques. In general, the conformation of adsorbed polystyrene is dependent upon competition among polymer molecules for the available surface sites. At low surface concentration, the adsorbed polymer molecules show a relatively low extension and flat conformation. As the adsorption time continues, or the surface concentration increases, the molecular conformation becomes more extended. Rates of adsorption and desorption provide additional information on the adsorbed molecular conformation. The rate of adsorption is dependent on the molecular weight, solvent, and substrate. For a given molecular weight, the rate of desorption is dependent on the initial adsorbance.

INTRODUCTION

The properties of polymers adsorbed at the solid-solution interface are relevant to a wide range of technical problems. For example, the addition of polymers to colloidal suspensions can significantly modify their stability. The formation of interparticle bridges by the adsorbed polymer can result in flocculation[1] important in water purification[2], while the addition and adsorption of large amounts of polymer can stabilize a suspension[3] and provide lubrication capacity.[4] The application of polymer films for microencapsulation and polymer adhesives similarly depends on both the amount and conformation of adsorbed polymer. The importance of the conformation of the adsorbed polymer in these effects has been reviewed by a number of investigators.[5,6]

Investigators in the field of adhesion have long speculated on the relationship between polymer adsorption and adhesion. The difficulty in establishing a direct relation between adhesion and polymer properties at interfaces probably results, in part, from the greatly different conditions under which the problems usually are studied[7], i.e., dilute solutions for polymer conformation and concentrated solution or bulk for adhesion. An indication of the importance of polymer adsorption in adhesion is shown by studies of polystyrene and poly(methyl methacrylate) composites with glass fibers.[8] The correlation between enhanced adhesion and increases in T_g implies that an increase in order due to adsorption has occurred. Thermodynamic models of polymer-filler interactions[9] relate the observed effects of thermal expansion, heat capacity, etc. to the change in conformation of adsorbed polymer compared to the same polymer in solution or in bulk. Finally, recent studies[10] have probed the thickness of the failure zone following a peel test utilizing infrared internal reflection spectroscopy in an attempt to correlate the microscopic and macroscopic parameters describing adhesion.

In this paper, we shall review the theory and experimental evidence pertaining to the conformation of adsorbed polystyrene.

THEORETICAL MODEL

In general, an adsorbed polymer is attached to the surface at a number of locations along the chain. Adsorbed segments can occur singly or in runs with attached portions separated by loops of unattached segments which extend away from the surface into the solution. The conformation of the adsorbed polymer molecule is a function of the number and arrangement of the attached segments, and the size and distribution of the loops and tails in the solution. The distribution of segments normal to the surface, the number of attached segments, and the average extension of the loops and tails provide measures of the adsorbed polymer conformation.

Statistical-mechanical treatments of polymer adsorption at a planar surface have been pursued extensively[11] using both analytical and Monte Carlo techniques. These procedures place polymer chain configurations in a one-to-one correspondence with random walk configurations on a lattice. While the analytical methods are limited to massless segments, and the Monte Carlo techniques are restricted to relatively short chains because of computational limitations, both provide results capable of experimental verification. The restriction to dilute solutions and non-interacting adsorbed molecules has been circumvented in recent theoretical treatments[12] of concentrated polymer solutions.

Using these statistical mechanical theories, predictions of the segment distribution for both loops and tails, the bound fraction, and the average number of loops for a polymer adsorbed on a single plate, have been made as a function of degree of polymerization N and segment-surface interaction energy. These studies show, at a given temperature, that the polymer segment density distribution depends critically on the surface interaction energy. For energies less than the transition energy (that energy resulting in an average of one attached segment per molecule), the distribution goes through a maximum and results in adsorbed molecular dimensions with an $N^{0.5}$ dependence. For energies greater than the transition value, the distribution decreases exponentially. At higher energies, most segments occur very close to the surface resulting in dimensions essentially independent of N. A decomposition of the total distribution into that arising for loops and tails shows that segments in the unadsorbed tails dominate the distribution at low interaction energies, while loops determine the distribution at higher energies. The adsorbance isotherms for many polymers indicate a high affinity for the surface, and probably a large attractive segment energy. In these real systems, the extension of the adsorbed segments could nevertheless depend on $N^{0.5}$, if there are sufficient interactions between the adsorbed molecules.

The fraction of segments attached to the surface (bound fraction) increases monotonically for all molecular weights as the surface interaction energy increases. For energies above the transition point, the average number of surface contacts is proportional to N and the bound fraction approaches a constant, large value. The constant bound fraction results in adsorbed polymer dimensions which are independent of N, with the molecule lying close to the surface.

At the transition point, the bound fraction varies as $N^{-0.5}$ with the actual number of contacts depending on $N^{0.5}$. To the left of the transition energy, the number of contacts is constant so the bound fraction approaches zero for large N.

Particularly relevant to the study of adhesion and flocculation are the theoretical results for an adsorbing polymer between two parallel plates. This problem has been attacked analytically by DiMarzio and Rubin[11], and numerically with Monte Carlo techniques for the excluded volume case by McCrackin[13]. In addition to a determination of the segment distribution and bound fraction, another important parameter, the number of polymer bridges between the plates, can be calculated. For a given separation between the plates, the average number of bridges per segment exhibits a maximum at the transition point as the interaction energy is varied. These bridges can sustain a tension resulting in a force between the plates which is repulsive in the desorption region and attractive in the adsorption region. The tension passes through a maximum as a function of the segment-surface interaction energy since, while the number of segments per bridge decreases at higher energies making the bridge more taut, the number of bridges decreases rapidly at high energies and dominates the effect. In the analytical development, the tension always lead to a collapse of the system and a zero plate separation. The Monte Carlo calculations, which include the excluded volume effect, also predict an attractive force at energies greater than the transition energy, no force at the transition point, and a repulsive force at lower energies which tend to push the plates apart. In this case, however, the plates do not collapse to a zero separation distance in the attractive region.

The experimental test of the theoretical models describing polymer adsorption has been slow in developing. While the situation is far from satisfactory, a number of techniques, including ellipsometry, infrared spectroscopy, viscosity, attenuated total reflection, and radiotracer rate studies have provided information on the adsorbed conformation and conformational changes. We will summarize the information provided by these techniques for polystyrene.

ELLIPSOMETRY

The measurement of the average extension of the loops and tails into the solution provides important information on the adsorbed polymer conformation. Ellipsometry, an <u>in situ</u> optical technique[14] measuring changes in the state of polarization of light reflected from a surface, can be used to characterize the adsorbing surface and a thin film overlying that surface. These measurements enable the simultaneous determination of both the thickness and refractive index of the film. Utilizing the refractive index increment for the polymer, these independent quantities can also yield the polymer concentration and adsorbance per unit area in the adsorbed film. A wide variety of materials, both metallic and dielectric, may be studied if the surface or interface is well defined and smooth enough to give specular reflection, and the system remains stable under the experimental conditions.

From the change in optical constants of the film covered surface compared to the bare surface, one can calculate[15] from first principles the thickness and refractive index of the film using the equations derived by Drude[16]. The sensitivity of the measurement is dependent primarily on the differences in refractive index of the solution and film, and the film and substrate.

The model upon which Drude equations are based is that of a homogeneous film of constant refractive index with discrete boundaries. However, the adsorbed polymer film, consisting of a mixture of polymer and solvent is not homogeneous, so it deviates from this model. The polymer concentration in the film, and hence the refractive index, would be expected to vary with distance from the surface, probably decreasing exponentially. The use of the Drude equations will result, therefore, in an "average" thickness value which will be greater than the mean square thickness. It is necessary to relate this calculated average for a homogeneous film model to the inhomogeneous adsorbed polymer film[17] to deduce molecular parameters.

Ellipsometry has been used to study the extension of the polystyrene molecule adsorbed near the theta temperature onto a number of metallic surfaces[18,19], including mercury[20], under similar experimental conditions. The experimental studies on polystyrene, except for the adsorption on the mercury surface, show an increase in the extension of the adsorbed film and adsorbance during the adsorption time period. These results on the solid metallic surface were interpreted to indicate that a polymer molecule is initially adsorbed in a rather flat conformational state. As more polystyrene molecules are adsorbed, some of the previously adsorbed molecules undergo a rearrangement and desorb some of their attached segments resulting in more extended loops and an increased thickness of the adsorbed film until an equilibrium is established[19,23]. As mentioned above, the polymer on the mercury surface behaved quite differently. The extension of the adsorbed polystyrene film remained constant during the adsorption time period on the liquid mercury substrate; during this time, the adsorbed polymer molecule seemed to attain its stable conformation relatively early in the adsorption period regardless of additional adsorption. This behavior seems to suggest that the initial conformation is relatively flat and that the initially adsorbed molecules do not become more extended as the surface population increases. The adsorbance, however, was similar on the mercury and solid surfaces.

The film thickness was also found to increase with increasing solution concentration, until an equilibrium or a plateau value was attained.[19] The interpretation of these results is nearly identical to that given for the change in the extension with time. The adsorbed polymer molecules are probably lying close to the surface with many attachments at low solution concentrations. At high solution concentrations, there is increased competition for adsorption sites resulting in fewer attachments and more loops extending into the solution.

The thickness of the adsorbed polymer layer at maximum adsorbance was found to be proportional to the square root of the molecular weight[19]. DiMarzio and McCrackin[22], from their one-dimensional study of adsorption under conditions of independent adsorption sites, predicted an increase in the extension from the surface that is proportional to the square root of the molecular weight.

From these ellipsometric results, it is evident that the conformation of an adsorbed molecule is dependent upon the competition for available sites as well as the interaction energy between the polymer molecule and surface.

INFRARED BOUND FRACTION

Another direct experimental measure of the conformation of adsorbed polystyrene is obtained using the infrared bound fraction technique developed by Fontana and Thomas[23]. The interaction of the chromophores of an adsorbed molecule with a surface frequently results in a shift of their characteristic spectral adsorption bands. In the case of polystyrene, the out-of-plane C-C ring vibration shifts from 697 to 701 cm^{-1}. If these bands were well separated and did not overlap, one could immediately utilize the optical density of the shifted band to determine the bound fraction of polystyrene, i.e., the fraction of the chromophore segments per adsorbed molecule directly in contact with the surface. The actual number of attached segments per molecule could then be calculated using the known molecular weight (and, thereby, the total number of segments per molecule). Relatively small values of the bound fraction for adsorbed polymer would, in general, imply few attachments with probably only small distortions of the solution conformation, while larger values would indicate many attachments with the surface and a relatively flat adsorbed conformation.

Usually, there is an overlap of the bands for the free and bound segments necessitating the use of difference techniques. In the original method described by Fontana and Thomas, the polymer is adsorbed from solution on a high surface area powder, the suspension centrifuged, and the infrared difference spectrum recorded for the resulting gel. Signals due to the solvent and unbound segments of the adsorbed polymer are compensated for by adjusting the path-length of the reference cell containing polymer solution of arbitrary concentration. To prevent the introduction of uncertainties in polymer concentration resulting from centrifugation, this method has been modified[24] to permit a direct _in situ_ analysis of the suspension. The optical density of the unshifted band of the adsorbing chromophore[25] obtained from the difference spectrum of polymer suspension and the same polymer solution can then be used to calculate the bound fraction. The extinction coefficient of the adsorbed segments at the frequency maximum of the unbound segments is determined by successive dilutions of the reference if matched cells are used, or by changing the pathlength of the reference cell. To normalize the data to a per molecule basis, the amount of adsorbed polymer is determined by solution depletion methods. This technique is, of course, limited to high surface area adsorbents, which are transparent in the infrared region used for analysis.

Using the modified bound fraction technique, Thies[26] has determined the conformation of polystyrene with $M_n = 1.05 \times 10^5$ adsorbed on silica from solutions in trichloroethylene. These studies were carried out as a function of solution concentration, time of adsorption, and the presence of a second polymer, poly(methyl methacrylate), competing for adsorption sites.

The bound fraction measured along the adsorption isotherm decreases from 0.28 at low concentration to 0.13 at the adsorbance plateau. This trend indicates a conformation more extended normal to the surface for the adsorbed polymer at high surface concentration. Increased competition for the available surface, therefore, decreases the number of attachments per molecule.

The effect of the time of adsorption following the formation of a suspension was determined by agitating the adsorption vessels for 18-185 hours. While the adsorbance did not change significantly for the conditions selected, the bound fraction decreased from 0.24 to 0.13 over this time period. This could be interpreted as a rearrangement toward a more extended conformation possible at very long adsorption times or as a mechanical degradation of the polymer.

Competition for adsorption sites can have a marked effect on the bound fraction of polystyrene as noted by the decreasing bound fraction with increasing adsorbance along the isotherm. Studies of the competitive adsorption of polystyrene and poly(methyl methacrylate)[26] show that the polystyrene adsorbance is depressed by the presence of poly(methyl methacrylate), and that previously adsorbed polystyrene can be displaced from the surface. The simultaneous adsorption of both polymers results in a small decrease in the polystyrene bound fraction at low equilibrium concentrations, while the value of the bound fraction is unaffected at higher concentrations for a given constant adsorbance of poly(methyl methacrylate). Although polystyrene cannot effectively compete with poly(methyl methacrylate) for adsorption sites, it does not, however, undergo drastic structural changes while occupying these sites in the presence of poly(methyl methacrylate). The inability of the styrene segments to effectively compete is supported by studies of the adsorption of copolymers of polystyrene and poly(methyl methacrylate)[27]. The incorporation of only 8 mole percent poly(methyl methacrylate) into polystyrene produces a decrease of 50% in the polystyrene bound fraction.

ATTENUATED TOTAL REFLECTION

Peyser and Stromberg[28] have determined the thickness of an adsorbed film of polystyrene adsorbed on crystalline quartz using attenuated total reflection in the ultra-violet region of the spectrum. In these experiments, polystyrene was adsorbed directly on the surface of a quartz internal reflection element. When light enters the element and reflects between the parallel faces, it penetrates beyond the interface as an envanescent, nonpropagating wave whose amplitude decreases exponentially with distance from the interface. At the wavelength of the polystyrene adsorption bands, approximately 250-270 nm, there is an attenuation of intensity observed which is a function of the wavelength of light, the internal angle of incidence, the adsorbed film thickness, and the indices of refraction of the adsorbing solution, the adsorbed film, and the internal reflection element. Applying the Drude equations[16], and a method of analysis similar to that utilized in ellipsometry[15], to the attenuation as the angle of incidence is varied, it is possible, in principle, to determine the adsorbed polymer segment distribution normal to the surface. However, for the experimental conditions used, only an average thickness and refractive index of the film could be determined. For polystyrene with M_v = 76,000 an average film thickness of 24.0 nm was measured, which agrees well with the ellipsometric thickness of 22.5 nm measured for the same polystyrene fraction.

VISCOSITY MEASUREMENT

Viscosity techniques have been used by a number of investigators to determine the extension from the surface of the adsorbed polymer molecule. Using this technique, the change in the time of flow of a given volume of liquid through a capillary is interpreted as a change in the diameter of the tube due to the adsorption of polymers. Early studies by Eirich and coworkers[29,30], who measured the flow rates of polystyrene solutions through sintered glass discs showed that the layer thickness was linearly dependent on the solution concentration and proportional to the square root of the polymer molecular weight. More recent studies by Priel and Silberberg[31] for flow through capillaries established that, for a given molecular weight, the extension increased with solution concentration. Viscosity measurements give a "hydrodynamic" thickness. The correlation between the hydrodynamic value and the molecular dimensions is unclear. The adsorbed film thicknesses are of the order of polymer dimensions in solution, suggesting that the adsorbed polymer has large loops with a small number of attachments to the surface.

RATES OF ADSORPTION

Competitive rates of adsorption provide information not only on the kinetics of the adsorption process, but also on the adsorbed polymer conformation. The overall rate of adsorption of a polymer molecule is probably comprised of two separate rates; the rate of initial attachment and the rate of reorientation of an attached molecule until it achieves its equilibrium conformation. The rates of desorption and exchange can also yield information on the adsorbed conformation since removal of a molecule from the surface is a function of the number of attached segments.

The rates of desorption and adsorption were studied[32] using polystyrene labeled with radioactive tritium which enables the direct determination of the amount of polymer adsorbed. The desorption results showed slow, but continuing desorption with from 40 to 80% of the initially adsorbed material remaining over periods of time as long as three weeks. These results qualitatively substantiate the conclusions reached from ellipsometry. The rates of desorption decreased with decreasing initial adsorbance values, indicating larger values of the bound fraction as the competition for available sites decreased. To keep the bound fraction constant, the rate of exchange of the adsorbed molecule with a solution of the same molecular weight and concentration was investigated. No difference was observed between the rate of exchange and that for desorption. If there is a large distribution in the value of the bound fraction among the adsorbed molecules, then those with the lowest number of attachments would be expected to desorb first. This is consistent with the observation that the rates of desorption at early times are greater than the rates at later times.

Recently, much more extensive adsorption rate studies[33,34] have been carried out onto chrome from cyclohexane at the theta point and from benzene solutions. The times required to attain equilibrium with cyclohexane as the solvent were approximately independent of solution concentration for the high molecular weight material in the concentration range studied. This indicates molecular rearrangement on the surface as the rate controlling step rather than diffusion to the surface. This concept is consistent with the interpretations obtained from ellipsometry studies.

In adsorption from a good solvent, benzene, a plateau region is attained more rapidly than for cyclohexane, from which it may be inferred that less molecular rearrangement on the surface occurs in the adsorption process. One explanation is that the molecules initially adsorbed are not occupying as many sites as they would in cyclohexane because the lower energy sites on the surface cannot be occupied due to higher polymer solvent interaction. In addition, because of this interaction, fewer molecules are adsorbed requiring less rearrangement of the early arrivals. This is reflected by lower adsorbance values than for adsorption from cyclohexane, as well as the shortened times to equilibrium. Similar solvent effects were shown by others[35] for the adsorption of polystyrene on a graphon surface.

Studies of the adsorption onto liquid mercury show a much more rapid attainment of a plateau which can be attributed in part to the accommodation of the polymer conformation by the liquid surface.

The tagged polymer also provided a means of studying the preferential adsorption of one molecular weight species with respect to another. When a high molecular weight polystyrene is mixed with a relatively low molecular weight material, the results show that, although both polymer fractions can compete with each other for surface sites, the higher molecular weight polymer is preferentially adsorbed. The preferential adsorption at equilibrium of the high molecular weight is consistent with the concept that the higher molecular weight materials would have a larger absolute number of attachments than lower molecular weight materials. Thus, the probability that a molecule remains on a surface or displaces a previously attached molecule increases with increasing molecular weight.

SUMMARY

While the experimental evidence describing the adsorbed conformation of polystyrene generally supports the statistical mechanical models that have been developed, the data are not extensive enough to enable a definitive evaluation of the various existing theories. The experimental data presented here provide a general picture of adsorbed polystyrene conformation. From the ellipsometric studies at a given solution concentration, the extension and adsorbance of the adsorbed molecule increased with time. Infrared bound fraction results showed that the fraction of segments attached to the surface decreased with increasing adsorbance. The viscosity measurements established that the adsorbed film thickness increased with increasing polymer solution concentration. Data on the rates of desorption showed that adsorbed polystyrene desorbed more rapidly from a higher adsorbance indicating a lower bound fraction. In these cases, where conformational changes are observed the equilibrium conformation is primarily determined by competition for surface sites which in turn depends upon the surface concentration.

An experimental determination of the adsorbed polymer segment distribution could provide critical comparisons with existing theories, while the recent theoretical studies on adsorption from concentrated polymer solutions will hopefully lead to experimental polymer conformation studies under these conditions. Additional adsorption studies, especially using concentrated polymer solutions, may well enable the prediction of the adhesive properties of polymers at interfaces and provide information on the mechanism of sorption that will allow selective modifications of these properties. In addition, further studies of adsorbed polymer conformation may also lead to a prediction of the stabilizing and flocculating characteristics of adsorbed polymers.

REFERENCES

1. V.K. LaMer and T.W. Healy, Rev. Pure Appl. Chem. 13, 112 (1963).
2. P.V. Freese and E. Hicks, NTIS No. P13-211 240, U.S. Government Printing Office, Washington, D.C. 20402.
3. G.J. Fleer and J. LyKlema, J. Colloid & Interface Sci. 46, 1 (1974).
4. B.J. Fontana, J. Phys. Chem. 67, 2360 (1963).
5. B. Vincent, Adv. Colloid Interface Sci. 4, 193 (1974).
6. F. Th. Hesselink and J. Th. G. Overbeek, J. Phys. Chem. 75, 2094 (1971).
7. Yu. S. Lipatov and L.M. Sergeeva, *Adsorption of Polymers*, pp. 158, 149, John Wiley and Sons, New York (1974).
8. Yu. S. Lipatov, T.E. Lipatova, Ya. P. Vasilienko and L.M. Sergeyeva, Polym. Sci. USSR 4, 920 (1963).
9. T.K. Kwei, J. Polym. Sci. Part A, 3, 3229 (1965).
10. E.H. Andrews and A.J. Kinloch, Proc. Roy. Soc. Lond. A, 332, 401 (1973).
11. For example, see E.A. DiMarzio and R.J. Rubin, J. Chem. Phys. 55, 4318 (1971).
12. R.J. Roe, J. Chem. Phys. 60, 4192 (1974).
13. F.L. McCrackin, ACS Polymer Preprints 11(2), 1246 (1970).
14. F.L. McCrackin, E. Passaglia, R.R. Stromberg and H.L. Steinberg, J. Res. Nat. Bur. Std. A67, 363 (1963).
15. F.L. McCrackin, "A Fortran Program for Analysis of Ellipsometer Measurements", NBS Technical Note 479, Washington, D.C. 20234.
16. P. Drude, Ann. Physik 272, 532 (1889); 272, 865 (1889); 275 (1890).
17. F.L. McCrackin and J.P. Colson, in *Ellipsometry in the Measurement of Surfaces and Thin Films* (Symposium Proceedings), NBS Misc. Publ. 256, Washington, D.C. 20234, p. 61 (1964).
18. R.R. Stromberg, E. Passaglia and D.J. Tutas, J. Res. Nat. Bur. Std. 67A, 431 (1963).
19. R.R. Stromberg, D.J. Tutas and E. Passaglia, J. Phys. Chem. 69, 3955 (1965).
20. R.R. Stromberg and L.E. Smith, J. Phys. Chem. 71, 2470 (1967).
21. E. Killmann and M. Kuzenko, Angew. Makromol. Chemie. 35, 39 (1974).

22. E.A. DiMarzio and F.L. McCrackin, J. Chem. Phys. $\underline{43}$, 539 (1965).
23. B.J. Fontana and J.R. Thomas, J. Phys. Chem. $\underline{65}$, 480 (1961).
24. C. Thies, P. Peyser, and R. Ullman, *Proceedings of the 4th International Congress on Surface Activity*, Brussels, 1964, Vol. 2, Gordon and Breach, New York, 1967, p. 1041.
25. B.W. Morrissey and R.R. Stromberg, J. Colloid Interface Sci. $\underline{46}$, 152 (1974).
26. C. Thies, J. Phys. Chem. $\underline{70}$, 3783 (1966).
27. J.M. Herd, A.J. Hopkins and G.J. Howard, J. Poly. Sci. Part C, $\underline{34}$, 211 (1971).
28. P. Peyser and R.R. Stromberg, J. Phys. Chem. $\underline{71}$, 2066 (1967).
29. F.W. Rowland and F.R. Eirich, J. Poly. Sci. $\underline{A1}$, 2401 (1966).
30. R. Rowland, E. Rothstein and F.R. Eirich, Ind. Eng. Chem. $\underline{57}$, 46 (1965).
31. Z. Priel and A. Silberberg, ACS Polymer Preprints $\underline{2}$, (2), 1405 (1970).
32. R.R. Stromberg, W.H. Grant and E. Passiglia, J. Res. Nat. Bur. Std. $\underline{68A}$, 391 (1964).
33. W.H. Grant, L.E. Smith and R.R. Stromberg, in press.
34. W.H. Grant, L.E. Smith and R.R. Stromberg, to be published.
35. M.J. Schick, Polymer Letters $\underline{7}$, 495 (1969).

Recent Advances in Wetting and Adhesion

W. A. Zisman

Laboratory for Chemical Physics

Naval Research Laboratory

Washington, D.C. 20375

    A synopsis is given of the controversy about the wetting of pure gold and other noble metals by water and final conclusions are summarized. Adsorption of water as a monolayer or as a thicker film on the surface of high energy solids such as glass, quartz, sapphire, and metals considerably lowers the surface energy of such solids; resulting energy values are independent of the chemical nature of the surfaces. This work makes it evident that extreme care is needed in drying gases and surfaces used in research in surface and colloid science, and in practical applications. Wetting studies of liquid ceramics and high-melting metals and alloys require high temperature studies; the critical surface tension of wetting, γ_c, decreases as the temperature rises, and an outline is given for the correct ways these results should be applied. Research with high polymers has taught that for good adhesion, complete wetting must first be obtained. The results point to new circumstances for preventing permeability to water, and new copolymers exist now which prevent such liquid penetration. Major advances for hot-melt adhesives are indicated.

CONCERNING WETTABILITY OF GOLD AND OTHER METALS

In 1955, Fox, Hare, and Zisman[1] published the generalization that all pure liquids spread spontaneously on high-energy surfaces unless the liquid is (a) "autophobic" or (b) is able to decompose on adsorption to release a more adsorbable product which converts the solid surface into one of "low-surface energy".

In 1964, White[2] published contradictory results on the water wettability of gold, and Fowkes[3] offered the suggestion that, since dispersion forces alone act from noble metal surfaces, the observation that water would spontaneously spread on them meant they were covered by surface oxide or some hydrophilic coating. In 1965 Erb[4], working with a closed distillation system in which a pure stream of water was continuously cycled over a pure gold surface, reported a steady state value of the contact angle of between 55° and 85° after several thousand hours of continuous still operation.

Bewig and Zisman[5] obtained complete wetting of pure gold and platinum by using pure water and by taking great care to eliminate all traces of organic impurities from the atmosphere. In 1966, White and Drobek[6] dropped their claims that gold oxide caused water wetting of gold and stated that presence of inorganic impurities such as alumina polishing agents made the gold surface more hydrophilic. Therefore, they used fine diamond paste suspended in kerosene as their polishing agent and followed that by heating the gold specimen in pure oxygen at 1000°C. By this procedure they obtained water contact angles on gold of 61°. Erb, whose earlier results were used for some theoretical calculations published by Thelen in 1967[7], reported in 1968[8] that his cyclic still method had produced values of from 61° to 65° which were the only reliable contact angles of water on gold. It should be noted here that water contact angles of 61° to 65° are the same as those reported much earlier by Bewig and Zisman[9] for water on gold and platinum surfaces which had been precoated deliberately and carefully with an adsorbed monolayer of n-hexane and also of benzene. They had pointed out in Ref. 5 the likely possibility that organic contaminants may have adsorbed on the gold surface in the earlier experiments of White[2] and Erb[4].

In 1970, a meticulous nuclear activation analysis investigation by Bernett and Zisman[10] demonstrated that pure gold surfaces wetted by pure water always exhibited a contact angle of zero in the absence of any polishing agent or particle contaminant detectable with an X-ray microanalyzer provided that the most scrupulous precautions were taken to prevent adsorption of traces of any organic compounds. Almost simultaneously Schrader[11], using ultra high vacuum methods and extremely pure gold demonstrated independently that gold was wetted by pure water exhibiting a zero contact angle. In 1974, Schrader[12] showed with his high vacuum system and other careful experimental techniques that pure copper and pure silver exhibited zero contact angles with oxygen-free water.

Therefore, the correctness of the early generalizations by Fox, Hare, and Zisman[1] had been fully sustained for gold, platinum, silver, and copper. This sequence of results makes it evident that the theoretical work of Fowkes[3] has not been supported for noble metals and other high-energy metal surfaces. The only open question now is how Fowkes' approach should be revised to accord with the facts established firmly by so many diverse experiments and investigations.

EFFECT OF ADSORBED WATER UPON THE SPREADING OF ORGANIC LIQUIDS ON SODA-LIME AND BOROSILICATE GLASS, QUARTZ, AND SAPPHIRE

Fox, Hare, and Zisman[1] had reported that at 20°C and 50% relative humidity (RH) pure, high-boiling, aliphatic, naphthenic, and aromatic liquid hydrocarbons spread spontaneously with a zero contact angle on polished clean quartz and synthetic sapphire (α-alumina). Such solid surfaces at this temperature and RH under equilibrium conditions are coated with only one monolayer of physically adsorbed water[13]. In marked contrast to the above observations, these organic liquids are not able to spread upon such solids coated with a duplex water film (hundreds of molecules thick) nor can they do so upon the clean surface of bulk water. The latter observation exemplifies the well-known generalization of Harkins and Feldman[14] that the initial spreading coefficients of high-boiling paraffinic and aromatic hydrocarbon liquids are always negative on bulk water at 20°C. Zisman[15] pointed out that all of these facts indicate that as a physically adsorbed film of water on hydrophilic solids like glass becomes more than one or two molecules thick, there must result a significant decrease in the spreading coefficient of each higher liquid alkane (such as hexadecane) and of white mineral oil as well as an accompanying increase in the equilibrium contact angles.

There resulted an investigation of the extent to which water adsorbed on various silicate glass and related surfaces affects the spreading and the equilibrium contact angle of various pure organic liquids[16a,16b]. Of especial interest was the effect of the transition of the physically adsorbed water film from a polymolecular to a monomolecular film.

Wettability in Controlled Atmospheres

Our experimental approach was to observe directly the spreading behavior and the equilibrium contact angle θ at 20°C of various pure organic liquids on a horizontal plane sheet of clean glass coated with water films of various thicknesses under conditions of controlled RH such as to preclude any surface contamination by organic material. The physically adsorbed water layer was varied in thickness from that of a bulk liquid, through that of a "duplex" polymolecular film which exhibited interference colors, and finally to that of an adsorbed monolayer at less than 1% RH. For this purpose an enclosed, compact, readily assembled and disassembled chamber made of Pyrex (borosilicate glass) was constructed so that the interior could be acid-cleaned and thoroughly washed before each experiment; thus, residual vapors or adsorbed films from one experiment could not contaminate clean surfaces introduced for a subsequent experiment. Grease and other organic materials were avoided by using carefully cleaned poly(tetrafluoroethylene) for stop-cock plugs and sleeves to seal the standard taper joints in the gas-purification train which delivered the purified gas to the specimen viewing chamber (Fig 1).

The wall curvature of the viewing chamber did not cause an error in the measurement of the contact angle, for the value of θ so obtained with sym-tetrabromoethane on poly(tetrafluoroethylene) was 79° ± 1°, in good agreement with previously published results[17]. A metered stream of nitrogen gas which had been freed of organic contaminants by passage through a series of long columns of Drierite, cocoanut charcoal, and activated chromatographic alumina was continuously delivered to the viewing chamber. A Teflon stop-cock permitted the purified gas to pass through (or around) a gas-washing bottle containing sufficient distilled water or aqueous solution of calcium chloride to provide the desired RH.

The soda-lime glass was cleaned by first washing with Tide solution, then immersing for 10 minutes at 80°C in an 0.5 wt-% solution of tetrasodium ethylene diamine tetraacetate dihydrate (Na_4EDTA) followed by rinsing with triply distilled water[16a]. Effective degreasing was obtained with the Na_4EDTA; this material has also been shown by the electron micrograph studies of Tichane and Carrier[18] to produce an essentially smooth surface on soda-lime glass when immersed in a hot, dilute solution for a short time. For cleaning the borosilicate glass, the method by Tichane[18] was used; this consisted of mild etching with 1% NaOH solution at 95°C, followed by cleaning with 5% HCl at 50°C, and final rinsing at room temperature with triply distilled water[16b]. Polished fused quartz specimen were cleaned by: immersion in a hot 50-50 vol. % nitric-sulfuric acid bath, rinsing with distilled water, heating three times in freshly distilled water to the boiling point, and final rinsing at 20° to 25°C with triply distilled water[16b]. The single-crystal, fused sapphire specimens were immersed in the hot nitric-sulfuric acid bath for 2 hours and then were rinsed with triply distilled water[16b]. The last step in the preparation of all specimens consisted of drying each for 10 minutes at 120°C in a grease-free oven and then placing it immediately in the controlled humidity chamber for observation of contact angles.

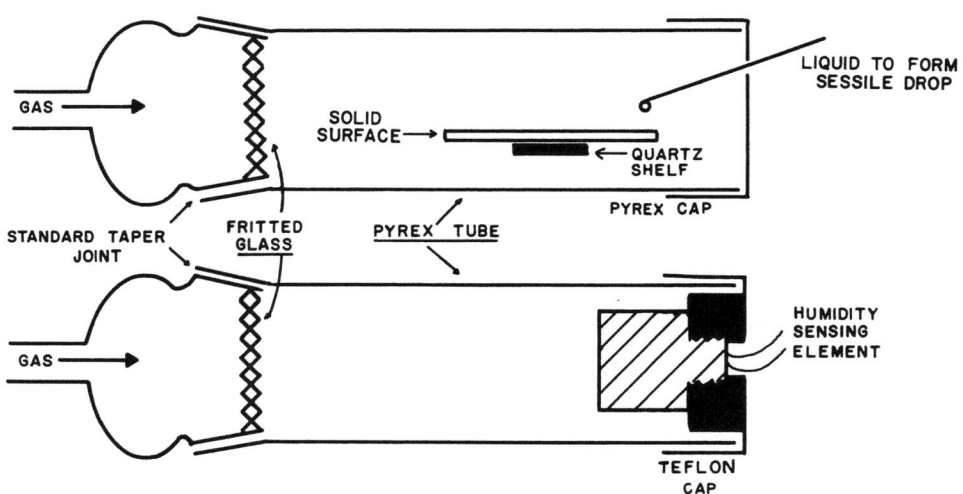

Fig. 1. Cross section of transparent viewing chamber used for measuring contact angles (upper) and monitoring relative humidity (lower).

Table I
Spreading Behavior on Glass of Pure Liquids Having Negative Spreading Coefficients on Bulk Water (All Data 20°C)

Liquid Sessile Drop	Surface Tension (γ_{LV}) (dynes/cm)	Contact Angle (θ) or Spreading Behavior of Drop			
		On Duplex Water Film	On Glass Equilibrated at		
			95% RH	63-53% RH	1% RH
Methylene iodide	50.8	37°	36°	31°	13°
Tetrabromoethane	47.5	31°	36°	9°	9°
1-Methylnaphthalene	38.7	22°	7°	—	<5°
iso-Propyl biphenyl	34.8	19°	16°	—	<5°
Dicyclohexyl	32.8	31°	21°	21°	Spread
p-Octadecyl toluene	31.5	29°	17°	—	<5°
iso-Propyl bicyclohexyl	30.9	33°	13°	—	<5°
Squalane	28.5	24°	Spread	—	Spread
n-Hexadecane	27.6	23°	Spread	Spread	Spread

Table I summarizes the results of our observations on soda-lime glass on the spreading and contact angles exhibited by the hydrocarbons investigated. The liquids are listed in the second column of the table in the order of decreasing surface tension (γ_{LV}). In the remaining columns are given the values of θ observed for each liquid on the indicated surfaces.

Of the nine liquids listed in Table I, the last seven are hydrocarbons of several structural types. When each of these seven hydrocarbons was placed on the surface of a very thin but visible layer (duplex film) of distilled water coating a clean glass slide, a nonspreading lens similar to that observed on bulk water was formed. When a hexadecane drop was placed on a glass slide equilibrated at 1% RH and the ambient RH was subsequently slowly raised, the rate of spreading decreased. When the drop of hexadecane was subjected to a rapidly increasing RH, the initial rate of spreading remained comparable to that observed under equilibrium conditions until the RH had reached about 50%; thereafter the rate of spreading decreased until at 92% RH, no further spreading was observed. When the amount of water present was progressively decreased, the drop remained nonspreading until the water film had evaporated at which time the hexadecane contact angle decreased rapidly, and the liquid spread.

Thus, the transition of hexadecane from spreading to nonspreading occured when the glass surface had become coated by a water film which was thicker than that developed on equilibration at 98.5% RH (i.e., probably not more than a few molecules thick), but definitely thinner than that required for the appearance of interference colors (hundreds of molecules thick).

The remaining hydrocarbons in Table I all exhibited nonzero contact angles on soda-lime glass at low RH and larger values at higher RH. When the thickness of the water layer on the glass increased to that of a duplex film, θ increased further. Conversely, if the water from the duplex layer was allowed to evaporate, θ remained large initially, then decreased as evaporation continued but not to zero, i.e., the drop did not spread indefinitely.

Remarkable alterations in the contact angle of tetrabromoethane on glass were observed when making large and sudden changes in RH by breathing moist air on the sessile drop. A drop of tetrabromoethane having an initial diameter of 8 mm and contact angle of 10° at 55% RH contracted promptly to a diameter of 2 mm when exposed to a sudden rise in RH. A subsequent decrease in RH caused the diameter to expand to nearly its original value. This procedure could be repeated as many as twenty times on the same drop without any detectable change in the results except that the diameter of the drop in its most extended condition decreased slightly with repeated expansions.

The rapid, reproducible, and reversible changes in θ of organic liquids on glass surfaces with changes in RH shown schematically in Fig. 2 are indicative of the mechanism involved. It has been established experimentally that these changes occur on hydrophilic surfaces and not on hydrophobic surfaces, that they cannot arise from alterations in the surface tension of the organic liquid, and that they occur more rapidly than is consistent with a mechanism of alteration of the liquid/glass interface by adsorption of water molecules migrating through the organic liquid. It is concluded, therefore, that the important changes caused by altering the RH take place at the periphery of the drop, i.e., at the line defining the triple interface. After an increase in RH, (a) water molecules from the atmosphere adsorb on the glass surface immediately adjacent to the organic liquid, (b) this water migrates laterally along the glass surface and displaces the organic liquid from a small peripheral area, (c) the newly exposed area adsorbs water molecules from the atmosphere, and (d) the sequence repeats rapidly until an equilibrium is reached. During reduction of the RH the process reverses; it is the loss of water molecules adsorbed just at the edge of the organic liquid which permits the drop to spread.

Since the soda-lime glass behaved as if it had a critical surface tension of wetting (γ_c), a plot of the cosine of the equilibrium contact angle (θ) observed at 95% RH against the surface tension (γ_{LV}) of the organic liquid resulted in a good straight line which fitted all graphical points except 1-methylnaphthalene (Fig. 3B); from the intercept on the line $\cos\theta = 1$, γ_c = 30 dyn/cm. The exceptional behavior of the 1-methylnaphthalene is understandable because it can adsorb with the aromatic ring horizontal in the water surface; this compound would therefore be expected to have some hydrogen-bonding ability. When the system was equilibrated at 1% RH, the data points for the identical liquids (Fig. 3A) were on, or very close to, the line $\cos\theta = 1$, except for methylene iodide. If a straight line is fitted to the graphical points for the halocarbons and is made parallel to the graph in Fig. 3B for 95% RH, the intercept on the line $\cos\theta = 1$ in Fig. 3A is 46 dyn/cm, a fair estimate

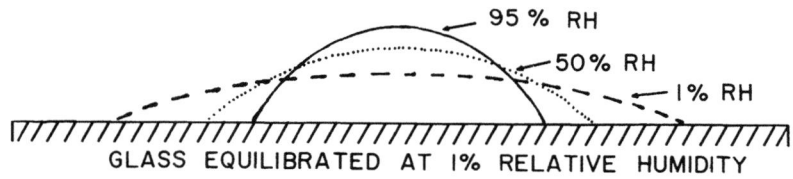

Fig. 2. Changing Ambient Relative Humidity.

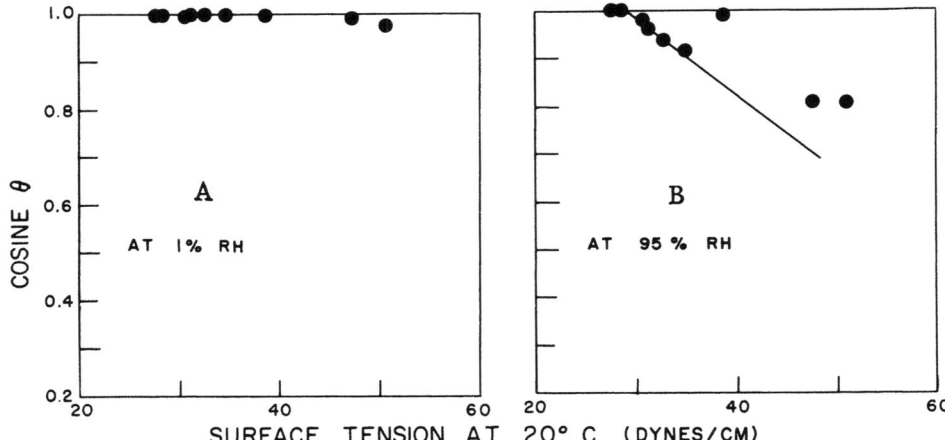

Fig. 3. Comparison of the wettability by selected hydrocarbon and halocarbon liquids of soda-lime glass surfaces at two different RH values.

of the value of γ_c for glass at 95% RH. It is remarkable that the increase in the amount of adsorbed water caused by raising the RH from 1% to 95% caused such a decrease in γ_c, namely, from 46 to 30 dyn/cm.

The closeness of the value of γ_c of 30 dyn/cm for a glass surface at high RH to that of 30 to 31 dyn/cm characteristic of smooth solids contaminated with a condensed film of an aliphatic hydrocarbon could imply that our results arose from contamination of the glass surface by adsorbed aliphatic materials. However, the careful experimental procedures excluded such possibilities[16a].

Soda-lime glass is a hard high-melting solid bound together with strong ionic forces; thus it would be expected to have a high surface energy. Although the critical surface tension of wetting (γ_c) is always less than the free surface energy ($\gamma_s°$), it would likewise be expected that γ_c of dry glass would be high. As is true of most solids, neither the surface tension nor the free surface energy of glass at ordinary temperatures is readily (or reliably) measured. Some of the experimental approaches tried in the past include microhardness measurements, comparisons of tensile strength in air vs in water, and the energy changes involved in crack propagation. The resulting room temperature values of surface energies or tensions have hung around 200 to 400 ergs/cm^2. Values of the surface tension of around 330 dyn/cm at 900°C were obtained by Davis and Bartell[19] using the pendant drop method on solidified drops of glass, and Parikh[20] obtained around 280 dyn/cm at between 500 and 700°C by fiber elongation methods. In the latter measurements on commercial soda-lime glass, the presence of water vapor was effective in lowering the measured surface tension even at such high temperatures. Bryant[21] has reported that the energy required for cleavage of mica in an ultrahigh vacuum or in helium, argon, or nitrogen atmospheres is 30 times greater than in air or water vapor atmospheres, which also demonstrates a high sensitivity of fracture strength and the surface energy to the adsorption of water vapor.

The reversibility of the contact angles and spreading properties of the various hydrocarbon and halocarbon liquids on glass following cycling of the RH indicated that in these experiments there was present an outer layer of physically adsorbed water which was readily removed. The conversion of the glass surface from one of high energy to one of much lower energy was believed to be the result primarily of the adsorption of an initial layer of chemically bound water; additional water adsorbed on top of that surface is physically adsorbed, and it is that layer whose variation in thickness is responsible for the sensitivity to RH observed in the contact angle behavior on glass of the liquids of Table I. The experiments revealed that the spreadability of such liquids on glass is very limited in the presence of water physically adsorbed on the accessible glass surface. Since such limitation at high RH is caused by the presence of one or more condensed monolayers of physically bound water on the surface of the glass, the effect on θ at 95% RH or higher of changing the chemical composition of the underlying glass would be expected to be minor.

Johnson and Dettre[22] and Shafrin and Zisman[23] demonstrated and reported by different methods that the clean surface of bulk water is a low-energy surface with respect to the spreading of non-hydrogen-bonding liquids, such as the alkanes. The latter reported that γ_c = 22.0 dyn/cm at 20°C. This result is to be compared with the conclusion that γ_c = 30 dyn/cm for glass coated with a physically adsorbed condensed film of water at 95% RH.

These data can be related by a plot of γ_c vs RH, where RH = p/p_o (p is the partial vapor pressure of the water in the atmosphere and p_o is the vapor pressure at saturation) (Fig. 4). At p/p_o = 1 the value of γ_c for bulk water[23] is 22.0 dyne/cm. The value of γ_c = 75 dyn/cm obtained by Olsen and Osteraas[24] is used for liquid metals on glass. By assuming that p/p_o for the water vapor is about 10^{-5}, the graph in Fig. 4 fits a straight line very well. Because γ_c for bulk water was about 8 dyn/cm less than for the adsorbed water at 98.5% RH, we concluded that the local field of force emanating from the surface of bulk water is weaker than that from the surface of glass coated with only one or two monolayers of adsorbed water. The surface field of force of dry glass is not suppressed completely by the contributions to that field resulting from the first several monolayers of water adsorbed on it. After many additional monolayers of water have been adsorbed on the glass, the surface field of force gradually ap-

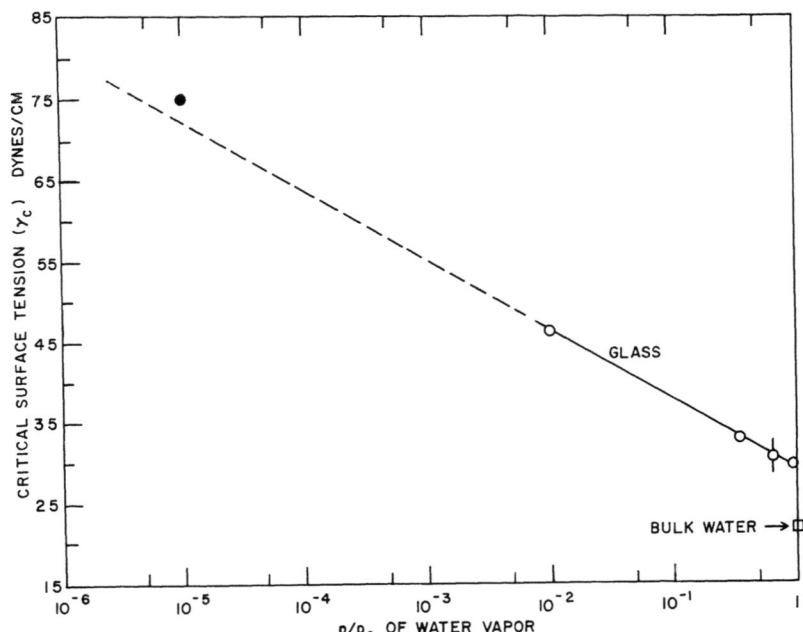

Fig. 4. The effect of the partial pressure of water vapor on the critical surface tension of glass.

proaches that of bulk water. Our presently available data do not allow us to estimate the thickness of the water layer needed to make the surface of glass behave like bulk water.

The other high-energy, hydrophilic, solid surfaces, fused quartz and synthetic α-aluminum oxide[16b], were known to be free from many problems arising from surface chemical reactions with water beyond the formation of a hydrated surface monolayer. Since it was necessary to measure contact angles on surfaces coated with a thin film of water, the liquids used for the sessile drops in these measurements had to be chosen so as to be neither hydrolyzable (such as some esters) nor able to spread spontaneously on water, as would be true of hydrogen-bonding liquids. Hence, the selection of the high-surface tension liquids was limited almost exclusively to halogenated hydrocarbons. The cos θ vs γ_{LV} plot resulted in a short straight line for the polar compounds whose intercept at cos θ = 1 was about 44 dyn/cm (Fig. 5a) for borosilicate glass. The value of γ_c for the hydrocarbons could not be determined at 0.6% RH in this manner, since all of them spread; however γ_c must have been greater than 38.7 dyn/cm, the highest value of γ_{LV} of the hydrocarbon liquids used. Table II reveals that the contact angle of every liquid investigated was larger at 95% RH than at 0.6% RH. In Fig. 5b are graphs of cos θ vs γ_{LV} at 95% RH for all the organic liquids. Three distinct straight lines were obtained: one for the homologous family of the n-alkanes whose intercept was at 26 dyn/cm; another for the various aromatic compounds; and a third for the halogenated n-alkanes. The latter two graphs showed more scattering of the graphical points from the straight line drawn since these compounds were

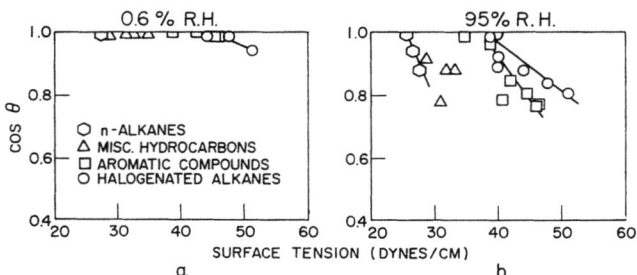

Fig. 5. Comparison of the wettability of borosilicate glass (a) at 0.6% RH and (b) at 95% RH.

not homologs. Although the slopes of the latter were different, the straight line for the halogenated alkanes intercepted the cos θ = 1 axis at 37 dyn/cm, and the line for the aromatics intercepted the axis at 36 dyn/cm.

Wettability data for quartz in Figs. 6a and 6b and for sapphire in Figs. 7a and 7b are very similar to those of pyrex. This investigation demonstrated that: (a) the clean, smooth surfaces of soda-lime glass, borosilicate glass, fused quartz, and fused synthetic sapphire (α-aluminum oxide) had each been converted to a surface of much lower γ_c, and therefore lower surface energy, by the adsorption of water molecules; and (b) the resulting critical surface tensions of wetting were nearly identical at a given relative humidity (RH) for all four types of high-energy, hydrophilic, nonmetallic surfaces. Since all four types of solid surfaces were exposed to the same controlled relative humidity and temperature, the conversion from high to low γ_c (and therefore lower surface energy, $\gamma_S°$) may have been the result of the physical adsorption of only a monolayer of water molecules onto each surface. We can conclude that the critical surface tension as well as the surface energy of clean, high-energy, hydrophilic solids after exposure to a humid atmosphere is dependent upon the surface concentration of water adsorbed on that surface, but that it is independent of the chemical nature of the underlying solid. At 20°C and the low relative humidity of 0.6% there cannot be as much as a close-packed monolayer of water physically adsorbed on each of these types of solid surfaces, yet it drastically decreased γ_c; additional water vapor adsorption in going from 0.6% RH to 95% RH formed little more than one condensed monolayer and

Table II
Wettability of Glasses at 0.6% RH and 95% RH at 20°C
(Specimens predried for 10 minutes at 120°C)

Liquid in sessile drops	Surface tension (dynes/cm) at 20°C	Contact angles (degrees)					
		0.6% RH			95% RH		
		Pyrex	Quartz	Sapphire	Pyrex	Quartz	Sapphire
Hydrocarbons							
α-Methyl naphthalene	38.7	5	<5	<5	14	6	16
Isopropyl biphenyl	34.8	7	5	<5	8	5	<5
Dicyclohexyl	32.8	spr.^a	<5	spr.	28	28	25
p-Octadecyl toluene	31.5	5	9	8	28	16	25
Isopropyl bicyclohexyl	30.9	<5	<5	<5	39	32	35
Squalane	28.6	<5	spr.	<5	22	23	19
n-Hexadecane	27.6	spr.	spr.	spr.	28	28	29
n-Tetradecane	26.7				17	16	17
n-Dodecane	25.4				spr.	<5	<5
n-Decane	23.9				spr.	spr.	spr.
Polar Liquids							
Methylene iodide	50.8	19	22	24	36	37	38
sym-Tetrabromoethane	47.5	8	17	18	33	35	33
Polyphenylthioether MCS 292	46.2				39	37	36
α-Iodonaphthalene	45.9	9	10	10	40	43	43
α-Bromonaphthalene	44.6	9	5	9	36	33	35
1,2,3-Tribromopropane	44.0	8	7	8	28	24	28
o-Dibromobenzene	42.0	<5	5	5	32	28	33
m-Dibromobenzene	40.4				38	33	27
1,3-Diiodopropane	40.0				23	25	18
1,2,3-Tribromobutane	39.9				27	18	24
1,3-Dibromopropane	38.9	spr.	spr.	spr.	10	8	6
1,1,2-Tribromopropane	38.9					11	9
1,2-Dibromo-1,1-dichloroethane	38.2					9	
Hexachloropropylene	38.1					<5	<5
1,2-Dibromoethane	37.8					12	7
Dibromomethane	37.4					10	6
Pentachloropropane	36.8					7	<5
1,2-Dibromopropane	33.0					spr.	
Bromotrichloromethane	30.3					spr.	spr.

^a Spr. = spreads spontaneously, i.e., contact angle is zero.

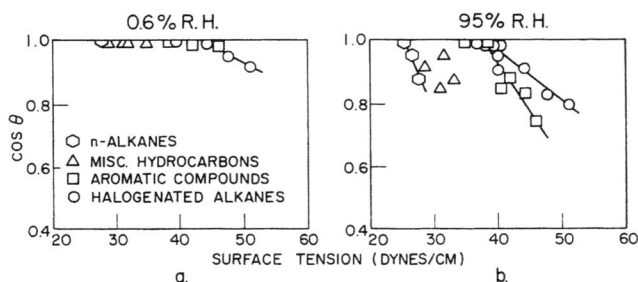

Fig. 6. Comparison of the wettability of quartz (a) at 0.6% RH and (b) at 95% RH.

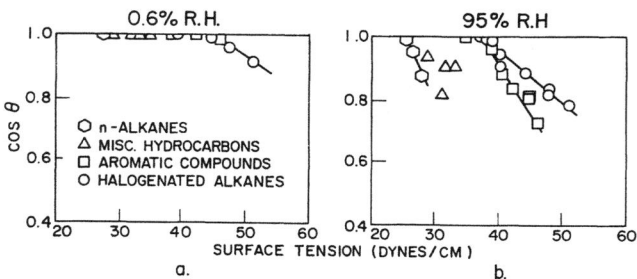

Fig. 7. Comparison of the wettability of sapphire (a) at 0.6% RH and (b) at 95% RH.

Table III
Comparison of Contact Angles at 0.6% RH and 20°C on Glasses after They Had Been Predried at Different Temperatures

Liquid	Surface tension (dynes/cm) at 20°C	Contact angle (degrees)								
		Pyrex dried at			Quartz dried at			Sapphire dried at		
		65°C	120°C	215°C	65°C	120°C	215°C	65°C	120°C	215°C
Hydrocarbons:										
Dicyclohexyl	32.8	spr.[a]	spr.	spr.	<5	<5	spr.	<5	spr.	spr.
Hexadecane	27.6	spr.	spr.	spr.	spr.	spr.	spr.	spr.	spr.	spr.
Polar Liquids:										
Methylene iodide	50.8	19	19	16	23	22	15	24	24	16
sym-Tetrabromoethane	47.5	12	8	10	12	17	9	12	18	12
Polyphenylthioether MCS 292	46.2				9		6	12		8
α-Iodonaphthalene	45.9	9	9	7	8	10	8	10	10	6
α-Bromonaphthalene	44.6	9	9	5	5	5	5	9	9	5
1,2,3-Tribromopropane	44.0	8	8	8	9	7	7	7	8	8

[a] Spr. = spreads spontaneously, i.e., $\theta = 0$.

Table IV
Comparison of Contact Angles at 95% RH and 20°C on Glasses after They Had Been Predried at Different Temperatures

Liquid	Surface tension (dynes/cm) at 20°C	Contact angle (degrees)								
		Pyrex dried at			Quartz dried at			Sapphire dried at		
		65°C	120°C	215°C	65°C	120°C	215°C	65°C	120°C	215°C
Hydrocarbons:										
Dicyclohexyl	32.8	27	28	27	24	28	27	26	25	25
n-Hexadecane	27.6	34	28	32	33	28	27	33	29	28
Polar Liquids:										
Methylene iodide	50.8	38	36	38	38	37	35	38	38	35
sym-Tetrabromoethane	47.5	34	33	32	32	35	28	35	33	32
Polyphenylthioether MCS 292	46.2		39			37			36	36
α-Iodonaphthalene	45.9	44	40	42	42	43	38	42	43	40
α-Bromonaphthalene	44.6	40	36	36	34	33	31	36	35	31

caused a further decrease in γ_c. This further reduction in γ_c was highly significant experimentally, although it was much smaller; thus γ_c at 0.6% RH was 45 dyn/cm and at 95% RH it was 25 dyn/cm.

Effect of Varying Baking Pretreatment Temperature

The results on the Pyrex, quartz, and sapphire surfaces were obtained after exposing each specimen to a drying temperature of 120°C for 10 minutes prior to subjecting it to the humid nitrogen atmosphere at 20°C. These conditions had originally been chosen to assure adequate removal of physically adsorbed water without causing excessive exposure of the surface to organic contamination. It has been known for many years that exposure of various types of glasses to increasing temperatures will remove successively decreasing amounts of adsorbed water[13].

In view of the importance revealed in the literature of effects at high temperature, the specimens were also dried at higher temperatures. Table III compares θ measured in nitrogen at 0.6% RH and 20°C after the specimens had been dried under one of the following conditions: 20 minutes at 65°C, 10 minutes at 120°C, or 10 minutes at 215°C. Table IV compares the wettability of each type of solid surface after the three different baking pretreatments at 95% RH. Since there were no significant differences attributable to any treatment, it can be assumed that any drying temperature from 65° to 215°C is sufficient to drive off all the physically adsorbed water molecules. Thus the contact angles observed at 0.6% and 95% RH necessarily must be caused only by the thin film of water molecules adsorbed at 20°C from the humid nitrogen atmosphere in which the solid had been immersed.

These results explain why many pure hydrocarbon and halogenated hydrocarbon liquids, excluding the autophobic liquids[25], are not able to spread spontaneously on such high-energy surfaces. We have shown the effect on γ_c of minute amounts of adsorbed water to be so great that even at less than 1% RH, high-energy surfaces behave like a low-energy surface toward these liquids. A monolayer of adsorbed water thus exerts an important influence upon the spreading and adhesion of organic liquids on high-energy solids. Unless the solid surface is completely free of adsorbed water molecules, spontaneous spreading of many organic liquids will be impeded, and consequently there will be decreased adhesion of the liquid to the solid. It should also be noted that the frequently used assumption that adsorbed water lowers the surface energy and so assists in crack propagation was given much reinforcement by our conclusions.

In view of the impracticality of attempting to remove all physically adsorbed water from glass, the existence of a critical surface

tension of glass of 30 dyn/cm at ordinary relative humidities and
temperatures is of interest in relation to many technological and
scientific problems. For example, it follows that spontaneous
spreading on glass at 20°C and ordinary relative humidities by any
nonhydrophilic liquid whose surface tension is more than 30 dyn/cm
cannot occur; hence, there is much difficulty in impregnating glass
fibers with such a polymeric hydrocarbon liquid[15].

Comparison with γ_c on Bulk Water

One would expect γ_c for any given solid or liquid surface to
be the same for all series of homologous liquids, providing the
interaction between the liquid and the solid surface arises from
London dispersion forces only. However, if the interaction includes
a significant contribution from nondispersion forces of adhesion,
such as those causing hydrogen bonding, γ_c should be displaced toward higher values, the amount of displacement being a measure of
the contribution of such nondispersion interfacial forces.

Pomerantz, Clinton and Zisman[26] had measured the extent to which
the unsaturated bonds in the aromatic hydrocarbon ring make such
compounds more hydrophilic than the analogous alkanes containing the
same number of carbon atoms; thus the benzene ring had to lie adsorbed
flat on the surface of bulk water and a hydrogen-bonding contribution
resulted from the interaction of the resonating double bonds with the
water molecules. In the investigation of the four high-energy surfaces, a water monolayer was always present on the solid substrate,
and hence the adhesional wetting energies for a homologous series of
aromatic liquids should result in a higher γ_c than those of a series
of n-alkanes. Our experimental results confirmed this conclusion,
as shown in Figs. 5b, 6b and 7b.

Shafrin and Zisman[23] have determined the critical surface tension of spreading on bulk water by plotting the initial spreading
coefficient (S_{ba}) on water of each organic liquid of a given series
vs. its surface tension (γ_{LV}); the intercept at $S_{ba} = 0$ of these
straight lines was then taken to be γ_c for this system. The value
of 21.7 dyn/cm at 20°C was thus calculated from the spreading of
the members of the pure n-alkane series of liquids. A higher value
of γ_c = 25 dyn/cm for water was obtained from the spreading of the
n-alkanes, and a still higher value of γ_c = 30 dyn/cm resulted for
the n-alkyl benzene series, when S_{ba} was plotted vs. the number of
carbon atoms in the alkyl chain.

Fig. 8 shows a plot of S_{ba} vs. γ_{LV} of several series of related
halogenated aromatic liquids on water; the values of S_{ba} are those
from publications of Harkins and co-workers[14] and are listed in Table
V. The intercept at $S_{ba} = 0$ for the halogenated benzenes leads to
a γ_c of 34 dyn/cm. Not enough points are available for the naphtha-

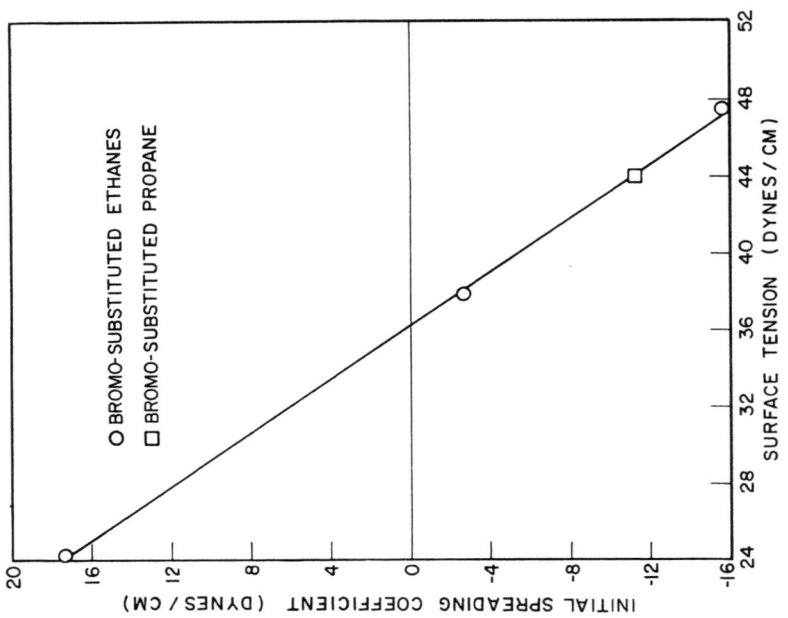

Fig. 9. Initial spreading coefficients of bromo-substituted alkanes vs. their surface tensions at 20°C.

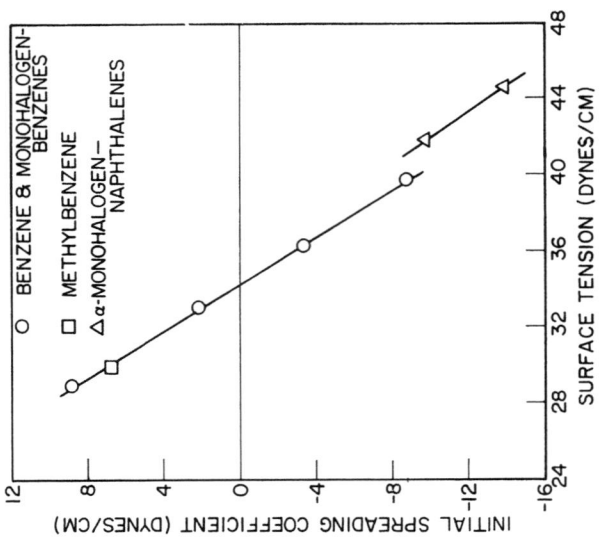

Fig. 8. Initial spreading coefficients of halogenated aromatics vs. their surface tensions at 20°C.

Table V

Surface-Chemical Properties of Aromatics and Brominated n-Alkanes with Respect to Bulk Water (All Data at 20°C)[a]

Liquid	(1) γ_b (dynes/cm)	(2) F_{ba} (dynes/cm)	(3) γ_{ab} (dynes/cm)	(4) W_{ba} (dynes/cm)	(5) σ_b (dynes/cm)	(6) ω_{ba} (dynes/cm)
Benzene	28.9	8.9	35.0	66.7	34.2	3280
Methylbenzene	29.9	6.8	36.1	66.6	38.8	3740
Chlorobenzene	33.1	2.3	37.4	68.6	38.5	3800
Bromobenzene	36.3	-3.3	39.8	69.3	42.0	4190
Iodobenzene	39.7	-8.8	41.9	70.6	45.5	4620
Bromoethane	24.2	17.4	31.2	65.8	21.5	2030
1,2-Dibromoethane	37.2	-2.5	37.5	73.1	28.7	3020
1,2,3-Tribromopropane	44.0	-11.1	39.9	76.9	44.8	4960
1,1,2,2-Tetrabromoethane	47.5	-15.7	41.0	79.3	43.7	4990

[a]Concepts and equations as developed in reference 26:
(1) b = organic liquid.
(2) Film pressure $F_{ba} = S_{ba}$ (Initial spreading coefficient); a = water.
(3) Interfacial tension from $\gamma_{ab} = (\gamma_a - \gamma_b) - F_{ba}$.
(4) Reversible work of adhesion from $W_{ba} = 2\gamma_b + F_{ba}$.
(5) σ_b = area occupied at the organic liquid/water interface by each molecule of organic liquid (from molecular models).
(6) Reversible work of adhesion per adsorbed molecule of organic liquid from

$$\omega_{ba} = \frac{W_{ba} \sigma_b (6.02 \times 10^{23})}{4.18 \times 10^7} \text{cal/gm-mole.}$$

lenes to define a corresponding γ_c; however, indications are that γ_c falls very close to that of the benzenes. Fig. 9 is a plot of S_{ba} vs γ_{LV} of several brominated n-alkanes; a straight line drawn through these points intercepts $S_{ba} = 0$ at γ_c of 36 dyn/cm. If one compares the intercepts in Figs. 8 and 9 at $S_{ba} = 0$, which correspond to the value of γ_c of bulk water for series of liquids of various molecular configurations, with the values of γ_c obtained in our study on moist, high-energy, hydrophilic solids for the same series of organic liquids, we find a good correlation (Table VI). In each case γ_c had the lowest value for the nonhydrogen-bonding, nonionic liquids (the n-alkanes). Also, γ_c for each series of organic liquids had a higher value on the high-energy solid surfaces covered with the condensed monomolecular layer of water adsorbed at 95% RH than on bulk water.

Table VI
Comparison of γ_c for Surfaces of Bulk Water (Calc) and of High Energy Glasses Equilibrated at 95% RH

Series of Liquids	γ_c (dynes/cm)	
	Bulk Water	Glasses at 95% RH
n-Alkanes	22[a]	25·5-26
Halogenated Aromatics	34[b]	36-37
Halogenated n-Alkanes	36[c]	37

[a] Reference 23; [b] Figure 8; [c] Figure 9

Since γ_c relative to water for several halogenated aromatic n-alkane series could be determined, some additional surface chemical properties for these same series were calculated. The concepts of these properties and the appropriate equations had been developed by Pomerantz, Clinton and Zisman[26]. Table V lists the initial film pressure (F_{ba}) (where $F_{ba} = S_{ba}$, "b" denotes the respective organic liquid being spread, and "a" the substrate water), the initial interfacial tension (γ_{ab}), the initial reversible work of adhesion (W_{ba}), the area occupied by each molecule of organic liquid at the organic liquid/water interface (σ_b) (as calculated from the molecular ball models), and the initial reversible work of adhesion per adsorbed molecule of organic liquid (ω_{ba}). When σ_b of the bromoethanes is plotted vs. the number (N) of substituted bromine atoms per molecule, a rectilinear graph is obtained which shows that the value of ω_{ba} contributed by each bromine atom substituted on such a molecule is about 990 cal/gm-mole. Nearly the same value of 910 cal/gm-mole is obtained when a bromine atom is substituted on a benzene molecule (Table V).

Unfortunately, no data were available on homologous families of chloro- or iodo-substituted ethanes which would permit calculation of the contribution of a chlorine or an iodine atom to the reversible work of adhesion. However, since ω_{ba} per bromine atom calculated from the difference for bromobenzene and benzene in Table V of between 910 and 990 cal/gm-mole agreed well with the graphed results from the homologous series of bromoethanes, we can assume that the values of ω_{ba} of about 500 cal/gm-mole for a chlorine atom and about 1350 cal/gm-mole for an iodine atom, as calculated for substituted benzenes (Table V) were good estimates and should agree with results that would be obtained for a series of chloroethanes or a series of iodoethanes.

EFFECT OF ADSORBED WATER ON THE CRITICAL SURFACE TENSION OF WETTING ON METAL SURFACES

Extending the investigation to additional high-energy solids, 12 metals of high purity (chromium, copper, gold, aluminum, iron, nickel, germanium, zirconium, niobium, tantalum, molybdenum, and tungsten) and one metal oxide were also studied.[27] The surface of every metal was carefully prepared to assure cleanliness and freedom from extraneous materials.

In Tables VII and VIII are listed the wetting liquids and the contact angle each liquid formed after the metal had been equilibrated for 2 hours at 0.6% RH and 95% RH, respectively. Contact angles of all liquids on every metal were considerably smaller at 0.6% RH than at 95% RH. Fig. 10 shows plots of cos θ vs. γ_{LV} for

Table VII
Wettability of Metals after a 2-Hour Equilibration at 0.6% RH
(All data obtained at 20°C in nitrogen)

Liquid sessile drop	Surface tension (dynes/cm)	Contact angle (degrees)										
		Cu	Au	Al	Ge	Zr	Nb	Ta	Cr	Mo	W	Fe	Ni
n-Hexadecane	27.6	Spr.	Spr.	Spr.	Spr.	5	6	5	<5	6	5	Spr.	Spr.
Methylene iodide	50.8	27	14	13	10	33	26	41	12	27	22	15	22
Tetrabromoethane	47.5	21	8	8	6	18	15	32	8	18	6	7	10
α-Iodonaphthalene	45.9	10	6	5	5	11	11	15	<5	12	6	6	10
α-Bromonaphthalene	44.6	12	<5	<5	<5	<5	5	8	<5	6	Spr.	<5	7
1,2,3-Tribromopropane	44.0	10	5	7	—	7	6	20	<5	12	5	<5	5
o-Dibromobenzene	42.0	—	<5	—	—	—	<5	—	—	—	—	<5	—

Spr = Spreads spontaneously over specimen surfaces.

Table VIII
Wettability of Metals after a 2-Hour Equilibration at 95% RH
(All data obtained at 20°C in nitrogen)

Liquid sessile drop	Surface tension (dynes/cm)	Contact angle (degrees)										
		Cu	Au	Al	Ge	Zr	Nb	Ta	Cr	Mo	W	Fe	Ni
Hydrocarbons													
α-Methylnaphthalene	38.7	7	<5	<5	22	13	12	15	8	12	10	6	10
Isopropyl biphenyl	34.8	7	Spr	Spr	12	<5	<5	<5	<5	<5	<5	6	9
Dicyclohexyl	32.8	8	Spr	16	15	18	17	14	15	12	16	9	12
Isopropyl bicyclohexyl	30.9	8	—	32	14	22	25	10	—	24	29	10	16
n-Hexadecane	27.6	Spr	Spr	Spr	12	20	19	5	18	10	23	Spr	12
Polar liquids													
Methylene iodide	50.8	32	22	39	37	33	33	44	32	36	30	25	39
Tetrabromoethane	47.5	26	16	35	30	30	28	35	30	27	27	26	34
α-Iodonaphthalene	45.9	30	17	36	37	34	35	33	35	29	29	27	35
α-Bromonaphthalene	44.6	25	18	32	30	30	32	29	31	27	29	22	26
1,2,3-Tribromopropane	44.0	22	12	20	27	25	22	30	19	25	15	17	26
o-Dibromobenzene	42.0	23	11	25	27	28	26	28	29	26	26	16	22
1,2,3-Tribromobutane	39.9	14	—	14	18	15	13	18	14	10	9	12	15

Spr = Spreads spontaneously over specimen surfaces.

each of the nonhydrophilic liquids on chromium observed at 0.6% RH and at 95% RH and 20°C and is representative of all other metals. Just as on the other high-energy surfaces[16a,16b], two distinct straight lines were obtained: one for the various aromatic compounds and another for the halogenated n-alkanes. The slopes of these straight lines were different, but the intercepts at the $\cos \theta = 1$ axis were very close to 38 dyn/cm for the former line and 36 dyn/cm for the latter. Since the hydrocarbon liquids (Table

WETTING AND ADHESION

VIII) were not homologous compounds, their graphical points did not define a straight line even though nonzero contact angles were obtained when γ_{LV} was above 27 dyn/cm.

No elaborate precautions were taken to prevent or control the extent of metal oxidation in the air at room temperature, but much care was taken to assure that a drop of grease-free distilled water would spread spontaneously on each polished specimen surface. Thus, each metal was free of even as little as one hundredth of a monolayer of adsorbed organic contamination when placed in the observation chamber. Since it was impossible to measure the thickness of oxide formed on the metal surface, an oxide of known thickness and structure, a single spinel crystal of pure Fe_3O_4, was studied in the same manner as the pure metals. The results were very much like those of the pure iron specimen; γ_c at 0.6% RH was 45 dyn/cm, and at 95% RH it was 38 dyn/cm (Table IX). These results are strong evidence that the thin, epitaxial oxide layer formed at room temperature on a pure iron surface during the experiments would not exhibit different results if the oxide were allowed to become thicker and assume

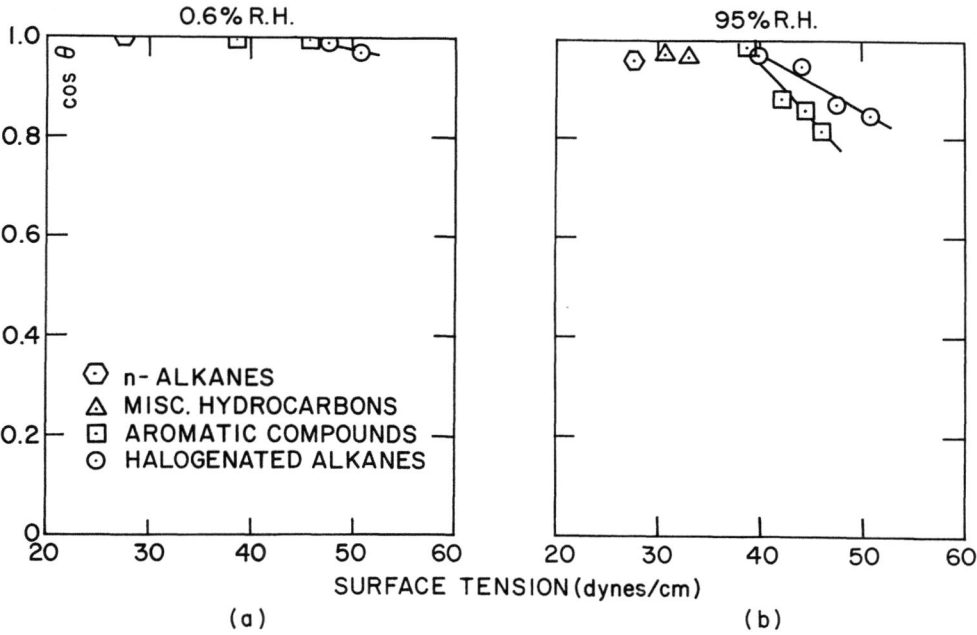

Fig. 10. Comparison of the wettability of chromium (a) at 0.6% RH and (b) at 95% RH.

Table IX

Comparison of Critical Surface Tension of Wetting for Twelve Metals and Fe_3O_4 (All data obtained at 20°C in nitrogen)

Surface	Atomic rad.[a]	Crystal structure	0.6% RH		95% RH					
					n-Alkanes		Aromatic Cpds.		Halogen. Alkanes	
			Slope[b]	γ_c (dynes/cm)	Slope	γ_c (dynes/cm)	Slope	γ_c (dynes/cm)	Slope	γ_c (dynes/cm)
Nickel	1.24	Cubic f.c.	−0.012	45	26	−0.024	38	−0.017	36	
Iron	1.26	Cubic b.c.	−0.007	46	>27	−0.014	37	−0.007	36	
Chromium	1.27	Cubic b.c.	−0.007	45		−0.022	38	−0.012	36	
Copper	1.28	Cubic f.c.	−0.014	44	>27	−0.016	38	−0.010	36	
Germanium	1.37	Diamond	−0.005	46	27	−0.018	35	−0.014	36	
Molybdenum	1.39	Cubic b.c.	−0.018	44	27	−0.016	37	−0.015	38	
Tungsten	1.39	Cubic b.c.	−0.012	45		−0.018	37	−0.012	37	
Aluminum	1.43	Cubic f.c.	−0.005	45	>27	−0.026	38	−0.017	37	
Gold	1.44	Cubic f.c.	−0.006	45	>27	−0.010	40	−0.013	40	
Niobium	1.46	Cubic b.c.	−0.018	44		−0.022	38	−0.013	38	
Tantalum	1.46	Cubic b.c.	−0.026	42	27	−0.022	37	−0.022	37	
Zirconium	1.60	Hexagonal	−0.028	45		−0.018	37	−0.013	37	
Fe_3O_4		Spinel	−0.026	45		−0.024	38	−0.019	38	

[a] R. T. Sanderson, "Chemical Periodicity," pp. 26-28. Reinhold, New York, 1960.
[b] Slope of the graph of cos θ vs. liquid surface tension (γ_{LV}).

a different crystal structure. Thus the thickness of oxide formed appears to be less relevant to the wetting properties of a hydrophilic metal surface than the amount of adsorbed water.

Since the effect of adsorbed water on the surfaces of soda-lime glass[16a], borosilicate glass, fused quartz, α-alumina[16b], and 12 metals and one metal oxide[27] had shown a lowering of γ_c for each of these surfaces to the same values of 45 dyn/cm at 0.6% RH and of 36 to 37 dyn/cm at 95% RH, we propose the generalization that the surface energy of any clean, smooth, hydrophilic surface, whether glass, metal, or metal oxide, after exposure to a humid atmosphere is mainly dependent upon the surface concentration of adsorbed water and is converted from a high to a low surface energy. Although the nature of the solid underlying the adsorbed water layer has little effect on γ_c, it does affect somewhat the slope of the graph of cos θ vs. γ_{LV}.

EFFECT OF TEMPERATURE ON WETTING OF HIGH- AND LOW-ENERGY SOLID SURFACES

The concept of γ_c was developed initially by Zisman and coworkers as an empirical parameter characterizing the wettability of a solid surface at constant temperature. The following presentation on the effect of temperature on γ_c is concerned with research published since 1964 when Zisman discussed that subject[28].

Much effort has been made in trying to extend the γ_c concept to characterize the wetting of inorganic and metallic high-energy solid surfaces by liquids of high surface tension (γ_{LV}) such as salt solutions, liquid nonmetallic and metallic elements, and liquid alloys. Thus, Olsen and Osteraas[24] reported values of γ_c for Pyrex glass, lead glass, fused silica, and soda-lime glasses. When concentrated aqueous solutions of K_2CO_3 or $CaCl_2$ were used, the cos θ vs. γ_{LV} plot revealed that γ_c = 73 dyn/cm at 20° and 20% RH. A similar value of γ_c = 75 ± 10 dyn/cm was obtained when low-melting metals were used at 20 to 156° under vacuum (Fig. 11, curve F). The authors therefore concluded that the values of γ_c obtained were characteristic of tightly bound surface water rather than of glass. Using the surface tension and contact angle data of Good et al[29] for mercury and gallium on baked, outgassed glass, Olsen and Osteraas[24] estimated values of γ_c = 260 dyn/cm from which they inferred that at least some of the adsorbed moisture had been removed from the glass. Olsen, Moravec, and Osteraas[30] used a variety of organic and inorganic liquids at 20° to obtain values of γ_c of 30, 30.5, and 31.5 dyn/cm for orthorhombic, monoclinic, and amorphous sulfur, respectively. Similar measurements on freshly cleaved surfaces of selenium and tellurium led to γ_c values of 32 and 35.5 dyn/cm. Finally, they also showed that γ_c of these elements plotted as a direct rectilinear function of the atomic radii. Eberhart[31] reported a rectilinear plot of cos θ vs. γ_{LV} for various liquid transition metals on sapphire at 1500° under vacuum, resulting in the estimated value of γ_c = 1050 ± 100 dyn/cm at 1500° (Fig. 11, curve A).

A straight-line relation between cos θ and γ_{LV} was reported by Rhee[32] for a series of related metal alloys of graded surface tensions on a high-energy surface. Using literature data of θ and γ_{LV}, he reported γ_c = 440 dyn/cm for sapphire at 1230° (curve B, Fig. 11) and γ_c = 230 dyn/cm for graphite (curve E) using θ data[33] at 600-650° and γ_{LV} values at 700-740°. Manning[33] also provided limited contact angle data for the same metal alloys on two additional high-energy surfaces, leading to γ_c = 340 dyn/cm for boron carbide (curve D) and approximately 350 dyn/cm for beryllium (curve E) for the same temperature range.

A critical surface tension for high-energy surfaces was also obtained by Rhee who employed a convenient and novel approach he had developed. By using contact angles of various liquid metals on solid ceramics[34-38], he determined the temperature dependence of θ for each metal over a limited temperature range generally corresponding to large values of θ. Since plots of cos θ vs. γ_{LV} for each liquid metal on each ceramic surface were found to be rectilinear, Rhee proposed that the intercept of that straight line extrapolated to the cos θ = 1 axis be used to characterize the critical surface tension of that solid substrate. Values obtained by this variable temperature method are designated by the symbol γ_{cs} corresponding to the particular temperature T_{cs} at which cos θ = 1 for a given wetting liquid. These values are thus distinguished from the values obtained isothermally and represented by the established symbol, γ_c.

The difficulties involved in evaluating the variable-temperature method are exemplified by the variation in Rhee's values of γ_{cs} for a single solid, depending on the wetting metal used[38]. Since some of the difficulty arose from the experimental restrictions dictated by the high temperatures involved in the study of high-energy surfaces, Shafrin and Zisman[39] sought to determine whether the γ_{cs} concept was valid for an experimentally more tractable system, namely one based on the wetting of a low-energy surface by an organic liquid. Because the range of temperatures characterizing the melting and boiling of organic liquids is experimentally accessible, no assumptions need be made concerning the nature of the θ vs. T relation near T_{cs}.

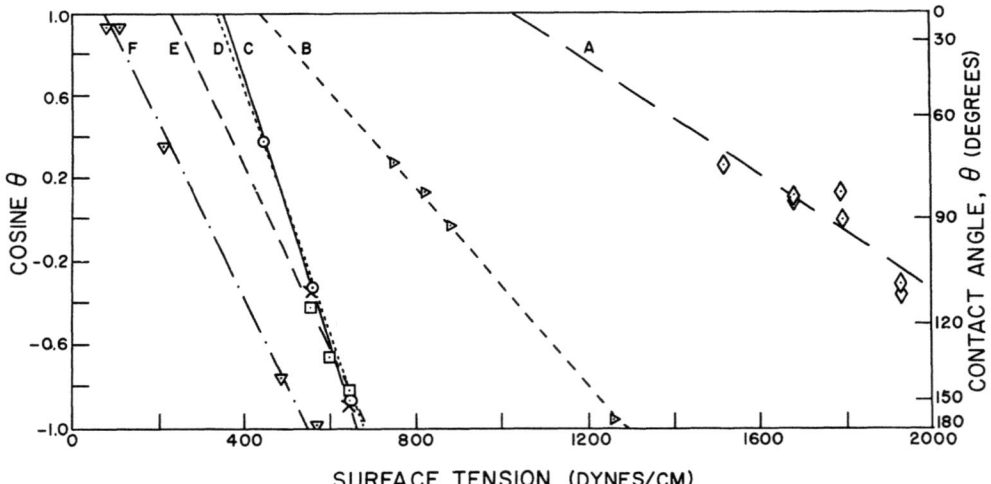

Fig. 11. Wetting of high-energy surfaces by liquid metals or alloys: (A) ◇, sapphire at 1500°[31]; (B) ▷, sapphire at 1230°[32]; (C) ✗, beryllium at 650-750°[33]; (D) ○, boron carbide at 650-750°[33]; (E) □, graphite at 650-750°[32]; (F) ▽, silica or pyrex at 20-156°[24].

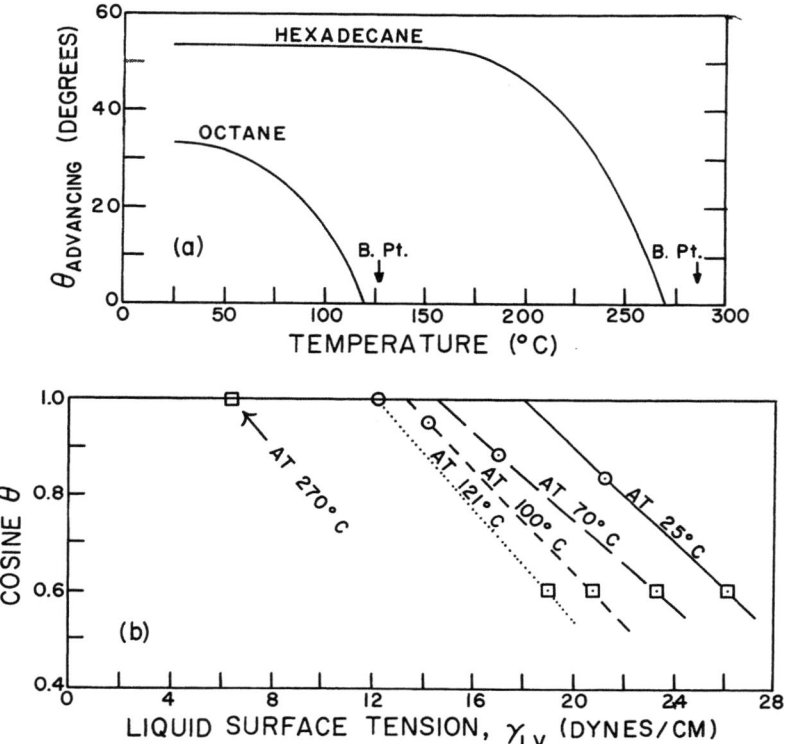

Fig. 12. Wettability of fluorinated ethylene-propylene copolymer by hexadecane (squares) and octane (circles): (a) advancing contact angles as a function of temperature[42,43]; and (b) cosine of the advancing contact angle as a function of the liquid surface tension at selected temperatures.

There is considerable information concerning the effect of temperature on θ for selected organic liquids on various solid surfaces, including both bulk polymers[40-43] and thin films of polymers (silicone-treated glass)[41]. The investigation of a fluorinated copolymer by Johnson and Dettre[42] had been pursued to temperatures such that sessile drops of a liquid which had not spread at lower temperatures began to spread. Plots of T vs. θ for two organic liquids proved to be curvilinear (Fig. 12a), in contrast with the rectilinear relation postulated by Rhee for the liquid metal/ceramic system. The evidence for nonrectilinear behavior precludes any attempts to extrapolate to θ = 0 the θ vs. T data from studies of low-energy surfaces which had been terminated before spreading occurred. Therefore, only reference (42) could be used to obtain values of T_{cs} and γ_{cs}.

Johnson and Dettre's observations[42,43] showed θ was essentially temperature-independent over an extended range of temperatures; at high temperatures, however, the θ vs. T graph developed a large negative slope and intercepted the θ = 0 axis at a T_{cs} only a few degrees below the boiling point of the respective wetting liquid (Figure 12a). Fig. 12b shows wetting isotherms based on θ values from reference (43) and the corresponding γ_{LV} values of Vogel[44] for two alkanes. Despite the limited number of liquids, values of γ_c obtained from the intercepts of the lower temperature isotherms with the cos θ = 1 axis compare favorably with literature values based on measurements of a large number of liquids at comparable temperatures. In addition, a plot of γ_c against temperature shows how consistent these isothermally derived values of γ_c (Fig. 13) are with the values of γ_{cs} derived from variable-temperature data for the individual alkanes.

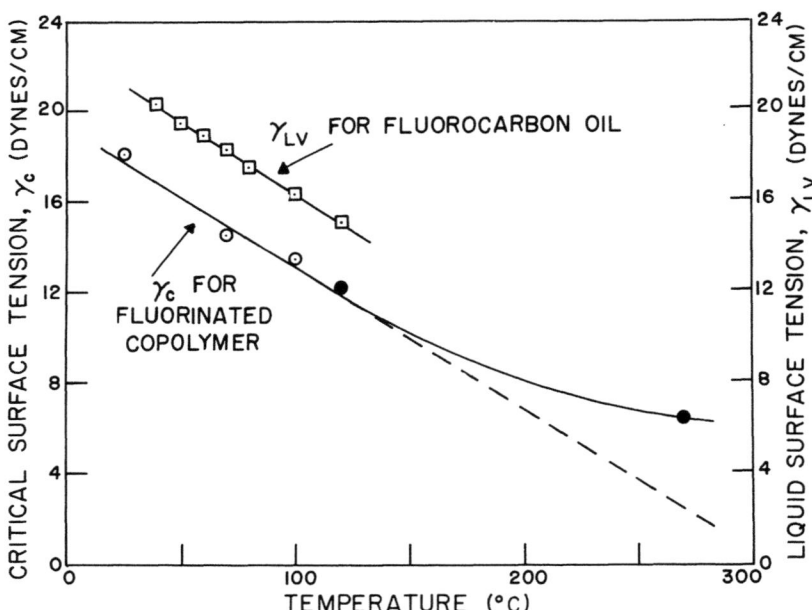

Fig. 13. Effect of temperature on the critical surface tension of a low-energy solid (γ_c, open circles; γ_{cs}, filled circles) and a comparison with the effect of temperature on the surface tension of a chemically related liquid.

The effect of temperature on γ_c of the fully fluorinated, chain-branched polymeric solid studied by Johnson and Dettre[42] strongly parallels that on the surface tension of a structurally and chemically related liquid. This is evident in Fig. 13 where the γ_c vs. T data are presented on the same energy scale as the surface tension data for a fully fluorinated, partially chain-branched oil of high viscosity at 25° and average composition $C_{21}F_{44}$.[45] A good straight line describes the γ_{LV} vs. T data up to the highest temperature reported (120°). Moreover, the slope of this graph (-0.064 dyn/cm deg) is close to that of the γ_c vs. T graph which agrees with the general observation of Petke and Ray[40] that the temperature coefficients of their γ_c values for several different polymers all fell within the values of $d\gamma_{LV}/dT$ for organic liquids.

These results are consistent with a model in which (a) γ_c is related monotonically to the surface energy (γ_S) of the low-energy solid; and (b) the change in γ_S with temperature for a solid of given chemical composition, configuration, and weak intermolecular force will not differ greatly from the change in the γ_{LV} of a liquid with similar chemical, steric, and interactive properties.

Fig. 14 presents a graphical summary of a large amount of literature data on the wetting of various low-energy surfaces[40-44] including both constant temperature and variable temperature approaches. The graphs emphasize how similar the effect of temperature on critical surface tension is for low-energy surfaces, whether the latter are primarily hydrocarbon-type or fluorocarbon-type polymers. As a result, the relative positions of the γ_c vs. T curves are maintained despite large increases in temperature so that, even at high temperatures, fluorine-rich surfaces remain far less wettable than nonfluorinated surfaces. The strong similarity in the several graphs of Fig. 14 and the conformity of curve K (derived from both γ_c and γ_{cs} data) to curve J for the same kind of polymer and to the remaining curves for the other polymers reinforces the conclusion that the value of γ_{cs} obtained by the variable-temperature method on a low-energy surface is, indeed, equivalent to γ_c for the temperature T_{cs}. This is evidence for the applicability to low-energy surfaces of what Rhee had postulated for high-energy surfaces, namely, that the value of the surface tension of a liquid at the temperature at which it just spreads on a given solid is equivalent to the critical surface tension of that solid at T_{cs}.

Thus, if Rhee's variable-temperature method is interpreted and used correctly, it can be applied to high-energy surfaces as Rhee has done for various metals and ceramics[34-38] or it can be applied to low energy surfaces. In either application, however, it is imperative that wetting behavior be observed at temperatures up to and including T_{cs}. The variable temperature method has particular value for systems where it would be difficult to find a sufficient number of nonspreading liquids to permit the construction of a con-

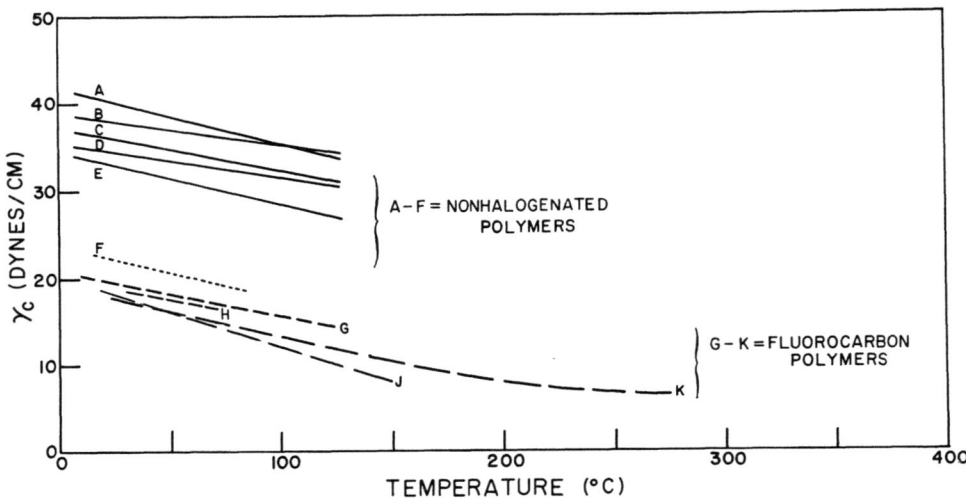

Fig. 14. Effect of temperature on the critical surface tension for spreading on low-energy surfaces: (A) poly(ethylene terephthalate); (B) polyoxymethylene; (C) polyethylene; (D) polycarbonate; (E) polystyrene; (F) silicone; (G) and (H) polytetrafluoroethylene; and (J) and (K) polytetrafluoroethylene-hexafluoropropylene copolymer.

ventional cos θ vs. γ_{LV} wetting isotherm, as in cryogenic studies.

Despite the attractiveness of the variable-temperature method, it has some important limitations. For example, it is often desirable for technical or theoretical reasons to determine the γ_c of a solid surface at some specific temperature; if the variable-temperature method is employed, it may be necessary to observe a variety of liquids to determine empirically which particular liquid has a value of T_{cs} in the temperature regime of interest. Moreover, because T_{cs} will generally differ from one liquid to another, it will be difficult to make direct comparisons of the effect of changes in the wetting liquid on γ_c at a single temperature. Reliance on a γ_c value obtained for only a single liquid increases the risk of allowing specific liquid/solid interactions to remain unrecognized. In the case of liquid metals in contact with the high surface energy ceramics, Rhee recognized this danger and took the precaution of examining the interface by various electron-optical techniques to see if phase interpenetration or compound formation had occurred. Low-energy surfaces would be less amenable to such examination. Yet, studies at constant temperature using a variety of organic and aqueous liquids indicate that specific liquid/solid interactions can profoundly affect wetting. Examples of such effects include hydrogen bonding[46] and possible changes in surface molecular conformation[47].

There are also limitations to the constant-temperature method. The critical surface tension of the solid may lie in a region where it is difficult to find liquids with sufficiently high surface tensions to be nonspreading or where the chemical compositions required to impart sufficiently high γ_{LV} values will be such that liquid/solid interactions cannot be avoided. This has been a serious difficulty, for example, in determining γ_c values much above 45 dyn/cm at room temperature for low-energy surfaces which are potential proton acceptors, since most organic liquids with $\gamma_{LV} > 45$ dyn/cm are capable of hydrogen-bond formation. In the case of high-energy surfaces, the limitation is more severe since the determination of γ_c for a specific, constant temperature would depend on the availability of a series of related liquids (such as liquid metal alloys) having the right range of γ_{LV} values to be useful in the temperature range of interest.

There are thus available two separate, but complementary, techniques for obtaining γ_c. It is safe to say that the analysis of contact angle behavior as a function of solid surface composition and liquid surface tension is now essentially consistent for a large variety of surfaces over a wide range of temperatures. Similarity of the temperature effect on γ_c can be shown for a variety of low-energy surfaces, but as yet cannot be done for high-energy surfaces because the results are not couched in a language which can presently be handled. The concepts of γ_c, γ_{cs}, and T_{cs}, and the recognition that γ_c is essentially a function of surface composition and a linear decreasing function of temperature appear adequate to provide a sound foundation and a framework on which to broaden the whole field concerned with the contact angle, spreading, and wetting.

TETRAFLUOROETHYLENE-PERFLUORO(PROPYL VINYL ETHER) COPOLYMERS AS WATER-RESISTANT, THIN-FILM ADHESIVES

The ideal adhesive for demanding structural applications would combine high-temperature strength with resistance to hydrolytic and oxidative attack. Moreover, it would not have the rigidity which makes ceramics and highly cross linked polymers susceptible to catastrophic failure upon crack initiation. An adhesive material which approaches this ideal state of affairs has been found in the form of random copolymers of tetrafluoroethylene with minor proportions of perfluoro(propyl vinyl ether), $CF_2=CF(OCF_2CF_2CF_3)$[48]. Unlike the

Table X

Changes in γ_c (dyn/cm) with Variations in Comonomer Content and Annealing Temperature

Code	Mol % perfluoro(propyl vinyl ether)	Melt flow no., g/10 min	γ_c (dyne/cm) at 20° as a function of heat treatment (24 hr)			
			None	200°	257°	284°
I	1.5	12	~18.9		16.6	
II	1.5	2	18.7	17.7	17.2	16.9
III	2.1	15	18.1		16.5	17.2
IV	2.6		17.7		16.2	16.8

well-known homopolymer of tetrafluoroethylene, PTFE, which has a softening point but no true melting point, these copolymers are true thermoplastics. We have found that they make poor substrates for ordinary adhesives, yet when they themselves are used as hot-melt adhesives, they adhere tenaciously to a wide diversity of surfaces. The key to this two-faced nature lies in the low surface energies of both the solid and the melt[49]. The critical surface tensions of wetting of these novel copolymers, along with comonomer content and melt flow number (an inverse function of viscosity) are shown in Table X. The copolymers were given various heat treatments, and Table X illustrates how γ_c decreases with increasing comonomer content under the same conditions of annealing as well as with higher annealing temperatures but constant comonomer content. The role of the comonomer in lowering the value of γ_c with increasing mole percent is involved with its ability to populate the surface with perfluoromethyl groups. It is interesting that in a copolymer of tetrafluoroethylene with hexafluoropropylene, 23 mol% of comonomer decreased γ_c by 1 dyn/cm below that of PTFE[50], whereas the same decrease was achieved with as little as 2.6 mol% of the perfluoro-(propyl vinyl ether) comonomer. The small range of comonomer content, 1.5-2.6 mol%, helps to account for the lack of definition in the relationship between γ_c and the copolymer composition. Increasing the levels of perfluoro(propyl vinyl ether) in the copolymer, if synthetically feasible, might yield increasingly lower energy surfaces, but probably at considerable expense to the high-temperature capability of these resins.

The molecular mobility brought about by the high temperatures of the annealing process expedites the migration of perfluoromethyl terminal side chains to the copolymer surface, a process which occurs as a result of the surface's thermodynamic drive to attain its lowest energy state. Molecular models indicate that the side chain would encounter considerable steric hindrance in shifting from one conformation to the other because bulky fluorine atoms immobilize the pendant side chain at both bonds to the ether oxygen. Such steric rigidity would account for the rather dramatic annealing conditions needed to affect γ_c. The lowering of γ_c by annealing to the extent observed here (8-8.5%, a sizable decrease in already low surface energy material) is, to our knowledge, larger by nearly an order of magnitude than what other experimenters have observed with other polymers. This simple technique, however, may possibly prove to be of general utility with polymers containing surface-active groups pendant to the main chain.

The low γ_c of the copolymers, from 16 to 19 dyn/cm, are the reason why common adhesives cannot wet them and, consequently, cannot form strong adhesive bonds with them. The low surface tensions of the molten copolymers, however, enable them to wet and spread over nearly any solid surface with which they come into contact, thereby fulfilling a necessary condition for good adhesion. The various chemical and physical attributes of these copolymers suggest corresponding benefits for adhesive applications, as summarized in Table XI. The specimens used in the study[51], supplied by du Pont de Nemours and Co., contained 1.5 mol% perfluoro(propyl vinyl ether) comonomer but differed in the viscosity of the melt (melt flow numbers 2, 6, and 12 g/10 min). Sheets of the copolymers were further reduced to a thickness of 0.006 in. and used as bonds by being sandwiched between carefully cleaned plates of the following materials: pyrex, sapphire, PTFE, 1100 aluminum, 2024 aluminum, and mild steel. The mild steel was used either untreated or preoxidized so as to impart a durable straw-to-blue colored oxide coating, since such an oxide had been used to promote maximum adhesion of electrostatically deposited PTFE on steel[52].

The results showed that the copolymers have strong potential as thin-film, hot-melt adhesives. For such applications three main points need to be discussed.

An important criterion of good adhesion is that the adhesive be able to wet the substrate. The copolymers studied vividly demonstrated their ability to wet all types of solid materials (provided the solid is sufficiently heat stable to accept bonding) when, in the molten state, they spread spontaneously on solid PTFE, a material of very low surface energy and thus unwettable by most liquids.

Table XI

Advantages of the Copolymers of Tetrafluoroethylene and Perfluoro(propyl vinyl ether) in Adhesion Applications as Suggested by Their Chemical and Physical Properties

Property	Typical Range of Values	Utility
Definite melting point [a]	300–310°C	Hot-melt adhesive
Low melt surface tension [a]	18 dyn/cm (estimated)	Able to wet nearly all substrates
Low critical surface tension [a]	16–19 dyn/cm	Resistant to penetration of liquids into the adhesive/adherend interface
Chemical inertness	Same as for PTFE	Stable to corrosive environments and weathering
High tensile strength [b] (ASTM D638-68)	4300 psi at 25°C 3400 psi at 100°C 1500 psi at 250°C	Structural joints with high-temperature capability
MIT flex endurance (7–8 mil compression-molded film)	50,000–500,000 cycles	Usable for joints subject to flexing and vibration

[a] Ref. (48)
[b] Heat-aging tests show no degradation of tensile properties after more than 2000 hr at 285°C.

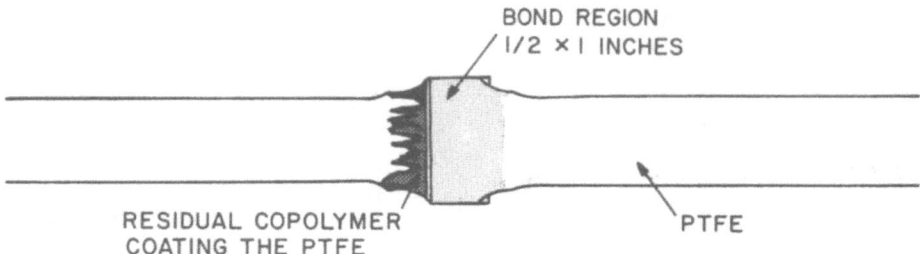

Fig. 15. A typical lap shear joint of PTFE and copolymer II after having been pulled on the Instron testing machine.

When the 1 in. wide, 0.5 in. overlap joints were pulled on an Instron testing machine at 0.05 in./min, the PTFE yielded indefinitely, yet the bond region suffered hardly any distortion (Fig. 15). Moreover, the ability of the copolymers to adhere without any sort of chemical or hydrogen bonding further underscored their wide range of applicability.

The copolymers, when used for bonding aluminum to aluminum or steel to steel, produced strong joints with tensile shear strengths up to 1600 psi[51] with reasonably good water resistance. The methods employed to prepare the metals for bonding were chosen mainly as convenient ways of getting flat, fairly smooth, and clean surfaces. Without a doubt, considerably greater adhesive strengths and improved water resistance could be achieved by a study of surface preparations and various other parameters of the bonding procedure. For example, chromate etching of aluminum, such as was used in this study, produces a porous outer layer of oxide which promotes adhesion through mechanical interlocking of adhesive and substrate[53]. Thus, the lower tensile shear strength of water-soaked aluminum joints is probably attributable to penetration of water through this porous outer layer into the bond interface. Water would disrupt the dispersion forces between adhesive and substrate, thereby re-

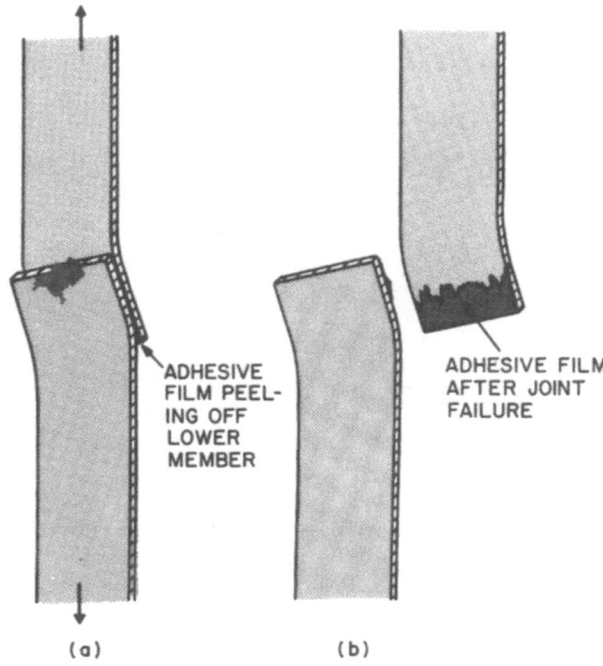

Fig. 16. Mode of failure of lap shear joints of 0.063 in. thick 1100 aluminum and 0.001 to 0.003 in. thick copolymer film: (a) during tensile loading; (b) after joint failure.

ducing their adhesion to mere mechanical interlocking of the two materials. The result, a larger degree of adhesive failure by peeling of the copolymer adhesive film (Fig. 16) would be reflected in lower tensile shear strength values. Similarly, the particular kind of oxide on a steel surface certainly has a heavy bearing upon the strength of the bond and its mode of failure. When adhesive joints were fabricated by bonding strips of 1100 aluminum to stainless steel plates with the copolymer, the various surfaces were probed with droplets of n-hexadecane after the aluminum strips were peeled off (Fig. 17). Large contact angles were a good indication of residual copolymer film. A film was detected by the droplets on the underside of the aluminum strip; although invisible to the eye, even under microscopic investigation (70X), it was also readily detectable by the very slick but dry surface. Such an invisibly thin film left on the peeled aluminum strip strongly suggests a mechanism of cohesive failure in the film just below the metal/copolymer interface. In general, optimization of the bonding parameters can be expected to be a result of reducing bond failure to pure cohesive failure, where bond

WETTING AND ADHESION

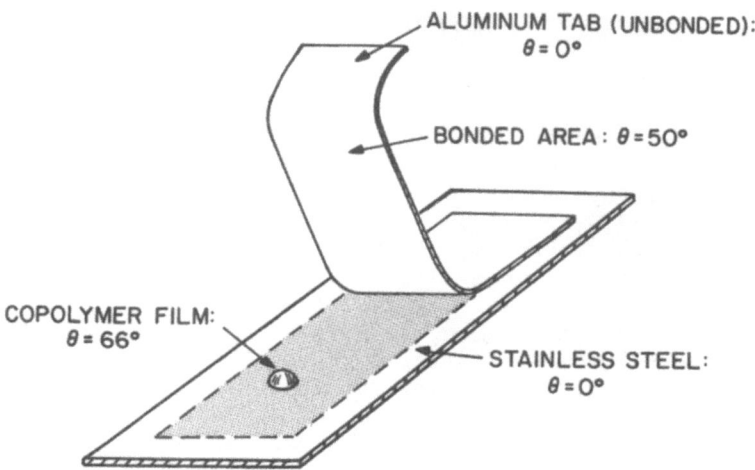

Fig. 17. A typical Al/copolymer/stainless steel peel specimen as probed with drops of n-hexadecane after peeling.

strength depends only on the inherent high mechanical strength of the copolymer adhesive itself.

The copolymers were essentially impervious to water. Bonded to another water-impervious material, they yielded totally water-resistant joints, as in the case when sapphire was bonded to sapphire with copolymer. Immersion in water for more than a month failed to weaken the joint, which was finally loosened only after several boiling water-cold water cycles. Specimens made with Pyrex instead of sapphire succumbed to the action of water within a few weeks of immersion, which was not a surprising result considering the readily hydrated nature of the silica glasses. Possibly water molecules penetrated into the adhesive/adherend interface by diffusion through bulk glass rather than by two-dimensional migration across the interface itself. Such evidence suggests that the key to totally water-resistant joints lies in both adhesive and adherend being impervious to water. As an operating principle this would dictate the design of surface preparation of the adherend as well as the basic selection of materials.

Initial results, then, show a great promise for these perfluorinated thermoplastic resins as hot-melt, thin-film adhesives. Their unique set of physical and chemical properties gives them a range of applicability few other polymers can begin to match. Work with them should continue with emphasis on their role in water-impervious joints. Hopefully such an investigation should lead to a better understanding needed for their optimum utilization.

REFERENCES

1. H.W. Fox, E.F. Hare, and W.A. Zisman, J. Phys. Chem. $\underline{59}$, 1097 (1955).
2. M.L. White, J. Phys. Chem. $\underline{68}$, 3083 (1964).
3. F.M. Fowkes, Ind. Eng. Chem. $\underline{56}$, 40 (1964).
4. R.A. Erb, J. Phys. Chem. $\underline{69}$, 1306 (1965).
5. K.W. Bewig and W.A. Zisman, J. Phys. Chem. $\underline{69}$, 4238 (1965).
6. M.L. White and J. Drobek, J. Phys. Chem. $\underline{70}$, 3432 (1966).
7. E. Thelen, J. Phys. Chem. $\underline{71}$, 1946 (1967).
8. R.A. Erb, J. Phys. Chem. $\underline{72}$, 2412 (1968).
9. K.W. Bewig and W.A. Zisman, J. Phys. Chem. $\underline{68}$, 1804 (1964).
10. M.K. Bernett and W.A. Zisman, J. Phys. Chem. $\underline{74}$, 2309 (1970).
11. M.E. Schrader, J. Phys. Chem. $\underline{74}$, 2313 (1970).
12. M.E. Schrader, J. Phys. Chem. $\underline{78}$, 87 (1974).
13. L. Holland, in *The Properties of Glass Surfaces*, Wiley, New York, 1964, Chapter 4.
14. W.D. Harkins and A. Feldman, J. Am. Chem. Soc. $\underline{44}$, 2665 (1922).
15. W.A. Zisman, Ind. Eng. Chem. $\underline{57}$, 26 (1965).
16a. E.G. Shafrin and W.A. Zisman, J. Amer. Ceram. Soc. $\underline{50}$, 478 (1967).
16b. M.K. Bernett and W.A. Zisman, J. Colloid Interface Sci. $\underline{29}$, 413 (1969).
17. H.W. Fox and W.A. Zisman, J. Colloid Sci. $\underline{5}$, 514 (1950).
18. R.M. Tichane and G.B. Carrier, J. Amer. Ceram. Soc. $\underline{44}$, 606 (1961).
19. J.K. Davis and F.E. Bartell, Anal. Chem. $\underline{20}$, 1182 (1948).
20. N.M. Parikh, J. Am. Ceram. Soc. $\underline{41}$, 18 (1958).
21. P.J. Bryant, Trans. National Vacuum Symposium $\underline{9}$, 311 (1962).
22. R.E. Johnson and R.H. Dettre, J. Colloid Interface Sci. $\underline{21}$, 610 (1966).
23. E.G. Shafrin and W.A. Zisman, J. Phys. Chem. $\underline{71}$, 1309 (1967).
24. D.A. Olsen and A.J. Osteraas, J. Phys. Chem. $\underline{68}$, 2730 (1964).
25. E.F. Hare and W.A. Zisman, J. Phys. Chem. $\underline{59}$, 335 (1955).
26. P. Pomerantz, W.C. Clinton and W.A. Zisman, J. Colloid Interface Sci. $\underline{24}$, 16 (1967).
27. M.K. Bernett and W.A. Zisman, J. Colloid Interface Sci. $\underline{28}$, 243 (1968).
28. W.A. Zisman, in Advan. Chem. Ser., No. $\underline{43}$, Am. Chem. Soc., Washington, D.C., 1964, p. 1.
29. R.J. Good, W.G. Givens and C.S. Tucek, in Advan. Chem. Ser., No. $\underline{43}$, Am. Chem. Soc., Washington, D.C., 1964, p. 211.
30. D.A. Olsen, R.W. Moravec and A.J. Osteraas, J. Phys. Chem. $\underline{71}$, 4464 (1967).
31. J.G. Eberhart, J. Phys. Chem. $\underline{71}$, 4125 (1967).
32. S.K. Rhee, J. Amer. Ceram. Soc. $\underline{54}$, 376 (1971).
33. C.R. Manning, Jr., and T.B. Gurganus, J. Amer. Ceram. Soc. $\underline{52}$, 115 (1969).
34. S.K. Rhee, J. Amer. Ceram. Soc. $\underline{53}$, 386 (1970).

35. S.K. Rhee, J. Amer. Ceram. Soc. 53, 426 (1970).
36. S.K. Rhee, J. Amer. Ceram. Soc. 53, 639 (1970).
37. S.K. Rhee, J. Amer. Ceram. Soc. 54, 332 (1971).
38. S.K. Rhee, J. Amer. Ceram. Soc. 55, 157 (1972).
39. E.G. Shafrin and W.A. Zisman, J. Phys. Chem. 76, 3259 (1972).
40. F.D. Petke and B.R. Ray, J. Colloid Interface Sci. 31, 216 (1969); erratum, ibid., 33, 195 (1970).
41. A.W. Neumann, G. Haage and D. Renzow, J. Colloid Interface Sci. 35, 379 (1971).
42. R.E. Johnson, Jr., and R.H. Dettre, J. Colloid Interface Sci. 20, 173 (1965).
43. R.E. Johnson, Jr., and R.H. Dettre, *Surface and Colloid Science*, Vol. 2, E. Matijević, Ed., Wiley-Interscience, New York, N.Y., 1969, p. 144.
44. A.I. Vogel, J. Chem. Soc. 133 (1946).
45. R.H. Dettre and R.E. Johnson, Jr., J. Phys. Chem. 71, 1529 (1967).
46. A.H. Ellison and W.A. Zisman, J. Phys. Chem. 58, 503 (1954).
47. W.A. Zisman, J. Paint Technology 44, 41 (1972).
48. J.F. Harris, Jr. and D.I. McCane, U.S. Patent 3,132,123 (May 5, 1964).
49. J.P. Reardon and W.A. Zisman, Macromolecules 7, 920 (1974).
50. M.K. Bernett and W.A. Zisman, J. Phys. Chem. 64, 1292 (1960).
51. J.P. Reardon and W.A. Zisman, Ind. Eng. Chem. Prod. Res. Develop. 13, 119 (1974).
52. V.G. FitzSimmons and W.A. Zisman, Ind. Eng. Chem. 50, 781 (1958).
53. K. Bright, B.W. Malpass and D.E. Packham, Brit. Polym. J. 3, 205 (1971).

Polymer Reactions on Solid Surfaces

K. Hamann, R. Laible and J. Horn

Forschungsinstitut für Pigmente und Lacke

D 7000 Stuttgart

Wiederholdstrasse 10/1, West Germany

Macromolecules can be covalently bonded on the surface of solids by the following methods, (1) initiation of polymerization by covalently bonded initiating groups, (2) copolymerization of covalently bonded vinylic groups and (3) chain termination reaction of polymers with reactive groups on the surface. Type (1) and (3) reactions are carried out with amorphous, pyrogene-produced silica. A radical graft polymerization of vinyl monomers is initiated by phenyl radicals generated by thermal decomposition of covalently bonded phenyl diazo groups (1), anionic living polymers react with superficial chlorsilyl groups by a chain terminating reaction (3). The amount of polymer coverage is dependent on reaction time, temperature and concentration of monomers and initiating diazo groups (1) and chain length, coil size and the degree of dissociation of the living polymers (3), respectively. The covalent nature of the bond between macromolecules and silica surface is exhibited by spectroscopic analysis (IR, mass spectra) and by adsorption-desorption experiments. Ten to one hundred times as much polymer can be covalently bonded on the silica surface by graft reactions than by irreversible adsorption reactions. The structure of the grafted polymer molecules can be described as ellipsoidically deformed, mutually penetrated random coils.

INTRODUCTION

Macromolecules can be covalently bonded onto solid surfaces by polymerization reactions carried out in the presence of the solids[1]. Principally, there are three types of polymer reactions:

1) Initiation of polymerization by covalently bonded initiating groups:

A cationic styrene polymerization can be initiated by acyl perchlorate groups covalently bonded on the surface of Cu-phthalocyanine pigments[2]:

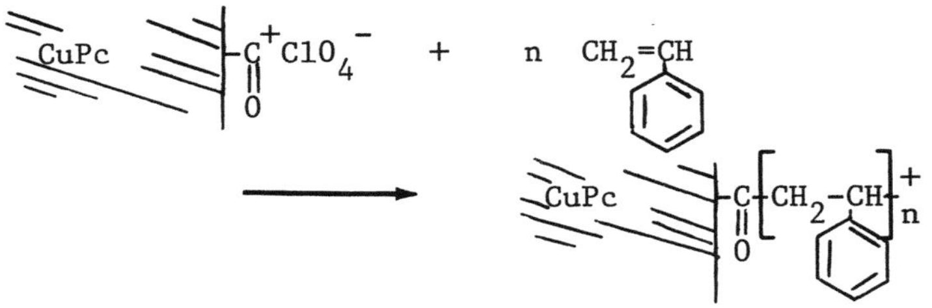

$$\text{CuPc}-\overset{+}{\underset{\underset{O}{\|}}{C}}ClO_4^- + n\ CH_2{=}CH(C_6H_5) \longrightarrow \text{CuPc}-\underset{\underset{O}{\|}}{C}{-}[CH_2{-}CH(C_6H_5)]_n^+$$

An anionic copolymerization of ethylene carbonate and phthalic anhydride (polyester synthesis) is initiated by carboxylate or sulfonate groups covalently bonded on the surface of Cu-phthalocyanine pigments or silica[3,4]:

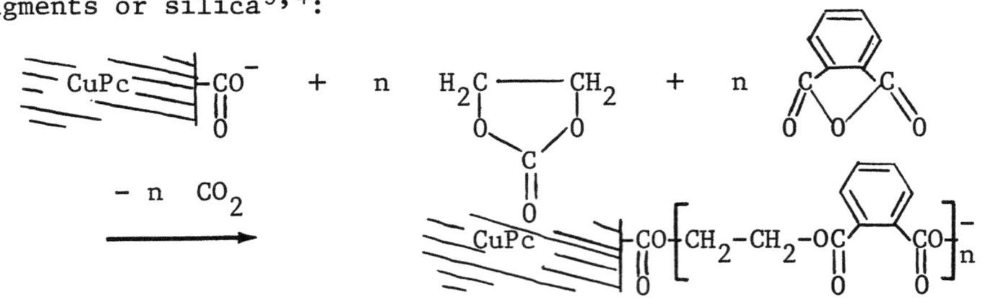

$$\text{CuPc}-\underset{\underset{O}{\|}}{C}O^- + n\ \underset{\underset{O}{\diagdown}\underset{C}{\underset{\|}{}}\underset{O}{\diagup}}{H_2C{-}CH_2} + n\ \text{(phthalic anhydride)} \xrightarrow{-n\ CO_2} \text{CuPc}-\underset{\underset{O}{\|}}{C}O{-}[CH_2{-}CH_2{-}OC(C_6H_4)CO{-}]_n^-$$

Polyamides are covalently bonded on the surface of silica by an anionic polymerization of N-carboxy-α-amino-acid anhydrides initiated by superficial amino-phenyl groups[5]:

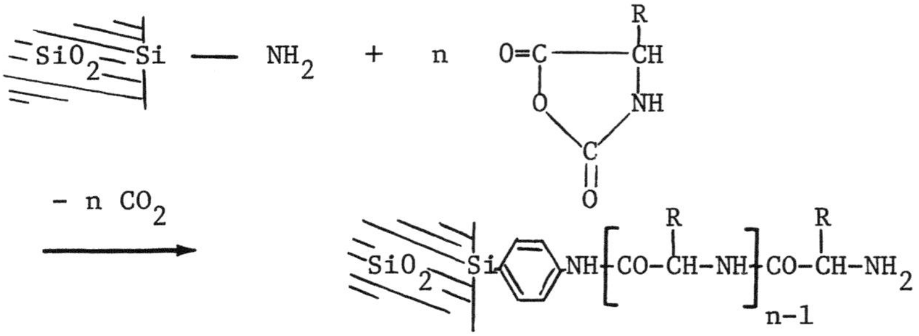

$$\text{SiO}_2{-}Si{-}NH_2 + n\ \underset{\underset{\underset{O}{\|}}{C}}{\underset{O}{\diagdown}\underset{NH}{\diagup}}O{=}C{-}\overset{R}{\underset{}{CH}} \xrightarrow{-n\ CO_2} \text{SiO}_2{-}Si{-}(C_6H_4){-}NH{-}[CO{-}\overset{R}{\underset{}{CH}}{-}NH]_{n-1}{-}CO{-}\overset{R}{\underset{}{CH}}{-}NH_2$$

2) Copolymerization of vinylic groups, bonded onto the
   surface with vinyl monomers:

Polar monomers of the acrylic acid type can be bonded on the surface of titanium dioxide by irreversible adsorption and then radically copolymerized with vinyl monomers[3]:

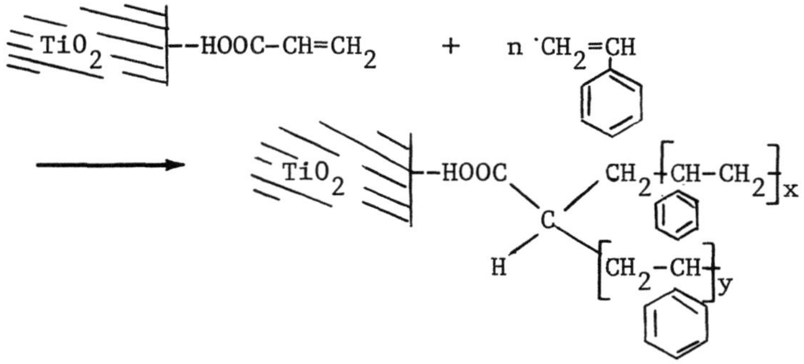

$$\equiv TiO_2 \equiv \!\!-\!\!HOOC\!-\!CH\!=\!CH_2 \;+\; n\, CH_2\!=\!CH\text{-}Ph$$

$$\longrightarrow \equiv TiO_2 \equiv \!\!-\!\!HOOC\!-\!C(H)(CH_2\text{-}[CH(Ph)\text{-}CH_2]_x)([CH_2\text{-}CH(Ph)]_y)$$

3) Chain termination reaction of polymers with reactive
   groups on the surface.

An example will be given below. In all reactions, solid particles are incorporated as end groups in the macromolecules.

In this paper, new type 1 and 3 reactions are presented, which are carried out on the surface of amorphous, pyrogene-produced silica (aerosil of Degussa, Frankfurt/M, West Germany):

a) <u>the radical graft polymerization of vinyl monomers,
   for example styrene, is initiated by phenyl radicals
   generated by the thermal decomposition of covalently
   bonded phenyl-diazo-thioether groups:</u>

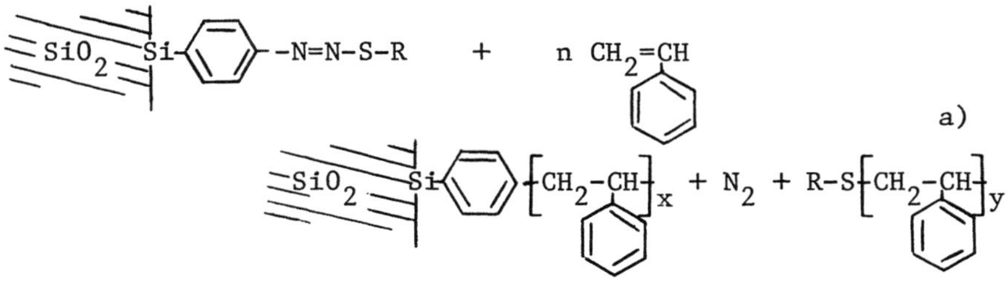

$$\equiv SiO_2 \equiv \!\!-\!\!Si\text{-}C_6H_4\text{-}N\!=\!N\text{-}S\text{-}R \;+\; n\, CH_2\!=\!CH\text{-}Ph$$

$$\equiv SiO_2 \equiv \!\!-\!\!Si\text{-}C_6H_4\text{-}[CH_2\text{-}CH(Ph)]_x \;+\; N_2 \;+\; R\text{-}S\text{-}[CH_2\text{-}CH(Ph)]_y \qquad a)$$

At the same time as the heterogeneous graft reaction, a homopolymerization takes place in the homogeneous phase initiated by the other radical partner.

b) <u>Anionic living polymers, such as polystyrene-Li and
   polyvinylpyridine-Li, react with chlorsilyl groups
   of the silica surface by a chain terminating reaction:</u>

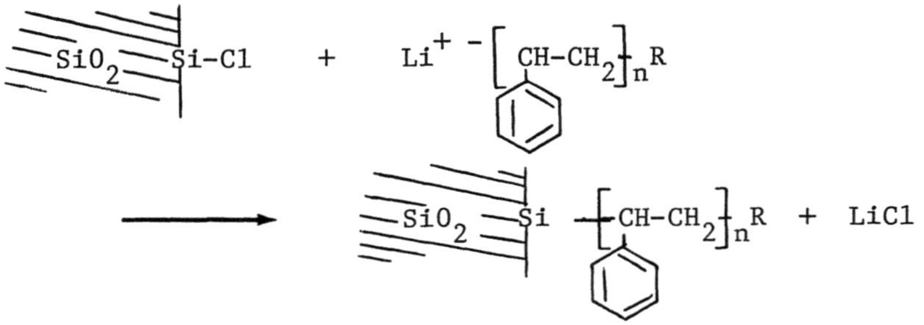

$$\equiv SiO_2 \equiv Si-Cl \;+\; Li^+ \; {}^-[CH-CH_2]_n R$$
(phenyl group on CH)

$$\longrightarrow \;\equiv SiO_2 \equiv Si-[CH-CH_2]_n R \;+\; LiCl$$

b)

EXPERIMENTAL

a) <u>Radical-initiated graft polymerization</u>: The initiator SiO_2-phenyl-diazo-(naphtalene-2)thioether is prepared by the following reactions on the surface of silica: chlorination of the superficial silanol groups and reaction with phenyl lithium, nitration of the covalently bonded phenyl groups, reduction of the nitrophenyl groups and diazotization of the aminophenyl groups and coupling of the diazonium salts with naphtalene thiol-(2)[5]. The polymerization is carried out in vacuo: the monomer (styrene) and the solvent (toluene) are added to the diazo thioether using ampoules and a break-seal technique.

The homopolymer of styrene which is formed in addition to the graft polystyrene is separated from the graft product by extraction with toluene. The amount of homopolymer is determined gravimetrically, and the amount of polystyrene grafted onto silica, by elementary analysis or loss on ignition.

b) <u>Anionic graft reaction</u>: Living polystyrene and polyvinylpyridine are prepared by a stoichiometric polymerization of the monomers with n-butyl lithium in tetrahydrofurane and added to chlorinated silica using the same experimental techniques as above. The graft reaction is carried out in vacuo. Excess polymeric anions are treated with acetic acid, and the homopolymers formed are separated from the graft products by extraction with a solvent mixture of tetrahydrofurane and diethylether (3:1)[6]. The amount of grafted polymer is determined as above.

RADICAL-INITIATED GRAFT POLYMERIZATION

Reaction Parameters

The formation of the graft and homopolymers is dependent on reaction time and temperature, monomer concentration and concentration of initiating diazo groups.

Figure 1 shows the conversion time-diagram for the graft and the homopolymerization. After a period of a relatively fast reaction, wherein the reaction rate of the homopolymerization is normally greater than that of the graft polymerization, the amount of grafted polymer reaches a final value, i.e. the graft polymerization is terminated, whereas the formation of the homopolymer continues slightly and comes to an end at longer reaction times. An increase of the conversion is caused by the elevation of the reaction temperature. From 50 to 70°C, the amount of polymer increases linearly with the reaction time.

The dependence of the conversion on the monomer concentration becomes evident in polymerization reactions which are carried out in presence of excess monomer and in a solvent (toluene), respectively. Over a wide range, from 0.875 over 1.75 to 8.75 mol/ml, the amount of polymers increases with increasing initial monomer concentration.

The influence of the concentration of initiating diazo groups is shown in two polymerization experiments. It can be seen that in the presence of varying amounts of silica containing the same number of initiating diazo groups per surface unit (surface density), the amount of polymers increases with increasing initial molar concentration of the diazo groups. The surface density of grafted macromolecules is held constant in this reaction. In the case of silicas with varying surface density of diazo groups, no direct proportionality between superficial concentration of diazo groups and amount of grafted polymers is to be observed. A comparable surface density of diazo groups results in approximately the same surface density of grafted macromolecules. On the other hand, a two-fold increase in the number of diazo groups results in a four times greater number of macromolecules.

Molecular Weights of the Polymers

The grafted macromolecules can be separated from the silica by hydrolysis with HF. The molecular weights of these and of the homopolymers are determined by viscosimetric methods, table 1.

The average molecular weights range from a few hundred thousand to over one million. These relatively high values are caused by a large ratio of monomer concentration and concentration of initiating diazo groups (0.01 to 0.03 mol-% diazo groups relatively to the monomer concentration, in bulk polymerizations). If this ratio is decreased by varying the monomer concentration, the molecular weights, for instance of the homopolymers, fall linearly with the initial monomer concentration.

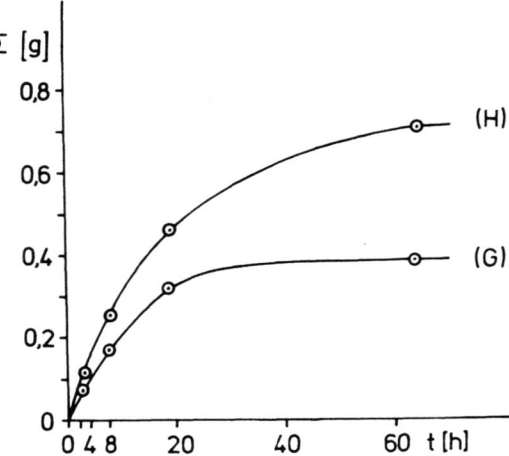

Fig. 1. Conversion-time diagram (radical-initiated graft polymerization) G-graft, H-homopolymerization.

Table 1

Molecular Weights of Graft and Homopolymers (radically initiated polyreaction) and Surface Densities of Diazo Groups (β_D) and Grafted Macromolecules (β_p).

β_D (Diazo / 100Å²)	θ_{too} (g/g SiO$_2$)	$\bar{M}_{graft}$	$\bar{M}_{homo}$	β_p (MM / 100Å²)	$\dfrac{\beta_D}{\beta_p}$
0.039	0.500	350,000	800,000	0.0050	7.8
0.030	0.557	560,000	810,000	0.0034	8.8
0.080	2.080	1,460,000	1,750,000	0.0051	16.5

A dependence of the molecular weights on the concentration of diazo groups, in which the molecular weight increases with decreasing initiator and constant monomer concentration, is not found. The molecular weights of graft and homopolymers are higher in the case of large surface density of diazo groups than in the case of lower surface density. It is evident that the molecular weight is not determined by the number of potentially initiating, but by the number of really initiating diazo groups. The ratio of the surface densities of diazo groups and of the grafted macromolecules indicates that approximately every eighth to sixteenth diazo group is used to initiate the polymerization reaction; see Table 1.

DISCUSSION

The kinetic study of the graft and of the homopolymerization shows that the reaction obeys the time law[1], which is known for radical polymerization reactions and in which a fraction f of the effective initiating radicals is inserted:

$$- d [M] /dt = v_{br} = k \cdot f \cdot [Diazo]^{0.5} \cdot [M] \qquad (1)$$

In the first period, the polymerization is of first order in relation to the monomer concentration and of 0.5th order in relation to the concentration of initiating radicals. The graft polymerization is initiated from the silica surface and takes place in the interfacial layer between the solid and the liquid phase. The homopolymerization is initiated by a thioaryl radical which diffuses from the solid surface to the liquid phase. The diazo groups bound to the surface decompose with unimolecular rate. The radicals are formed as pairs and must be separated for initiating the graft and the homopolymerization.

If the radicals combine in the solvent cage, the separation is not complete; further, the free access of the monomer molecules to the solid surface may be hindered sterically by a layer of grafted molecules: in both cases, only a fraction of the radicals is available for the initiating reaction. At a certain moment of the reaction, the radical concentration diminishes in a manner that the quasi-stationary equilibrium between initiating and terminating reaction is disturbed for the benefit of the latter, because initiating radicals are no more available. At this moment, the reaction rate of the graft polymerization becomes zero, while that of the homopolymerization falls rapidly, although monomer is present in sufficiently high concentration ("dead-end" polymerization).

ANIONIC GRAFT REACTION

Reaction Parameters

In the anionic grafting of macromolecules on SiO_2 surfaces, living polymers, having a narrow molecular weight distribution, react by a chain termination reaction with chlorsilyl groups on the surface—see formula b. When a sufficient excess of reacting carbanions is available, the amount of grafted polymer or, corresponding, the surface density of macromolecules, is independent of the concentration of polymer but, in contrast, dependent on the reaction time and temperature, as well as the chain length, the size of the molecule and the degree of dissociation of the macromolecules. The time dependence of the bonding of the polymers to the SiO_2 surface is shown in the conversion-time diagram (Figure 2) for polystyrene. The amount of polymer coverage increases strongly and rapidly at the beginning of the reaction. After about five hours the amount of coverage increases only slowly further and reaches a constant value after about 40 hours. This constant value cannot be increased by increasing the reaction time.

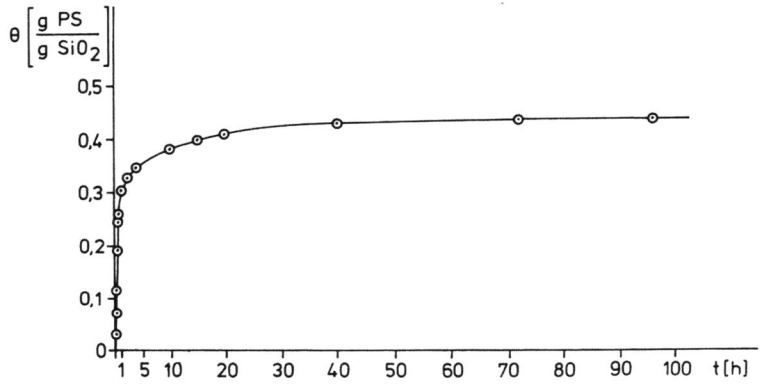

Fig. 2. Conversion-time diagram (anionic graft reaction).

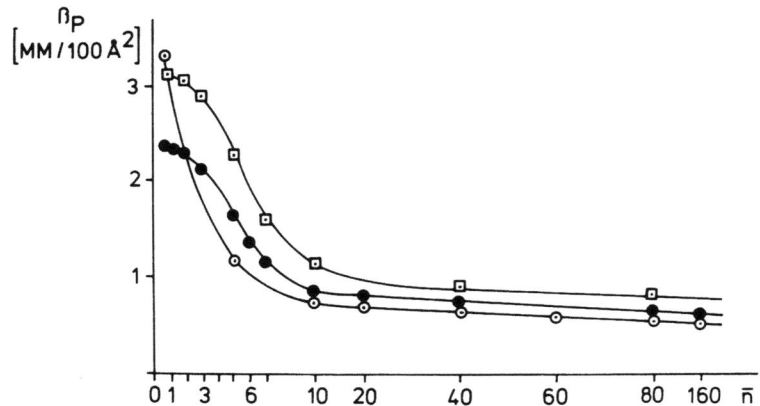

Fig. 3. Surface density of the grafted macromolecules dependent on chain length (⊙ polystyrene 20°C, ● polyvinylpyridine 20°C, ⊡ polyvinylpyridine 45°C).

The influence of counterions of the living polymers can be investigated for polystyrene in the solvent hexane in which macroions and counterions are available in the form of ion pairs. It is seen that the amount of polymer coverage decreases continuously with increasing ion-diameter in the sequence Li, Na, K. This means that the reactivity of the ion pairs in a graft reaction follows the same course as that for the addition of further monomer (as is well-known for anionic polymerization).

The dependence of the surface density of the grafted macromolecules on the chain length of the reacting polymers can be shown for polystyrene and also for polyvinylpyridine. It is difficult to form oligomers through an anionic polymerization initiated by means of n-butyl lithium, in the case of polystyrene; phenyllithium serves as a model of a monomer unit. In the case of polyvinylpyridine, polymer homologues may be formed having precise short (n=1-10) or long chain lengths. The surface density of the grafted macromolecules is greatest for very short chains and decreases with increasing chain length, see Figure 3. The surface density is about 2.4 molecules per 100$Å^2$ for monomers and dimers of vinylpyridine when the reaction temperature is 20°C. At the higher reaction temperature of 45°C, the surface density is 3.1 macromolecules per 100$Å^2$, and thus approaches the surface density of hydroxyl groups on the silica surface, which are converted to chlorsilyl groups for the graft reaction. With increasing chain length, up to about n=10, the surface density of the macromolecules falls and reaches a value of about 0.6 to 0.8 macromolecules per 100$Å^2$, this is also true for polystyrene. Thus, the amount of coverage, which is determined by the number of reaction sites and by the molecular weight, increases continuously.

DISCUSSION

A kinetic study of the graft reaction shows that the time-dependent increase of the amount of grafted polymer, θ, is of first order in relation to the chlorsilyl groups still available, see Equation 2; the polymer concentration is inserted as a constant since macrocarbanions are present in excess:

+)
$$-(d\theta/dt) = k \cdot (\theta_\infty - \theta) \cdot c_{pol} \qquad (2)$$

+)
  (∞ : amount of grafted polymer at infinite reaction time)

This means that it is not the fast reaction of the macrocarbanions with the chlorsilyl groups on the surface which determines the reaction rate, but the diffusion of the reactive polymer molecules through the solvent and the adsorption of the macromolecules on the surface. A structural re-orientation of the macromolecules, by means of which the reactive site is brought sufficiently close to the chlorsilyl groups, may be regarded as very fast and not rate determining. The time-dependent equation 2 holds experimentally only for the period of rapid increase in the amount of grafted polymer at the beginning of the reaction. A deviation from the time law means that there is no longer free access to the chlorsilyl groups. The macromolecules approaching the surface must diffuse through a layer of already-grafted molecules; the reaction rate decreases and finally falls to zero.

STRUCTURE OF THE GRAFTED MACROMOLECULES

To obtain some idea of the structure of macromolecules covalently bound to the surface of a solid, we may compare the space requirement of the macromolecule as a random coiled molecule in solution with that on the surface. For this purpose, the root-mean-square end-to-end distance in the relevant solvent is obtained from the molecular weight of the macromolecule and hence the largest cross section, F_L, of the molecule in this solvent may be calculated. From the surface density and the molecular weight, it is possible also to calculate the cross-section, F_0, which is available to the molecule on the SiO_2 surface. The ratio F_0/F_L allows us to make conclusions about the shape of the molecule on the surface. For values of $F_0/F_L = 1$, the molecule lies on the surface in the same form that it has in solution (statistically, an ellipsoidal tangle); for $F_0/F_L \to 0$ the macromolecules stand upright in brush form on the surface.

For radical grafted polystyrene (solvent toluene) the ratio F_0/F_L is dependent on the molecular weight, and lies between 0.150 and 0.035. For anionically grafted polystyrene (solvent THF), these values decrease with increasing molecular weight from 0.330 to 0.022, see Table 2. Values of F_0/F_L smaller than one indicate a contraction of the coil on the surface, or a mutual penetration of the coils or an extension of the coil in the molecule axis perpendicular to the surface. It is therefore suggested that the structure of the grafted molecules is a deformed ellipsoid, partly contracted and partly interpenetrated with other random coils, see Fig. 4.

COVALENT NATURE OF THE BOND BETWEEN MACROMOLECULES AND THE SOLID SURFACES

The covalent bond of radical-grafted polymers to the SiO_2 surface may only be demonstrated indirectly using I.R.-spectroscopy. The band at 1430 cm^{-1} which is characteristic for the Si-C(phenyl)

Table 2

Ratio of the Space Requirement of the Macromolecules in Solution (F_L) and Grafted Onto the Silica Surface (F_O)

	Reaction	$\bar{M}$ or $\bar{n}$	F_O/F_L
a)	radical	350,000	0.148
		560,000	0.132
		1,460,000	0.035
b)	anionic	10	0.266
		20	0.143
		40	0.077
		80	0.042
		160	0.022

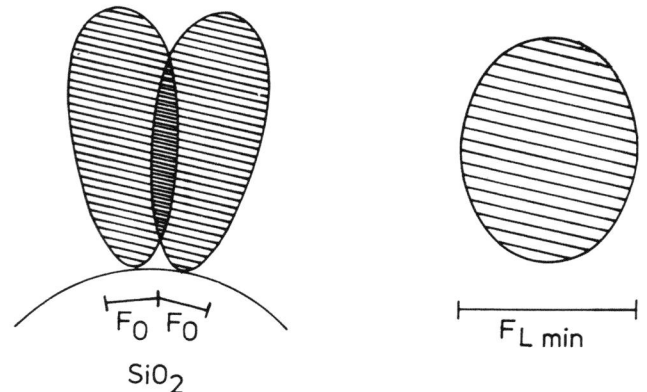

Fig. 4. Structure of the macromolecules grafted onto silica and in solution (schematically).

groups is hidden in the spectrum of the grafted silica by bands of the polystyrene, but is sharp in the spectrum of the phenylated silica. According to the reaction mechanism of the initiating reactions, the grafted polymer molecules are covalently bonded to the surface via these phenyl groups.

Direct evidence for the Si-C-bond between anionically grafted polymer molecules and the silica surface may be obtained through a hydrolytic investigation. Silica grafted with polystyrene is hydrolyzed with (i) HF and (ii) with HF/D_2O; simultaneously (iii) a mixture of homopolystyrene and silica is also treated with HF/D_2O. Afterwards, the isolated polymers are pyrolyzed and the resulting fragments examined using mass spectroscopy. A simple positive monodeuterostyrene fragment resulting from a covalent Si-C-bond and having a M/e value of 105 has only been found in case (ii).

Adsorption-desorption experiments show clearly, further, that approximately ten to one hundred times as much polymer can be bound to the surface by means of polyreactions (a) and (b) than can be irreversibly adsorbed, see Table 3.

Table 3

Adsorption of Polystyrene (from toluene) Onto Silicas, Compared with Radical and Anionic Graft Reaction

adsorbents	polymer coverage (g/g SiO_2)		polymer coverage (g/g SiO_2) by polyreactions
	before / after adsorption	after desorption	
SiO_2-⟨phenyl⟩	0.059	0.262	0.070	a) radical
SiO_2-⟨phenyl⟩-NH_2	0.040	0.250	0.049	$0.1 < \theta < 2.000$
SiO_2-PS-cov.	0.421	0.626	0.422	
SiO_2	0	0.340	0.030	
SiO_2-PS-cov.	0.240	0.345	0.240	b) anionic
SiO_2-PS-cov.	0.900	0.990	0.900	$0.1 < \theta < 1.800$
SiO_2-PS-cov.	1.600	1.680	1.600	

Polystyrene is adsorbed onto silica, phenylated silica, aminophenyl-SiO_2 and silica grafted with polystyrene. It is then step by step desorbed using toluene, until the amount of polymer coverage remains constant (4-5 desorption steps). In the cases of silica, phenylated silica and aminophenyl-SiO_2, approximately 0.03 and 0.01 gms of polystyrene per gram SiO_2, respectively, is irreversibly bound, while in the case of the grafted silica the adsorbed polystyrene may be quantitatively separated by means of desorption. The amount of polymers covalently bonded to the surface through reactions (a) and (b) reaches up to 2000 mg per gram of solid, and does not change with repeated desorption.

REFERENCES

1. R. Kroker, M. Schneider and K. Hamann, Progr. Org. Coatings 1, 23 (1972).
2. R. Kroker and K. Hamann, Angew. Makromol. Chem. 13, 1 (1970).
3. K. Nollen, V. Kaden and K. Hamann, Angew. Makromol. Chem. 6, 1 (1969).
4. E. Dietz, N. Fery and K. Hamann, Angew. Makromol. Chem. 35, 115 (1974).
5. N. Fery, R. Laible and K. Hamann, Angew. Makromol. Chem. 34, 81 (1973).
6. J. Horn, R. Hoene and K. Hamann, Makromol. Chem. (1975), Supplement Band 329, June 1975.

The Role of Interfacial Processes in Adhesion

Robert J. Good

Department of Chemical Engineering

State University of New York at Buffalo

Buffalo, New York 14214

The interfacial processes that influence adhesion are examined. These are the microscopic processes (wetting, adsorption and charge transfer) and the macroscopic processes (elastic energy storage, the elimination of voids, and dissipation of energy during separation). Some novel components of this treatment are: A discussion of the thermodynamic nature of an interface under tension; the heat capacity of the interfacial region; the wetting of a polymer by a liquid that swells it; the storage of elastic energy in the region of the interface; the range of conditions under which craze material will contain two-component fibrils. A condition which will lead to high, practical, interfacial adhesion strength is that craze fibrils should contain strands of both polymeric species, bridging interfacial crazes. If grafting and chemisorption are absent, this condition should exist only for a narrow range of matching of properties of the two phases.

INTRODUCTION

This paper will evaluate the relative importance of the principal interfacial processes in adhesion. These processes are of two kinds: The microscopic, involving the formation of intimate molecular contact between phases, and the macroscopic or mechanical, involving the transfer of work or force from one phase to another, in tension or in shear. This subdivision of the processes corresponds to the set of complementary definitions of the work, adhesion, which has recently been proposed[1].

The microscopic processes are:
- Wetting
- Adsorption
- Charge transfer

The macroscopic processes and properties are:
- Load-bearing (energy storage) by the interface itself
- Wetting as a mechanism of elimination of voids
- Electrostatic contribution to crack-closing force
- Dissipation of energy in the separation process.

Before these processes can be discussed, it is essential to clarify what the term, "interface" means, physically.

DEFINITION OF "INTERFACE"

An interface is a region where there is a steep gradient in the local properties of the system. This may be a gradient in intensive properties such as chemical composition, as when two immiscible phases are in molecular contact; or it may be a gradient in structure, as with the interface between two phases which have the same chemical composition, but differ in density or arrangement of atoms or molecules. There will, in general, be a gradient across the interface, of other properties such as elastic modulus, Poisson's ratio, refractive index, etc.

Chemists commonly envision interfaces as existing between isotropic phases, and the properties that have the steep gradients as being scalar functions. More generally, the properties which have the steep gradients across an interface may, themselves, be anisotropic, and hence require vectorial descriptions. This is particularly true for polymeric solids.

For chemical purposes, it commonly fulfills first-order operational requirements, to consider a polycrystalline material as a single phase. Thus, the chemical potential of iron is effectively constant throughout a solid piece of iron, at equilibrium. Grain boundaries can often be considered to be of second-order importance, although in etching processes, or in corroding systems, they can be of primary importance.

When a polycrystalline solid is under stress, the anisotropic character of many physical properties will be manifested, often quite strongly. This is obviously true in crystalline or cross-linked polymer systems. (It is also true of an amorphous polymer above its T_g, when in a nonequilibrium state of deformation.) As a direct consequence, a crystal grain boundary exhibits the <u>mechanical</u> character of an interface, for there is a steep gradient of one or more <u>vector</u> properties, across the interface. Such a gradient can be described in tensor notation.

The condition of "adhesion" can be defined[1] by the macroscopic specification that the stress tensor is continuous, across the interface. This is a mathematical generalization of the specification[2], that force or work, in tension or shear, be transmissible from one phase to the other. In molecular terms, the condition of adhesion can be defined as the existence, and persistence during stress or deformation, of molecular contact between the adherends, and the persistence of appreciable intermolecular or interatomic attractions (which may or may not include chemical bonds) across the interface.

A gradient of properties is considered to be "steep" if the distance, λ_{12}, over which the gradient exists is small compared to dimensions that can be observed conveniently in the laboratory. Historically, observation was specified to be by light microscopy, and so the practical condition was $\lambda_{12} < 10^{-5}$ cm, for a region of space to be considered an interface. (Interfacial gradients less steep than this can be observed in special cases, such as a two-phase system near a critical point; but when λ_{12} approaches macroscopically observable dimensions, the word <u>phase</u> loses its meaning.) Regardless of ease of observation, λ_{12} must be small compared to the dimensions of the two phases. Otherwise, the system would be better described as an adsorbed layer on a condensed phase, rather than as a system of two condensed phases.

The steepest concentration gradient that can exist is one that extends over a space equal to the pertinent equilibrium interatomic distance in the system. For an interface between two metals, this would mean $\lambda_{12} \simeq 2$ to 3 Å; for two immiscible polymers, λ_{12} would be in the vicinity of 5 to 10 Å. It is not strictly proper to speak of a "gradient" existing over distances any smaller than this, because on the scale of one angstrom, ordinary matter cannot be considered to be homogeneous, even if its average density of composition is constant right up to a Gibbsian dividing surface[3]. So this phraseology must be interpreted with a molecular picture: The "steepest possible gradient" exists when there is no molecular mixing, in the layers on either side of a geometric surface, over and above the average degree of mixing in the bulk phases.

In most real systems at temperatures above 0°K, the gradient of composition is not as steep as the maximum. A surface of a condensed phase is normally in a dynamic state, with molecules entering and leaving it with frequency that can easily be as high as 10^5 sec^{-1}. (The degree of such activity in a surface is much lower than this, for high-melting solids.) For a liquid-vapor interface, there is a fractional concentration of "holes" in the surface layer that is of the order of 0.5, according to estimates that have been made for the purpose of explaining surface entropy[4,5]. The thickness of the overlap region between two unlike polymers is not as yet known; recent estimates have ranged from about 3 Å[6] for two polymers in the absence of grafting, to 11 Å for the inter-phase region in a block-polymer system[7].

The principal thermodynamic properties of interfaces have been widely discussed. However, one property that is pertinent to adhesion has not been treated adequately, up to the present. This is the heat capacity per unit area of a layer of small but arbitrary thickness. (We do not refer, here, to the Gibbsian surface excess heat capacity of an interface between two condensed phases, which is, in general, small.) Formally, we define this property as follows: $C_a(\lambda)$ is the heat capacity per unit interfacial area, for a layer of thickness λ.

$$C_a(\lambda) \equiv C\rho\lambda \tag{1}$$

where C is the mean specific heat and ρ the density. If $\lambda = \lambda_{12}$, C_a is the total heat capacity associated with the region where there is a concentration gradient. The specific heats of common polymers are in the range, 1 to 1.5 joules/gm; so for $\lambda = 10$ Å, $C_a(10)$ is in the range 1 to 1.5 x 10^{-7} joules/cm^2. We will examine, later, the case where λ is the thickness of the region that undergoes plastic deformation during separation.

THE MICROSCOPIC (i.e. MOLECULAR OR ATOMIC) INTERFACIAL PROCESSES

A. Wetting

The direct measure of the wetting of a solid by a liquid is the contact angle, θ. It was pointed out a few years ago[8], and it is now well recognized, that optimal wetting (which results in maximum contact in the atomic level) is a necessary condition for the production of a system with high strength of adhesion. Phenomeologically, it is known that for any smooth solid, there exists a "critical surface tension for wetting", γ_c, such that any liquid having surface tension less than this value is found to wet the solid with zero contact angle. Fox and Zisman[9] proposed a method of determining γ_c, by plotting $\cos \theta$ vs γ_ℓ for an appropriate series of liquids on a given solid, and extrapolating to zero contact angle. Good and Girifalco[10,11] proposed two alternate methods of handling contact angle data, which will be described below. These methods have the advantage over the earlier technique, in that there is no <u>a priori</u> basis for making a <u>linear</u> extrapolation of $\cos \theta$ vs

INTERFACIAL PROCESSES

γ_ℓ. (Indeed, it has been pointed out[12] that the observed data actually fall on a smooth but notably nonlinear curve.)

The theory of Good and Girifalco is easily derived[10,13]. We will review the theoretical basis here, in order to introduce a contribution to the theory of contact angles which has not been published heretofore.

Consider the free energy of cohesion of phase i, ΔG_i^c, and the free energies of adhesion, ΔG_{12}^a, between phases 1 and 2.

$$\Delta G_i^c \equiv -2\gamma_i \qquad (2)$$

$$\Delta G_{12}^a \equiv \gamma_{12} - \gamma_1 - \gamma_2 \qquad (3)$$

We then define the thermodynamic function Φ_{12}:

$$\Phi_{12} \equiv \frac{\Delta G_{12}^a}{\sqrt{\Delta G_1^c \, G_2^c}} \qquad (4a)$$

$$\equiv \frac{\gamma_1 + \gamma_2 - \gamma_{12}}{2\sqrt{\gamma_1 \gamma_2}} \qquad (4b)$$

(The sign of the square root, in Equations (4a) and (4b) is chosen so that Φ is positive.)

Equation 4, being a thermodynamic relation, adds nothing to science that is not implicit in Equations (2) and (3). Where new scientific knowledge appears in the method of evaluating Φ from molecular, (i.e. non-thermodynamic) properties. It has been shown[10,11] that Φ_{12} can be evaluated, as

$$\Phi_{12} = \Phi_{12}(\alpha_1, \alpha_2, \mu_1, \mu_2, I_1, I_2) \qquad (5)$$

where α is molecular polarizability, μ is dipole moment, and I is ionization energy. For polymers, α and μ are the polarizability and dipole moment per segment. An equation was reported[11,13] by which Φ_{12} can be computed with excellent accuracy, and which contains no adjustable parameters:

$$\Phi_{12} = \left[\frac{\dfrac{1.5\, \alpha_1 \alpha_2 I_1 I_2}{I_1 + I_2} + 0.33(\alpha_1 \mu_2^2 + \alpha_2 \mu_1^2) + \dfrac{0.29\, \mu_1^2 \mu_2^2}{kT}}{\left\{ 0.75 \alpha_1^2 I_1 + 0.66 \alpha_1 \mu_1^2 + \dfrac{0.29 \mu_1^4}{kT} \right\}^{1/2}} \right]$$

$$\left[\frac{1}{\left\{ 0.75 \alpha_2^2 I_2 + 0.66 \alpha_2 \mu_2^2 + \dfrac{0.29 \mu_2^4}{kT} \right\}^{1/2}} \right] \quad (6)$$

Figure 1 shows a graph of this function, for water vs. organic liquids. The range of α from 5 to 10 x 10^{-24} cm^3, includes most common organic compounds and polymers.

In the interests of convenient computations, we now introduce a simple approximation to Φ, which is only a little less accurate. For water vs organic liquids, as a function of the dipole moment μ_2 of the organic compound,

$$\Phi = 0.55, \qquad \text{for } 0 \leq \mu_2 \leq 0.7$$
$$\Phi = 0.38 + 0.24\mu_2 \qquad \text{for } 0.7 \leq \mu_2 \leq 2.6 \quad (7)$$
$$\Phi = 1, \qquad \text{for } 2.6 \leq \mu_2 \leq 4$$

This applies to non-hydrogen-bonding organic compounds. For water vs organic compounds that form strong hydrogen bonds, e.g. alcohols, $\Phi \simeq 1$. The π electrons of nonpolar aromatic compounds take part in the formation of weak hydrogen bonds; and for benzene, toluene, etc., $\Phi \simeq 0.70$. (This number cannot be predicted <u>a priori</u> from molecular properties, because the strength of the hydrogen bond between water and the π electrons of an aromatic ring is now known with any accuracy.) For water vs nitro-aromatic compounds, $\Phi \simeq 0.8$. Aromatic molecules are discussed in detail, in Reference (13).

Equation (4) can be rearranged into the form:

$$\gamma_{12} = \gamma_1 + \gamma_2 - 2\Phi_{12} \sqrt{\gamma_1 \gamma_2} \quad (8)$$

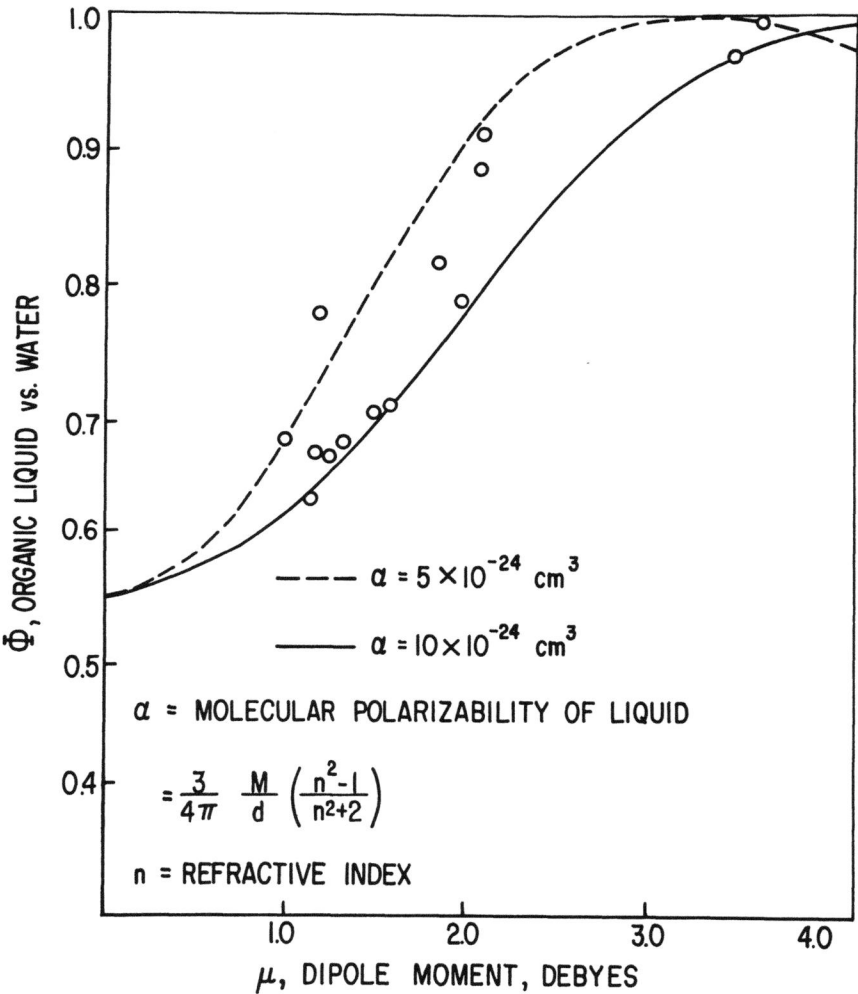

Fig. 1. Interaction parameter Φ vs. dipole moment of organic compound, μ, for water vs. organic liquids. Circles: experimental values of Φ, from interfacial tension data, using Equation 4b. Lines: computed values of Φ, for water vs. model organic compounds with variable dipole moments, using Equation 6.

Estimates of γ_{12} have been made[13] using Equation (6), that are accurate within one dyne/cm for water vs saturated hydrocarbons, and within 5 dynes/cm for water vs polar compounds other than nitroaromatics.

It must be emphasized that γ_1 and γ_2, in Equations (2,3,4 and 8), are the surface tensions of the pure liquids, not the mutually saturated solutions. Physically, this is a consequence of assuming that, in a <u>binary</u> liquid-liquid system, there is not any appreciable surface excess of either component at the interface. This is reasonable for a binary system; but in a ternary system, it is not permissible to make any such assumption; c.f. Melrose[14].

It has been shown[13,15] that, when mutual solubilities of the two phase are high, γ_1 and γ_2 in Equations (4) and (8) must be replaced by the concentration-weighted average surface tensions:

$$g_a \equiv \phi_1^a \gamma_1 + \phi_2^a \phi_2$$
$$g_b \equiv \phi_1^b \gamma_1 + \phi_2^b \gamma_2 \quad (9)$$

Here the ϕ's are volume-fraction concentrations of substances 1 and 2, in the saturated phases a and b. Then Equation (8) becomes:

$$\gamma_{ab} = g_a + g_b - 2\Phi_{ab}\sqrt{g_a g_b} \quad (10)$$

This expression is important in the consideration of solubility effects in wetting.

When a pure liquid is placed on a solid, the relations given above may be applied--with the amendment, of course, that γ for a solid-vapor or solid-liquid interface must be interpreted as a surface free energy, not a surface tension. (Surface tension and surface free energy are identical only for fluid interfaces.)

$$\gamma_{sv} - \gamma_{sl} = \gamma_{lv} \cos\theta \quad (11)$$

γ_{sv} is the surface free energy of the solid in the saturated vapor of the wetting liquid, and γ_{lv} is the surface tension of the liquid saturated with molecules from the solid. The film pressure, π_e, is defined as

$$\pi_e \equiv \gamma_s - \gamma_{sv} \quad (12)$$

It has recently been shown theoretically[16] that π_e will be negligible for a low-energy surface such as polymethylene, in contact with a liquid such as water, which forms a non-zero contact angle on it. If the surface is a poorly-prepared one, or an intentionally oxidized one, the value of the surface free energy, γ_s^* may be larger than γ_s. In such a case, the value of π_e may be appreciable[16].

Combining Equations (8) and (11), and neglecting π_e, we obtain;

$$\cos\theta = -1 + 2\Phi\sqrt{\frac{\gamma_s}{\gamma_{\ell v}}} \qquad (13)$$

$$\gamma_s = \frac{\gamma_{\ell v}(1+\cos\theta)^2}{4\Phi^2} \qquad (14)$$

$$\gamma_{s\ell} = \gamma_{\ell v}\left[\frac{(1+\cos\theta)^2}{4\Phi^2} - \cos\theta\right] \qquad (15)$$

Young's equation is limited to the range of values of γ_s, γ_ℓ and $\gamma_{s\ell}$ such that

$$\frac{\gamma_{sv} - \gamma_{s\ell}}{\gamma_{\ell v}} \lesseqgtr 1 \qquad (16)$$

The limiting case, in which $\cos\theta \to 1$, is important. Let $\gamma_{\ell v} \to \gamma_c$ as $\cos\theta \to 1$. Then

$$\gamma_s = \frac{\gamma_c}{\Phi_c^2} \qquad (17)$$

For the purposes of Equation (17), γ_c is the constant that is characteristic of the combination of the solid with a real "γ_c liquid", if one exists. If an extrapolation is employed in determining γ_c (without use of theoretical values of Φ) it should be an extrapolation of a plot of $\cos\theta$ vs. $1/\sqrt{\gamma_\ell}$, which is much more linear than the Fox-Zisman extrapolation. Then five cases must be distinguished:

(1) $\Phi = 1$ for all members of the series of liquids, vs the solid. (This is rare).

(2) Φ is the same, for the solid vs all members of the series employed in the extrapolation. (This is probably a good approximation when the liquids are saturated hydrocarbons, with most low-energy solids.) This value of Φ should be computed and used in Equation (17). Very often, it is close to unity.

(3) Φ varies systematically in the series. The Φ_c is the appropriately extrapolated value of Φ. There is reason to believe that $\Phi_c \simeq 1$ in a large number of cases. See Neumann, et al.[17] for a very useful treatment that is rigorously applicable with such systems, and a good approximation with others.

(4) Φ does not vary systematically with γ_ℓ. This condition can be expected to hold when the test liquids do not form a strictly homologous series. It is probably true of many sets of measurements on polar polymers that have been reported, using liquids with $\gamma > 30$. In this case it is difficult to choose what value of Φ_c to use, and so estimates of γ_s from γ_c extrapolations are not as reliable as in cases (1) to (3). A convenient method of estimating γ_s with the aid of a single contact angle measurement has been developed by Neumann et al., based on the existence of a correlation between Φ and $\gamma_{s\ell}$ that is much better than that between Φ and γ_ℓ. See Reference (17) for details.

(5) $\Phi > 1$ as a consequence of specific interactions. Examples are, hydrogen bonding between nitrile groups in one polymer and the hydrogens of CHCl groups in another polymer. This kind of interaction in low molecular weight systems leads to complete miscibility. When one phase is polymeric, it leads to complete miscibility of the polymer with the liquid, unless solubility is restricted on account of crosslinking, crystallinity, loading with reinforcing filler, etc. For two amorphous, uncrosslinked polymers, it can lead to complete compatibility. In principle, it should be possible to estimate Φ from the heat of mixing. The release of heat, when the solvent-free polymer (e.g. anhydrous cellulose) is swelled by the solvent, should be found to correlate with the existence of a value of Φ greater than one, and hence (see below) with low or zero contact angle of a liquid on a polymeric solid which is swelled by the liquid. A secondary effect may appear, in such systems, that the solvent may interact strongly enough with the polymer segments than it would be negatively adsorbed at the swelled polymer-vapor interface. (This should not occur at the polymer-liquid interface.) Since, in such a system, $\gamma_{s\ell}$ would be extremely low, or even negative, the contact angle should be zero. See below.

Clearly, γ_s is a more fundamental property of the solid than is γ_c; and the physical significance of γ_c (as Zisman has repeatedly pointed out) is as "a measure" of the surface free energy of the solid, not necessarily identical with γ_s.

When the solid is a polymer that is swelled by the wetting liquid but is not dissolved by it, Equations (9) and (10) must be employed. Swelling is appreciable only when the cohesive energy densities of the liquid and the solid are nearly the same; so Φ_{ab} will be approximately unity (or greater). We replace γ_s (the γ of the unswelled polymer) with g_s, defined by

$$g_s \equiv \phi_s^s \gamma_s + \phi_\ell^s \gamma_\ell$$

Since $\phi_s^\ell \approx 0$, $g_\ell = \gamma_\ell$. Here ϕ_s^s is the volume fraction of polymer in the swelled-polymer phase. We consider first the case in which $\Phi = 1$. Equation 13 is replaced by

$$\cos\theta = -1 + 2\sqrt{\frac{g_s}{\gamma_\ell}} \qquad (18a)$$

$$= -1 + 2\sqrt{\phi_\ell^s + \phi_s^s \frac{\gamma_s}{\gamma_\ell}} \qquad (19a)$$

As can be seen from these equations, θ will be lower for the liquid on the swelled polymer, than on the same polymer if it were not swelled, or if it were swelled to a lower degree. But swelling alone will not lower θ to zero, if $\gamma_\ell > \gamma_s$.

To be more general, we now allow for the possibility that $\Phi > 1$, as in case 5, described above. Then we write,

$$\cos\theta = -1 + 2\Phi_{ab}\sqrt{\frac{g_s}{\gamma_\ell}} \qquad (18b)$$

$$= -1 + 2\Phi_{ab}\sqrt{\phi_\ell^s + \phi_s^s \frac{\gamma_s}{\gamma_\ell}} \qquad (19b)$$

Now, unlike the previous case, it is evident that θ may be zero. As noted above, it is, in principle, possible to predict whether or not Φ will be greater than one, due to specific interactions.

Hysteresis of contact angles is a practical complication; all the above discussion ignored hysteresis. On very smooth, homogeneous, rigid solids, with slow movement of the liquid-vapor front, hysteresis is small. It is not yet known whether a truly hysteresis-free system of a liquid on a bulk polymer, with θ > 0, exists. The lowest hysteresis observed by the author with a polymeric solid is of the order of 2 degrees. (A method of silicone treatment for glass, which yields a surface on which hysteresis is reported to be zero, has been described by Neumann et. al.[18] and by Hertzberg et.al.[19].) It is known experimentally (and explained theoretically) that hysteresis increases with increasing roughness and heterogeneity[20-25]. On a smooth, binary surface, it is believed that the advancing contact angle, θ_a, is characteristic of the low-energy component, and the retreating angle, θ_r, is characteristic of the higher-energy component of the solid. The sensitivity of estimates of γ_s to uncertainties in θ, e.g. due to hysteresis, may be determined by differentiating equation (14) with respect to θ. The result is

$$\frac{\delta\gamma_s}{\delta\theta} = \frac{-\gamma_{\ell v}\sin\theta(1 + \cos\theta)}{2\Phi^2} \qquad (20)$$

For a typical organic liquid on an organic polymer, a 1° uncertainty θ corresponds to an uncertainty that may be as large as about 3 ergs/cm^2 in γ_s.

With regard to adhesion, the advancing contact angle is of far greater practical importance than the receding angle. This is quite obvious, because the wetting phase advances over unwetted areas, in making molecular contact.

B. Adsorption

If there is chemisorption of the molecules of the liquid onto the solid, then the contact angle of the liquid will, in general, be zero. The so-called autophobic systems[26] comprise the exception to this rule. For such systems, the strong adsorption does not contribute to strong adhesion, because the bulk liquid (or adhesive) does not adhere strongly to the chemisorbed layer. Indeed, the formation of an autophobic film may contribute to the prevention of strong adhesion. In polymeric adhesive systems, the chemisorbing group is often a unit of (or side chain on) the polymer structure of the (initially) liquid phase. This is a condition highly conducive to strength in the adhering system. In addition to contributing to macroscopic energy-dissipating processes (which will be discussed in the last section of this paper) chemisorption bonds tend to resist environmental attack, e.g. by water or heat, and to be far more permanent than the bonds due to so-called "physical" forces. The beneficial influence of chemisorption is, thus, to be expected on theoretical grounds; and (as is well known) it is found to be extremely beneficial in practice.

Physical adsorption is generally present, whether chemisorption occurs or not. Physical adsorption involves interaction energies that are weaker than those of chemisorption, by one to two orders of magnitude; hence the ease of displacement of physically adsorbed material, by more-strongly adsorbed species; and hence also the greater ease of thermal detachment. Yet the _forces_ of physical adsorption are, very commonly, strong enough to produce a zero contact angle for the liquid on the solid. All that is required is that the energy of attraction of the solid for the molecules of the liquid be stronger than the energy of attraction of the liquid molecules for each other. See Reference (16). For example, physical adsorption is always present in a fluorocarbon-hydrocarbon system, but the contact angle depends on which component is solid: Any fluorocarbon will adsorb, in a thick multilayer, on a hydrocarbon polymer; and the contact angle is zero. But a higher n-alkane such as decane will adsorb only to a very small fraction of monolayer coverage on a flat, homogeneous surface of a fluorocarbon polymer[16]; and the contact angle will be high. Thus, θ for decane on teflon TFE is observed to be about 40°[9,27]. The contact angle _per se_ is not a _direct_ thermodynamic measure of the interaction between two phases; that is given by the free energy of adhesion, Equation (3), and the Young-Dupre equation,

$$\Delta G^a = -\gamma_\ell (1 + \cos\theta) \qquad (21)$$

Physical adsorption can lead to a large energy of adsorption per polymer molecule, provided the molecule is adsorbed at the solid surface at several points along the chain. Thus, an energy of adsorption of, say 1 kcal per mole of segments, times 50 segment sites of adsorption, corresponds to 50 kcal per mole, which is energetically equivalent to a chemisorption bond. We will discuss, below, the relation between the thermodynamic work of adhesion and the mechanical strength of an adhering system.

C. Charge Transfer

Charge transfer across an interface is a microscopic process, which must occur in essentially every two-phase system. This is the process of contact electrification, and it is closely related to the most ancient of electric phenomena, tribo-electricity. (It is not to be confused with ionization of atoms or molecules.) Once the charge separation has been accomplished, there will be an electrostatic contribution to the mechanical strength of the system. However, it is difficult to measure the charge transfer without separating the two phases; and so there has been little success, to date, in quantitative measurement or prediction of the charge separation or of its direct contribution to adhesion. As a result, discussion of this component of adhesion must be somewhat speculative. The remarks here will be confined to the role of the interface; and in this regard, matters are fairly clear.

The interface is the locus of transfer of charge. On a metal an oxide may act as an insulating layer, vs. electronic charge transfer; but we are not concerned with such effects. First, it is well known that for two metals to adhere at all, the oxide must first be removed, e.g. with the aid of a flux, in soldering. (If the oxide is not removed, then if any adhesion is present, it is between the two metals and the oxide, i.e. with the oxide acting as an adhesive or cement.) Second, adhesion of a polymer or other nonmetal to a metal, <u>in practical systems</u>, is never direct, but through the oxide that is always present. The oxide may, of course, be modified or replaced by a chemical film, e.g. a phosphate on iron. All the concepts that are discussed here regarding the oxide apply to the "chemical film". Thus, the charge transfer between phases that are separated by an oxide layer is of little importance.

After the charge separation has been accomplished, there will be an electrostatic contribution to the force across the interface.

Just how that force manifests itself will depend strongly on the mechanical properties of the two phases and, possibly, on their electrical conductivities. We will return to this question as part of the discussion of the macroscopic processes.

MACROSCOPIC PROCESSES

A. Load-bearing by the Ideal Interface

The storage of elastic energy is the measure of the ideal load-bearing contribution of any region of an adhering system. Since the thickness, γ, of the interfacial region is small relative to the thickness of[12] either phase, the absolute amount of energy that can be stored is small. If the energy of the chemical bonds across the interface is, say, the geometric mean of the bond energies in the bulk phases, then the contribution to the load-bearing of the system, due to the interfacial bonds, will be intermediate between the contributions made by comparable volumes in the one phase and in the other. If the cohesive forces in each of the two phases are strong (e.g. covalent bonds) and if the interfacial forces are weak (e.g. van der Waals force) then the interfacial contribution will be small.

The qualitative estimate just given, and quantitative estimates of this type which have been made in the past[28] (including those based on the free energy of adhesion, Equation 21) are generally irrelevant. This is because they are based on a totally unrealizable idealization: Two rigid solids, perfectly parallel, with a flaw-free interface, and no stress concentrations arising during loading, such as shear stresses at the boundary of the interface due to differences in elastic modulus and Poisson ratio. There is not even a distant approach to any of these four abstractions, in real systems. The departures of practical adhering systems are so serious that it is to be expected (and it is found in practice) that experimental results would be several orders of magnitude below the predicted values. (See Reference (29).) The discrepancy resembles, both in magnitude and in physical origin, the discrepancy between the observed tensile strength of a single-phase solid and that calculated for an ideal single crystal. It will be remembered that that failure of prediction led to the invention of dislocations, and to the Griffith crack theory.

Interfacial voids or cracks, and specimen boundaries, as stress concentrators, have been given extensive consideration by others. We will concentrate, below, on the prevention of voids, by achieving optimum wetting; on their inactivation, by an increment to crack-closing force; and on the dissipative processes in separation, particularly crazing.

INTERFACIAL PROCESSES

B. Wetting and the Elimination of Interfacial Voids

Wetting as a microscopic process has already been discussed. Wetting is also a macroscopic process, because it involves bulk flow of one phase over the other. It is the process by which interfacial voids, and particularly re-entrant voids, are eliminated or prevented. (This is a viewpoint that has been advocated by Zisman for nearly two decades[8,30].) This elimination of voids can be accomplished effectively only if one phase is applied to the other as a liquid, or passes through a fluid stage during processing (e.g., with heat or pressure) and if the advancing contact angle of the liquid on the solid is very low or zero.

We may now point out a mechanical aspect of the void-filling functions. It may be noted[1] that continuity of the stress tensor, between the two solid phases, is a necessary condition for strong adhesion. When interfacial voids are filled, the regions which would have been weak because of stress concentrations around voids are replaced with regions where there is no discontinuity of the stress tensor. The void-filling function, performed by an adhesive, is often more important than its possession of <u>high</u> shear or tensile strength.

It has been known for some years that there is a correlation between the thermodynamic work of adhesion and the mechanical strength of adhesive-bonded systems. See the paper by Mittal[31], in this volume, where an extensive review of the subject is given. All attempts at numerical prediction of mechanical strength of adhering systems from the thermodynamic work of adhesion have led to numbers that are orders of magnitude away from the observed values. From these facts, we may conclude that a <u>direct</u> causal relation between mechanical strength and thermodynamic properties (surface free energies and contact angles) does not exist. Rather, both the mechanical strength and the thermodynamic properties are causally related to a set of other variables, which must be the molecular properties of the phases. The list of pertinent properties must include polarity, segment size, chain stiffness, chain branching, etc.--those properties that control rheological and solubility behavior of polymers.

C. Electrostatic Contributions to Crack-Closing Force

The electrostatic force, which develops following charge transfer, obviously must make a contribution to the load-bearing capacity of an ideal interface. Cf. sections III C and IV A, above. In this regard, it is roughly equivalent to an augmentation of the elastic modulus. But another mode of action must be present in non-ideal (i.e., real) systems which, because it relates to local stress amplification, should be much more important than the contribution to the (never-observed) ideal strength. Also, this mode of action is more truly attributable to the interface.

If a crack or void exists at an interface, then the electrostatic force across the void will be opposite to any tensile stress that is applied. Acting across a gap, electrostriction does not have the rapid decay of force with increasing distance that is characteristic of interatomic or intermolecular forces. If the charges are not conducted away from the facing crack surfaces, the electrostatic force of attraction provides a crack-closing force, and hence renders such flaws less dangerous. The load that is necessary to start a flaw growing will be larger, on account of the charge transfer, than it would be in the absence of this contact electrification. This may be mathematically equivalent, in the Griffith-Irwin crack theory[32,33], to an increment to the term. Thus, for fracture of a homogeneous solid, we write

$$\sigma_f \simeq \sqrt{\frac{\mathcal{G} E}{\ell}} \qquad (22)$$

where ℓ is crack length, E is elastic modulus, and $\mathcal{M}$ is the energy requirement per unit length of crack extension. For a two-phase solid with crack in or near the interface, Good has pointed out[34] that E must be replaced by an effective <u>mean</u> modulus, $\bar{E}$, and $\mathcal{M}$ by a similar quantity $\bar{\mathcal{G}}$. When there is a potential difference, V, between the phases, then we must add to $\bar{\mathcal{G}}$ some function of V. We have not yet derived the form of this function, but clearly it increases monotonically with V. It may also depend on crack shape. We write, then,

$$\sigma_f \simeq \sqrt{\frac{\bar{E}[\bar{\mathcal{G}} + f(V)]}{\ell}} \qquad (23)$$

Experimental evidence for this effect has not yet been obtained, so this treatment must be regarded as conjectural at the present time.

D. Energy Dissipation in the Separation Process

Finally, we examine the role that interfacial processes play in the dissipation of energy, during a real separation. It is the dynamic, dissipative processes in crack extension that provide resistance to separation. When the energy requirements for these dissipative processes are too large (under a specified loading) then crack-extension does not proceed, and it is said that the limit of strength of the system has not been exceeded. When the loading system, and the elastic energy stored in the two solid phases, can supply sufficient energy to fill the requirements of the dissipative processes, then these proceed, and some preexisting void starts to grow, as a propagating crack. (If there is no preexisting void, one

INTERFACIAL PROCESSES

may be nucleated out of dislocations.)

The dissipative processes in adhesive separation[2] and fracture[35,36] release heat. It is well known[37] that the layer in which local mechanical deformation takes place is relatively thick. The mechanism which causes the layer to be thick may be regarded as a thermal one (particularly for polymers) as may be seen from the following argument.

The fracture energy, $\mathcal{H}$, for a brittle polymer such as polystyrene, is reported to be of the order of 10^3 joules/m^2 [38], or 0.1 joule/cm^2. This is three orders of magnitude larger than the energy required to break the covalent bonds in unit cross section area of polymer. (Assuming 25 A^2 per molecule--which corresponds to close-packed, fully-oriented polymethylene--and 80 kcal per mole of carbon-carbon bonds broken, the required energy input is, at most, about 2000 ergs/cm^2. For an unoriented, linear polymer, with molecular cross section larger than that of $(CH_2)_n$, the energy would be nearer 500 ergs/cm^2 [39]). The excess energy over that required to break covalent bonds must be dissipated as heat, and must raise the local temperature.

The mean temperature rise can be estimated as follows: Assume the processes to be locally adiabatic within the time regime of the fracture process; and assume the final temperature to be uniform over the layer whose thickness is λ. The temperature rise is[2]

$$\Delta T(\lambda) = \mathcal{H}/C_a(\lambda) \qquad (24)$$

We have already estimated the heat capacity, $C_a(10A)$, to be around 1.5×10^{-7} joules/cm^2; so for an interfacial region only 10A thick, the temperature rise would be over $10^5 °C$. This is far above the pyrolysis temperature of any organic compound, and indeed, it is above the melting point of any known solid. If we assume $\Delta T = 150°$, i.e. that the temperature will rise to some region above T_g, then we find that $\lambda \sim 4 \times 10^{-4}$ cm. This is about the right order of magnitude, for interference colors on fracture surfaces of a polystyrene and PMMA[39] indicate the formation of a drastically disturbed zone, the thickness of which is of the order of the wavelength of visible light.

We may conclude that the irrecoverable plastic work, in fracture, cannot be confined to the molecular region of the interface, but it must be dissipated in regions that are the order of 10^4 A thick for common glassy polymers.

We can further explain the temperature rise, in terms of the model in which a craze forms at the tip of a crack in a glassy polymer under stress, and advances ahead of it[40]. (The extensive work of Kambour and others, in the last decade or so, has elucidated the nature of crazes. It is believed that in a craze, there exist micro-voids, and that the region is bridged by minute, strong fibrils or strips of polymer, drawn out of the bulk phase.) With static loading below the stress that will cause failure, this occurs isothermally, and with a stable crack and craze which advance very slowly; the visco-elastic description of this process must be <u>creep</u>. When the stress is increased, from zero up to the critical load to initiate this slow propagation, there is a dissipation of energy due to the creep process. Because of the finite thermal conductivity of the solid, the temperature rises locally, by no more than a very small fraction of a degree. Cf. the Gruntfest theory of thermal feedback in fracture[2,35].

At a higher level of loading, the temperature rise is larger, but still small enough that the mechanical properties of the polymer are not appreciably affected. So the rate of growth of the craze, and the crack, remains low. With increasing stress, however, a level is reached where the heat evolved in craze formation cannot be conducted away fast enough for the temperature rise to remain infinitessimal. Then the local rise in temperature at and near the crack tip becomes large enough that there is a local increase in the plasticity of the polymer, as chains become more mobile. This leads to a local increase in the rate of the deformation or crazing process, which causes a further increase of the rate of heat release, and increase in local plasticity; and so on until the local temperature reaches some value which is effectively infinite, e.g. above the T_g of the polymer. Compare the estimate of ΔT given in the preceding paragraph.

The role of the interface in such a thermal feedback process, in a two-phase system, is to provide a mechanism by which the local deformation process can proceed in a band that includes the transition region between the phases, (e.g. an interfacial craze[42]) or cross from one phase to the other. Suppose, for one extreme set of circumstances, that the intermolecular forces between the phases are of comparable strength to the forces within either phase. This means chemisorption, or polymer grafting, with about the same number of strong bonds per unit area as in the bulk. Then it is obvious how the local stress and deformation are transferred across the interface.

At the other extreme, suppose that the intermolecular forces between phases are a great deal weaker than the cohesive forces of either phase. An example is, the two phases being highly crosslinked and there being no chemisorption or grafting across the interface, or interpenetration. Then there will be no mechanism by which a strong, stable fibril or strip of polymer can be formed with one end in one phase and the other in the second phase, and a transition of composition in the middle. The local strength of that transition region will be negligible. So an interfacial craze will not form.

In an intermediate case, consider two linear polymers with about the same T_g, and cohesive energy densities that are not very different, and only physical forces acting across the interface. Consider a chain of one species lying, for a considerable part of its length, in the interface. When this chain is pulled, it exerts about an equal drag on chains of the other species as it does on chains of its own species. Therefore, in the formation of the fibrils that bridge a craze, there will be a good probability that there will be chains of both species drawn into the fibrils. In effect, there will be a region of interface longitudinally down the middle of many of the fibrils. Hence it is possible for an interfacial craze to be stable. This will, in effect, provide strength to the interface, to resist crack propagation, by requiring considerable dissipation of energy in craze formation.

Consider one more intermediate case, in which the cohesive energy densities are close but the T_g's are quite different. Then the craze matter for an interfacial craze will be drawn exclusively from the phase with the lower T_g. There will be orientational strengthening of the fibrils, but essentially no participation of the chains from the phase with the higher T_g, in the fibres. Hence, an interfacial craze would be, essentially, all derived from the phase with the lower T_g. The craze would contribute a negligible increment of strength, or resistance to fracture, to the system. The craze fibrils would terminate with only a very small cross section in butt contact with the phase with the higher T_g; and such contacts would be very easily severed.

Thus, if there is any contribution of the interface to the enhancement of dissipative processes in interfacial separation in systems containing polymers, it is by way of composite fibrils, in interfacial crazes. The conditions of matching properties as between phases, in order for this to occur, must be quite critical. So, in the absence of chemisorption, grafting, solubility or notable interpenetration, it is to be expected that these conditions of matching will be met only rarely.

SUMMARY

We have examined the processes that occur at interfaces, in adhesion. At the atomic level, these are the processes by which molecular or atomic contact is made between phases, covalent and ionic bonds are formed, and charges are transferred. The existence of molecular contact is one necessary condition for strength, in an adhering system. A second necessary condition is that the interface be free from voids or cracks, and that cracks do not nucleate under load or environmental attack or as a result of residual stresses in the system. If interfacial cracks or voids are present (as they often are) or if they nucleate, then a necessary condition for strength is that the energy requirement for initiating crack propagation be large.

An electrostatic component of crack-closing force is postulated, which (in the stages before the development of actual discharges) acts as a conservative, i.e. non-dissipative, component of the resistance to separation. The most important dissipative interfacial process, in systems in which one or both phases is a glassy polymer, is interfacial craze formation. Some details of this process are discussed. It is shown that a particular kind of matching of properties of the two phases can lead to craze fibrils which accomplish strong bridging between phases by being, themselves, two-phase fibres with the phase boundary parallel to the fibre axis. In the absence of such close matching of polymer properties, or of covalent bonding across the interface, interfacial crazing processes will not contribute a resistance to crack-propagation that is comparable to that made by crazing in bulk.

REFERENCES

1. R. J. Good, J. Adhesion (submitted for publication).
2. R. J. Good, *Aspects of Adhesion*, 6, D. J. Alner, Ed., University of London Press, 1971, p. 24.
3. J. W. Gibbs, *Scientific Papers, Vol. I, Thermodynamics*, Dover Publishing Co., N.Y. (1961).
4. A. Skapski, J. Chem. Phys. 16, 386 (1948).
5. R. J. Good, J. Phys. Chem. 61, 810 (1957).
6. P. J. Flory, B. E. Eichinger and R. A. Orwoll, Macromolecules 1, 287 (1968).
7. D. H. Kaelble, Akron Summit Polymer Conference, 1974.
8. W. A. Zisman, Ind. Eng. Chem. 55, 18 (1963).
9. H. W. Fox, W. A. Zisman, J. Coll. Sci. 7, 428 (1952).
10. R. J. Good and L. A. Girifalco, J. Phys. Chem. 64, 561 (1960).
11. R. J. Good, in *Contact Angle, Wettability and Adhesion*, Advan. Chem. Series No. 43, Am. Chem. Soc., p. 74 (1964).
12. R. E. Johnson, Jr., in *Surface and Colloid Science*, E. Matjevic, ed., Wiley, N.Y., Vol. 2, p. 85 (1969).
13. R. J. Good and E. Elbing, Ind. Eng. Chem. 62, No. 3, p. 54 (1970).
14. J. C. Melrose, in *Contact Angle, Wettability and Adhesion*, Advan. Chem. Series No. 43, Am. Chem. Soc., p. 159 (1964).
15. R. J. Good, in *Environment-Sensitive Mechanical Behavior of Materials*, Westwood and Stoloff, eds., Gordon and Breach Publishing Co., p. 541 (1967).
16. R. J. Good, *Spreading Pressure and Contact Angle*, in press, J. Coll. & Interface Sci.

17. A. W. Neumann, R. J. Good, C. J. Hope and M. Sejpal, J. Coll. and Interface Sci. 49, 291 (1974).
18. A. W. Neumann and D. Renzow, Z. Phys. Chem. (Frankfurt am Main) 68, 11 (1969).
19. W. J. Hertzberg, J. E. Marian and T. Vermeulen, J. Coll. and Interface Sci. 33, 1 (1970).
20. A. W. Neumann and R. J. Good, J. Coll. and Interface Sci. 38, 341 (1972).
21. J. D. Eick, R. J. Good and A. W. Neumann, submitted for publication.
22. R. J. Good, J. Am. Chem. Soc. 74, 5041 (1953).
23. R. E. Johnson, Jr., and R. H. Dettre, J. Phys. Chem. 68, 1744 (1964).
24. R. E. Johnson, Jr., and R. H. Dettre, Advan. Chem. Ser. 43, 112 (1964).
25. R. H. Dettre and R. E. Johnson, Jr., *ibid*, p. 136.
26. E. F. Hare and W. A. Zisman, J. Phys. Chem. 59, 335 (1955).
27. A. W. Neumann and W. Tanner, J. Coll. and Interface Sci. 34, 1 (1970).
28. J. J. Bikerman, *The Science of Adhesive Joints*, 2nd Edition, Academic Press, N.Y. (1968).
29. R. J. Good, in *Treatise on Adhesion*, R. L. Patrick, ed., Marcel Dekker, N.Y., Vol. I, p. 9 (1967).
30. W. A. Zisman, Advan. Chem. Ser. 43, p. 1 (1964); R. E. Baier, E. G. Shafrin and W. A. Zisman, Science 162, 1360 (1968); W. A. Zisman, in *Adhesion and Cohesion*, P. Weiss, ed., Elsevier, p. 176 (1962).
31. K. L. Mittal, This Volume.
32. A. A. Griffith, Phil. Trans. Roy. Soc., London, 221, 163 (1920).
33. G. R. Irwin, in *Fracturing of Metals*, Am. Soc. for Metals, p. 147 (1947).
34. R. J. Good, J. Adhesion 4, 133 (1972), also in *Recent Advances in Adhesion*, L. H. Lee, ed. Gordon and Breach, New York (1973).
35. A. A. Wells, Trans. Inst. Welding 15-16, pp. 34r-56r (1952-3).
36. I. J. Gruntfest, in *Fracture of Solids*, D. C. Drucker and J. J. Gilman, eds., Interscience, N.Y. (1963).
37. G. R. Irwin, in *Treatise on Adhesion*, R.L. Patrick, ed., Marcel Dekker, N.Y., Vol. 1, p. 233 (1967).
38. E. H. Andrews, in *The Physics of Glassy Polymers*, R. N. Hayward, ed., Wiley (1973).
39. J. P. Berry, J. Polymer Sci. 50, 107, 313 (1961).
40. S. B. Newman and I. Wolock, in *Adhesion and Cohesion*, P. Weiss, ed., Elsevier, p. 218 (1962).
41. R. P. Kambour, J. Polymer Sci. A2, 4159 (1964); A3, 1731 (1965).
42. R. E. Robertson, J. Adhesion 4, 1 (1972).

Surface Chemical Criteria for Maximum Adhesion and Their Verification Against the Experimentally Measured Adhesive Strength Values

K. L. Mittal

IBM Corporation

MHV Systems Products Assurance

Poughkeepsie, N. Y. 12602

Different surface-chemical criteria -- thermo-dynamic work of adhesion, interfacial tension, penetration, compatibility, and spreading co-efficient -- which are relevant to the adhesive strengths between the adhesives and low-energy substrates are discussed. The conditions optimizing these criteria are elaborated. Subsequently, predictions from these criteria are tested against the available adhesive strength values. In the case of incomplete wetting ($\theta > 0°$), there is an apparent direct correlation between the work of adhesion and the reported adhesive strengths; but such relationship is not valid when the adhesive completely wets the substrates. A careful analysis of the existing literature has revealed that the interfacial tension is the most important surface energetic parameter -- both in incomplete and complete wetting situations -- in determining the adhesive strengths. The lower the interfacial tension, the higher the adhesive strength. Various forms of γ_c's and their relationship to the surface free energies of substrates is discussed.

INTRODUCTION

The term "adhesion" (unmodified) simply signifies the sticking together of two materials, say, A and B; whereas "practical adhesion" (more commonly termed as adhesive strength, joint strength, and mechanical strength) measures the force required to effect separation between two adhering materials. On the other hand, thermodynamic or reversible work of adhesion (commonly denoted by W_A) is the change in free energy when the materials are brought in contact, and it is the same as the amount of work expended under reversible or equilibrium conditions to disrupt the interface.

Throughout the history of adhesion science and technology, researchers have hypothesized that the practical adhesion and the thermodynamic adhesion should somehow be correlatable, but there is no unanimity on this issue. It is universally agreed that the practical adhesion cannot be equated with the thermodynamic adhesion; at the most, one can expect a direct correlation between the two. This is so because (i) during the breaking of the joint irreversible processes such as inelastic deformations take place with the consequent dissipation of energy and (ii) W_A refers to a defect free interface whereas in real situations there are present a large number of flaws between the adhesive and the substrate. So there is no argument concerning the statement that one cannot predict the magnitude of practical adhesion on the basis of a knowledge of W_A.

As regards a possible relationship between the practical adhesion and W_A, the literature is replete with controversial and discordant findings. There are ample instances[1-14] which testify to the existence of a relationship between the measured adhesive strength and the work of adhesion expressed as $\gamma_{lv}(1 + \cos\theta)$, where γ_{lv} is the surface free energy of liquid adhesive and θ is the contact angle. In some cases, investigators have plotted the practical adhesion vs. γ_c, the critical surface tension of wetting of the substrate; but such plots can be converted into plots of practical adhesion vs. $\gamma_{lv}(1 + \cos\theta)$ with similar results, as shown later. However, it has been reported that the measured adhesive strength is not directly related to $\gamma_{lv}(1 + \cos\theta)$[15] or γ_c[16] of the substrates.

Moreover, in some cases, workers have reported that the maximum practical adhesion is attained under the conditions of equality of solubility parameters[17], and of surface free energies[18-19] between the substrate and the adhesive. In such cases, as will be shown in the Discussion section, the maximum practical adhesion does not correspond to maximum W_A. Wu[20] has suggested that the spreading coefficient is more important than W_A in determining the final adhesive strength values.

Recently Gent and colleagues[21-22] in their tensile rupture
and peel separation studies involving viscoelastic adhesives have
found that the measured peel strength is a product of two terms[22]:
an equilibrium work of detachment given by thermodynamic considerations and a numerical factor representing the enhancement of
strength when the adhesive is imperfectly elastic. In case of
viscoelastic adhesives, the second term is both large and strongly
dependent on the rate of peeling and the temperature of separation.
Also Gent and Kinloch[21] found that the failure energy per unit
area of interface at extremely low rates (10^{-20} cm/sec) of failure
is not much larger than would be expected on thermodynamic grounds.
So, based upon their findings, it is not surprising at all to find
a correlation between the practical adhesion and W_A.

Adhesives are applied in a fluid state so as to facilitate
the interfacial contact and also the penetration in the pores and
crevices in the adherends. In this paper, the terms "adherend" and
"substrate" will be used interchangeably. When a joint fails, if
the substrate remains intact, the failure should take place either
by yield or fracture of the adhesive. Cherry and Holmes[15] have
suggested that the mechanical properties of a polymeric adhesive
depend upon flaws in the adhesive which can facilitate flow in the
polymer when failure is by yield or which can propagate as cracks
when the failure is by fracture. Since the strength of a joint is
therefore a function of the strength of the adhesive, a meaningful
design of the experiments to test the correlation between the
practical adhesion and the thermodynamic adhesion should involve
the same adhesive; the adherends should be either different surfaces differing in their suface free engeries, or the surface of a
substrate can be modified by suitable deposition or other techniques.

There are various theories of adhesion[23,24] -- adsorption or
wetting theory[25-27], electrostatic[28-29], diffusion[30-31], weak
boundary layer[32-33]. The chemical aspects of adhesion have been
discussed by Dean[34] and recently, Fowkes[35] has suggested that
acid-base interactions at the adhesive-adherend interface could
play a very important role in adhesion phenomenon. Bikerman has
accentuated the non-existence of adhesive failure in a "proper
joint", i.e., according to his views, true interfacial failure
practically never occurs and what is taken for interfacial failure
is actually separation in a weak boundary layer[32,33,36]. If the
universality of his ideas is accepted, then it makes no sense even
to talk about the role of thermodynamics or surface energetics in
controlling the adhesive strength of a joint. However, cases are
chronicled in the literature where a true interfacial failure has
been observed[37], and its possibility has been suggested on theoretical grounds[38,39]. Good[38], has presented a critique of the

Bikerman "dogma" and has concluded that his principle is not universally true. The mechanical argument given by Good shows that the product, $E_i \mathscr{G}_i$, and the strength of the intermolecular forces across the interface, determine the locus of failure-initiation. Here E_i is the elastic modulus for phase 1 or 2, and $\mathscr{G}_i$ the corresponding energy dissipation per unit extension of the locus. If E_1-E_2 and $\mathscr{G}_1-\mathscr{G}_2$ have opposite sign, and if the interfacial forces are weak (e.g. dispersion forces) then true interfacial failure is to be expected.

Voyutskii[40] has also levelled objections at the thermodynamic approach to the strength of adhesive joints; his comments and their rebuttal are well summarized by Crocker[37].

No one would deny the importance of wetting in adhesion, as the lower the contact angle, more the interfacial area in contact which should result into improved adhesion. Sharpe and Schonhorn[27] in their earlier studies advocated the opinion that the liquid adhesive would have to spread (i.e., the surface tension of the liquid adhesive must be smaller than the surface tension of the solid substrate)[41] on the substrate in order to obtain good adhesion, or, in other words, if the adhesive spreads (i.e., makes a contact angle of zero) on the adherend, this should lead to good adhesion. However, there are instances where the spreadability condition was fulfilled, but the adhesion was poor with interfacial failure[17]. Moreover, the later results of Schonhorn and Ryan[42] were inexplicable in terms of wettability criteria and were explained on the basis of weak boundary layers. Therefore, it may seem that wetting is a necessary but not sufficient condition for high adhesive strengths[11,43,44].

It is evident from the above discussion that there is no consensus among the adhesiologists on the role of thermodynamics or surface energetics in dictating the practical adhesion of adhesively bonded joints. This subject has been treated in the literature in a fragmentary manner, and different views have been propagated by their respective proponents.

In the present paper, I wish to (i) review critically the literature pertaining to the relationship between the practical adhesion and thermodynamic adhesion, (ii) discuss various surface chemical criteria--W_A, interfacial free energy, penetration of the adhesive--and the conditions which optimize these criteria, (iii) test these criteria and conditions against the existing values of adhesive strengths, (iv) finally deduce the most important surface chemical criterion germane to the practical adhesion.

As the values of surface free energies of the solid substrates are sine qua non in the present analysis, so a subsection is devoted to the various ways of obtaining these values.

THEORETICAL CONSIDERATIONS

Young's Equation

For the case of a drop of liquid resting on the surface of a solid, Young's equation is expressed as

$$\gamma_{sv} - \gamma_{sl} = \gamma_{lv} \cos \theta \quad (1)$$

The above equation appears quite simple and straight forward but there are present conceptual and experimental difficulties, and it has been a popular equation both for defense and attack. The most frequently raised questions apropos of this equation are: (i) should the various γ's be expressed as surface tensions or surface free energies (ii) does it have thermodynamic basis (iii) is the solid in equilibrium, as any tensile stress existing in the surface of the solid would hardly be a system in equilibrium (iv) is it applicable at $\theta = 0°$.

It has been stressed that the various γ's be expressed in terms of surface tensions[45], but Good[46] has derived the above equation from considerations of free surface energies and so the various γ's should be properly expressed in terms of surface free energies. I do not intend to delve into the polemics of Young's equation, and throughout this paper all γ's—both for solids and liquids—will be expressed as surface free energies.

Thermodynamic Work of Adhesion or Gibbs Free Energy of Adhesion, W_A

In deriving the expression for W_A, two cases are discernible:

Case I—Consider an interfacial area of $1 cm^2$ is separated so as to create a liquid surface of specific surface free energy γ_{lv} ergs/cm^2, and a solid surface covered with an adsorbed monolayer of liquid in equilibrium with the vapor of the liquid of specific surface free energy γ_{sv} ergs/cm^2. In this process $1 cm^2$ of the interfacial area disappears and the thermodynamic work of adhesion is given by

$$W_A^1 = \gamma_{sv} + \gamma_{lv} - \gamma_{sl} \quad (2)$$

Case II—If the surface is saturated in vacuum (instead of the vapors of the liquid) then the film free surfaces of the solid and liquid give rise to the thermodynamic work of adhesion as follows:

$$W_A = \gamma_s + \gamma_l - \gamma_{sl} \quad (3)$$

However, $\gamma_l = \gamma_{lv}$, but $\gamma_s - \gamma_{sv} = \pi_e$, the equilibrium spreading pressure of the film.

Equations (2) and (3) can be related by the expression

$$W_A \equiv W_A^1 + \pi_e \tag{4}$$

Where π_e is given by

$$\pi_e = \gamma_s - \gamma_{sv} = RT \int_0^{P_o} \Gamma \, d(\ln P) \tag{5}$$

where Γ is the surface excess and P_o is the saturated vapor pressure of the liquid.

In the case of high energy surfaces, π_e has been determined[47,48] and is quite appreciable and in certain cases is comparable to W_A^1. As regards low energy surfaces, earlier views[41] were expressed that liquids having γ_{lv} much greater than γ_c have π_e essentially zero, and it becomes more important as γ_{lv} approaches γ_c. On the other hand, according to Melrose[49], π_e cannot be negligible unless the contact angle is greater than 90°, and in some cases π_e could be $\sim$ 10 ergs/cm^2.

The difficulty in obtaining reliable magnitudes of π_e is due to the paucity of adsorption isotherms of various liquids on low energy solids.

Most recently, Good[50] has analysed this issue in detail and according to him π_e should be negligible on a smooth, homogeneous surface of a low energy solid such as Teflon, for most liquids that form non-zero contact angle and particularly for those with high heats of vaporization. For the purpose of this paper, π_e may be neglected for simplicity without altering the arguments presented later and in many places, γ_s and γ_{sv} will be used interchangeably.

Young-Dupre Equation

It can easily be seen that equations (1) and (2) can be combined for cases in which Young's equation is valid, i.e., when

$$\cos\theta = \frac{\gamma_{sv} - \gamma_{sl}}{\gamma_{lv}} \leq 1$$

SURFACE-CHEMICAL CRITERIA

More properly, $\cos\theta \to 1$ as $\theta \to 0°$. The case of $\theta = 0°$ represents a limiting case, and when spreading takes place, Young's equation becomes inapplicable; however as pointed out earlier there is no general agreement on this issue. In any case, the above combination leads to Young-Dupre equation, which is one of the most basic equations of surface chemistry.

$$W_A^1 = \gamma_{1v} (1 + \cos\theta) \tag{6}$$

Various Forms of γ_c and Their Meanings

Before discussing the conditions dictating optimum thermodynamic adhesion, it is in order to introduce the concept of γ_c, the critical surface tension of wetting. It is obvious from equation (2) that for the determination of W_A^1, one must know the values of γ_{sv} and γ_{sl} (γ_{1v} can easily be measured), and these are not easily accessible because of many experimental difficulties associated with measuring solid surface free energies. However, in cases where equation (1) is valid, the difference of $\gamma_{sv} - \gamma_{sl}$ can be determined by measuring θ.

The determination of γ_s or γ_{sv} has been a long sought desideratum and Zisman pioneered the concept of γ_c[41], with the hope that γ_c could be related to γ_s or γ_{sv}. Zisman etal.[51] working first with the homologous series of n-alkanes showed that the contact angle, θ, on low energy solids decreases with decreasing surface free energy of the liquid. Therefore, the cosine of the angle increases and this increase was found to be linear for a given solid. The intercept of this linear plot at $\cos\theta = 1$ was termed the "critical surface tension of wetting", γ_c; γ_c represents the surface tension of the liquid which makes a contact angle of zero on the solid. Also, for $\gamma_{1v} \leq \gamma_c$ the contact angle observed was zero. Subsequently, it was observed that for a non-homologous series of liquids[52-53], the graphical points of the plot of $\cos\theta$ vs γ_{1v} did not lie on a straight line, rather they tend to collect in a narrow rectilinear band. In such circumstances, Zisman has used the lower lines of the band to represent γ_c, but in certain cases the width of the band could easily be ~ 10 dyne/cm.

It is clear from above that γ_c is not a characteristic of the solid surface rather it alters with the liquid series used. This dependence of γ_c upon the liquid series was pointed out by Fox and Zisman[53] for paraffin and n-hexatricontane as early as 1952. Although, very frequently, γ_c's of Zisman are used in the literature, but in the light of recent results, one should be discerning in the choice of proper γ_c. Dann[54] measured γ_c's of nine polymer surfaces with four series of polar liquids and his values differed considerably from the accepted values of Zisman. Recently, Kitazaki and Hata[19] have suggested four different γ_c's: (i) γ_c of Zisman; (ii) γ_c^A obtained using non-polar liquids such as n-alkanes and di-n-alkylether; (iii) γ_c^B obtained using polar liquids such as halogenated compounds and esters; and (iv) γ_c^C obtained with the use of hydrogen-bonding liquids such as water, glycerol, and formamide. It is important to notice that using the results of Zisman et al., they have observed a difference of 11 dyne/cm between the various γ_c's ($\gamma_c^A = 21$, $\gamma_c^B = 12$, $\gamma_c^C = 10$ dyne/cm) for n-hexatricontane.

In any event, obviously γ_c of Zisman is only one form of at least four γ_c's and neither γ_c is equal to the surface free energy of the solid surface. The values of various γ_c's for a variety of polymers are presented in Table I.

Relationship Between Various γ_c's and the Surface Free Energy of Solids

Below will be discussed four different approaches regarding the relationship between γ_c's and γ_s or γ_{sv}.

(i) Gray's ideas. According to Gray[45,55-57], the surface free energy of a solid (actually, he uses the term "surface tension") can be determined from the contact angle made on the solid by a liquid possessing zero interfacial free energy with the solid. Such a "liquid" is the solid itself, rendered fluid in some way. The solid and liquid surface tensions then become identical (γ_s) and equation (1) leads to

$$\gamma_s = \gamma_s \cos\theta \quad \text{(Putting } \gamma_{sv} = \gamma_{lv} = \gamma_s \text{ and } \gamma_{sl} = 0\text{)}$$

In other words, a solid spreading on itself will have an exactly zero contact angle. Furthermore, Gray stresses that if there is a mono-functional relationship between θ and γ_{lv} then any liquid with surface free energy equal to γ_s will just spread in the same way. According to Gray, extensive experiments by Zisman and collaborators suggest that there is in many cases a strict monofunctional relationship between θ and γ_{lv}. Furthermore, he states that the observed difference in γ_c (in case of polyethylene, for example) with different homologous series may be due to the modification of γ_s by the adsorbed vapors of different kinds of liquids and not to departure from a single line with γ_s constant. Gray extends his reasoning further and states that if the general monofunctional relation is assumed, then γ_s is clearly identical with γ_c and the technique of plotting $\cos\theta = 1$ is a valid method for the measurement of solid surface free energy.

Table I. Various γ_c's and Surface Free Energies of a Variety of Polymers

Polymeric Solid	γ_c^+	γ_c^{A++}	γ_c^{B++}	γ_c^{C++}	γ_s^{d+++}	γ_s^*	γ_s^{**}	γ_s^{***}
Poly(tetrafluoroethylene)	18.5	19.3	21.4	14.3	19.5	19.1	19.6	21.5
n-Hexatriacontane	21	21	12	10	21.0	19.1	20.8	-
Poly(vinylidene fluoride)	25	26	39.1	40.0	-	30.3	-	40.2
Poly(vinyl fluoride)	28	-	43.2	44.2	-	36.7	-	43.5
Poly(ethylene)	31	-	-	-	35.0	33.2	34.0	35.6
Poly(styrene)	33	-	43.0	33	43.3	42.0	43.3	40.6
Poly(vinyl chloride)	39	-	43.9	39	-	41.5	-	44.0
Poly(vinylidene chloride)	40	-	44.0	40	-	45.0	-	45.8
Poly(ethylene terephthalate)	43	-	43.4	43.5	44.0	41.3	44.0	43.8
Poly(hexamethylene adipamide) (nylon 6-6)	46	-	42.5	46	44.9	47.0	44.9	-

+ from reference 41
++ from reference 19
+++ This represents the dispersion component of surface free energy, taken from reference 66
* Calculated using extended Fowkes equation to two components, taken from reference 68
** from reference 64
*** Calculated using extended Fowkes equation to three components, taken from reference 19

A few comments should be made about Gray's ideas: (i) Gray does not discuss the variability of γ_c with different liquid series and this is very important in the light of recent results as pointed out earlier. According to him γ_s is the same as γ_c; so, the choice of γ_c should dictate the values for γ_s. (ii) He assumes that at $\theta = 0°$, γ_{sl} is zero, which may not be true as discussed later.

(ii) **Good and Girifalco's ideas**[58]. According to these workers, γ_{sl} is expressed as

$$\gamma_{sl} = \gamma_s + \gamma_{lv} - 2\Phi\sqrt{\gamma_s \gamma_{lv}} \qquad (7)$$

and substituting equation (7) in equation (1) leads to

$$1 + \cos\theta = 2\Phi\sqrt{\frac{\gamma_s}{\gamma_{lv}}} \qquad (8)$$

For $\theta = 0$, $\gamma_{lv} = \gamma_c$ and the following relation holds

$$\left(\gamma_s = \frac{\gamma_c}{\Phi^2}\right) \qquad (9)$$

If $\Phi = 1$, then from equation (9), $\gamma_s = \gamma_c$ and according to Gardon[59], since Φ is probably close to unity in Zisman systems, γ_c should be only slightly lower than γ_s. The difference being probably not more than 33%. Unlike Gray, Gardon[59] has mentioned the dependence of γ_c on the liquids used in its determination, but has not discussed it further. The dependence of Φ on the fractional polarities of the interacting phases has been discussed by Gardon[60]. According to Good[61], if only dispersion forces are involved in both phases, or if dipole - dipole forces predominate in both phases, or if both phases are metals, the portion of Φ determined by the nature of the interactions should be close to unity. For the details of calculating Φ, the reader should refer to Good and Elbing[62], and Good and Hope[63]. It should be added that as Φ depends upon the nature of the solid and liquid involved; so apparently, if Φ is known accurately then any of the γ_c's may lead to the same value of γ_s. The rationale being that both γ_c and Φ are functions of the nature of the solids and liquids, and most probably the effect could be such as to render a constant value of γ_s. Good[62] has discussed the dependence of γ_c on Φ.

(iii) **Kitazaki and Hata's ideas**. According to them[19], putting $\gamma_{lv} = \gamma_c$ and $\gamma_{sl} = \gamma_{sl}^*$ at $\cos\theta = 1$ in equation (1) one obtains

$$\gamma_c = \gamma_s - \gamma_{sl}^* \quad (\gamma_s = \gamma_{sv} \text{ if } \pi_e \text{ is neglected}) \qquad (10)$$

SURFACE-CHEMICAL CRITERIA

or in other words, they do not assume $\gamma_{sl} = 0$ at $\cos\theta = 1$. Furthermore it is obvious from equation (10) that the conditions minimizing $\gamma_{sl}*$ depend upon the solid and the liquid. Generally speaking, the combination of similar polarity will give a small value of $\gamma_{sl}*$, accordingly a large value of γ_c for a given solid. In essence, the value of γ_c determined under conditions of minimum $\gamma_{sl}*$ (ideally zero) should be the closest approximation to γ_s. Hata gives example of γ_c's on Nylon-66 to support this conclusion. Nylon-66 gives γ_c^C of 46 dyne/cm with hydrogen bonding liquids; γ_c^B of 42.5 dyne/cm with halogenated liquids; and nonpolar liquid spread on Nylon-66 and entirely wet it. Their ideas can be summarized as: (i) γ_c^A, γ_c^B and γ_c^C should be the best approximation to the surface free energy of non-polar, polar and hydrogen bonding solids respectively. (ii) No γ_c is equal to γ_s; the only possibility is to select the γ_c which comes closest to γ_s.

(iv) <u>Rhee's views</u>. Very recently, Rhee[64] by using Zisman's formulation has derived an expression between γ_c and γ_{sv} of solid surfaces. He combines the Young's equation

$$\gamma_{sv} - \gamma_{sl} = \gamma_{lv} \cos\theta \tag{1}$$

with

$$\cos\theta = 1 + b(\gamma_c - \gamma_{lv}) \tag{11}$$

(where equation (11) represents the Zisman relationship between $\cos\theta$ and γ_{lv}).

to express γ_{sl} as

$$\gamma_{sl} = \gamma_{sv} - \left[\gamma_c + \frac{1}{b}(1 - \cos\theta)\right]\cos\theta = \frac{1}{b}\left[\cos\theta - \frac{(b\gamma_c + 1)}{2}\right]^2 + \left[b\gamma_{sv} - \frac{(b\gamma_c + 1)^2}{4}\right] \tag{12}$$

If γ_{sl} is assumed to become zero at the minimum, the last term in equation (12) has to become zero, i.e.,

$$b\gamma_{sv} - \frac{(b\gamma_c + 1)^2}{4} = 0 \tag{13}$$

or from equation (13)

$$\gamma_{sv} = \frac{(b\gamma_c + 1)^2}{4} \tag{14}$$

A very special case of equation (14) is when minimum for γ_{sl} occurs at $\cos\theta = 1$; in such a case

$$\frac{1}{b} = \gamma_c \qquad (15)$$

If the condition imposed by equation (15) is satisfied then equation (14) becomes

$$\gamma_{sv} = \gamma_c = \frac{1}{b}$$

Rhee claims that equation (14) gives consistent values of γ_{sv} for solid surfaces irrespective of the liquid series used. However, it should be noted that equations (14) and (15) are based on an empirical approximation (equation (11)) which is not valid over a wide range of solid liquid combinations.

In addition to the determination of γ_{sv} from γ_c's, the solid surface free energies have been determined employing contact angle and other approaches[65-72]. Table I summarizes γ_{sv} for a variety of organic solids.

Various Surface Chemical Criteria For Optimum Adhesion

After presenting the relevant theoretical details, concept of γ_c and their relationship to solid surface free energy, it is now appropriate to examine the various surface chemical conditions which predict optimum adhesion. Thermodynamic work of adhesion, W_A^1, is the commonly used criterion for optimizing adhesion, but, as will be discussed in the following paragraphs, there are other quantities which should also be considered.

As can be seen from equation (2) that the thermodynamic work of adhesion is given by

$$W_A^1 = \gamma_{sv} + \gamma_{lv} - \gamma_{sl} \qquad (2)$$

It is obvious from equation (2) that for maximum W_A^1:

(i) γ_{sv} should be large

(ii) γ_{lv} should be large

(iii) γ_{sl} should be small.

In the field of adhesives, for a given solid adherend, most commonly, it is desired to determine the optimum surface free energy of the liquid adhesive which can lead to optimum adhesion.

SURFACE-CHEMICAL CRITERIA

So in the following pages, the optimum values of γ_{lv}, the surface free energy of the liquid adhesive, for a variety of surface-chemical criteria will be determined.

<u>Condition based upon contact angle.</u> Equation (6) reads

$$W_A^1 = \gamma_{lv} (1 + \cos \theta) \tag{6}$$

at complete wetting, $\theta = 0$

therefore $W_A^1 = 2\gamma_{lv}$ (16)

According to equation (16) once θ is zero, W_A^1 is independent of the nature of the solid substrate and is simply twice the surface free energy of the liquids used. Actually, in such cases π_e becomes appreciable and one should include this term to calculate the actual thermodynamic work of adhesion, which is W_A.

Furthermore, equation (16) shows that (i) for maximum W_A^1, γ_{lv} should be maximum consistent with zero contact angle (ii) γ_{lv} should be equal to γ_s, and not less than γ_s because for an adhesive of $\gamma_{lv} < \gamma_s$, W_A^1 will be lower compared with its value for $\gamma_{lv} = \gamma_s$.

<u>Condition based upon maximum W_A^1.</u> Zisman[41] has obtained the condition for maximum W_A^1 in the following manner.

$$\cos \theta = 1 + b (\gamma_c - \gamma_{lv}) \tag{17}$$

Equation (17) represents the linear relation between $\cos \theta$ and γ_{lv}

Now $W_A^1 = \gamma_{lv} (1 + \cos \theta)$ (6)

Eliminating $\cos \theta$ between equations (17) and (6), we obtain

$$W_A^1 = (2 + b\gamma_c) \gamma_{lv} - b\gamma_{lv}^2 \tag{18}$$

Equation (18) represents the expression for a parabola with the concave side towards the surface free energies axis. From equation (18), the value of γ_{lv} giving maximum W_A^1 can be obtained by setting

$$\frac{\partial W_A^1}{\partial \gamma_{lv}} = (2 + b\gamma_c) - 2\gamma_{lv} b = 0 \tag{19}$$

or

$$\gamma_{lv} = \frac{1}{b} + \frac{\gamma_c}{2} \tag{20}$$

and using this value of γ_{lv} in equation (18), the maximum value of W_A^1 is given by

$$W_A^1 = \frac{1}{b} + \gamma_c + \frac{b\gamma_c^2}{4} \qquad (21)$$

So according to this discussion, maximum W_A^1 for a given solid should be given by a liquid adhesive whose surface free energy is $\frac{1}{b} + \frac{\gamma_c}{2}$

In Table II, are summarized γ_{lv} values according to equation (20) for a variety of polymers. For example, for polyethylene γ_c of Zisman is 31 and the optimum γ_{lv} is 55.5 ergs/cm.^{2} As discussed earlier, for spreading $\gamma_{lv} < \gamma_c$, so for a γ_{lv} of 55.5 ergs/cm.^{2} there will be an appreciable contact angle θ which is undesirable for getting good adhesive strength.

<u>Condition based upon Kitazaki and Hata's views</u>. Kitazaki and Hata[19] have considered the following surface chemical criteria for optimum adhesion:

(a) Wetting pressure $\Delta F_i = \gamma_s - \gamma_{sl} = \gamma_{lv} \cos \theta$

(b) Work of adhesion $W_A^1 = \gamma_{lv} (1 + \cos \theta)$

(c) Interfacial free energy $\gamma_{sl} = \frac{(\sqrt{\gamma_s} - \sqrt{\gamma_l})^2}{1 - 0.015 \sqrt{\gamma_s \gamma_l}}$

and they have concluded that the condition of $\gamma_s = \gamma_{lv}$ is the one most conducive to obtaining optimum adhesion. $\gamma_s = \gamma_{lv}$ leads to (i) maximum ΔF_i (ii) minimum γ_{sl} and (iii) equality of solubility parameters. It should be noted that the term "wetting pressure" as used by them is more commonly known as "adhesion tension."

It is interesting to note that the condition for the optimum adhesion tension can be derived in the following manner, assuming the validity of Zisman relationship.

$$\cos \theta = 1 + b (\gamma_c - \gamma_{lv}) \qquad (17)$$

Adhesion tension, $\gamma_{lv} \cos \theta$, becomes

$$\gamma_{lv} \cos \theta = \gamma_{lv} (1 + b (\gamma_c - \gamma_{lv})) \qquad (22)$$

putting

$$\frac{\partial (\gamma_{lv} \cos \theta)}{\partial \gamma_{lv}} = 0 \text{ leads to } \gamma_{lv} = \frac{1}{2b} + \frac{\gamma_c}{2} \qquad (23)$$

Table II. Values of Optimum Surface Free Energy of Liquids (Adhesives) Calculated For a Variety of Conditions Predicting Optimum Thermodynamic Adhesion for Various Polymers

Polymeric Solid	γ_c	γ_s	b	$\frac{1}{b} + \frac{\gamma_c}{2}$	$\frac{1}{2b} + \frac{\gamma_c}{2}$
Poly(tetrafluoroethylene)	18.5	21.5	0.036*	37.05	23.15
Poly(ethylene)	31	35.6	0.025*	55.5	35.5
Poly(styrene)	33	40.6	0.028*	52.2	34.35
Poly(vinyl chloride)	39	44.0	0.0355**	46.7	32.6
Poly(vinylidene chloride)	40	45.8	0.027**	57.0	38.5
Poly(ethylene terephthalate)	43	43.8	0.030*	54.8	38.15
Poly(hexamethylene adipamide)	46	–	0.027*	60.0	41.50

*Calculated from the plots of γ_{lv} vs. $\cos\theta$ in reference 73

**Calculated from the plots of γ_{lv} vs. $\cos\theta$ in reference 74

Or in other words, if the maximum adhesion tension is taken as the criterion for optimum adhesion, then for a given solid, optimum γ_{1v} is given by equation (23). It should be noticed that equation (23) is different from equation (20) derived earlier.

In Table II, are listed γ_{1v}'s according to equation (23) for a variety of polymers.

<u>Condition based upon Gray's analysis for minimum γ_{s1} using Zisman's relationship</u>. As described in the earlier section, Gray[45,57], apparently, identifies γ_s with γ_c of Zisman and by the same token Zisman relationship

$$\cos \theta = 1 + b(\gamma_c - \gamma_{1v}) \tag{17}$$

becomes

$$\cos \theta = 1 + b(\gamma_s - \gamma_{1v}) \tag{24}$$

From Young's equation

$$\cos \theta = \frac{\gamma_{sv} - \gamma_{s1}}{\gamma_{1v}} \tag{1}$$

So eliminating $\cos \theta$ from equations (24) and (1) we obtain

$$\gamma_{s1} = b\gamma_{1v}^2 - \gamma_{1v}(1 + b\gamma_s) + \gamma_s \quad (\text{assuming } \gamma_s = \gamma_{sv}) \tag{24a}$$

putting

$$\frac{\partial \gamma_{s1}}{\partial \gamma_{1v}} = 0, \text{ one obtains } \gamma_{1v} = \frac{1}{2b} + \frac{\gamma_s}{2} \text{ for minimum } \gamma_{s1} \tag{25}$$

Furthermore, there are two values of γ_{1v} for which $\gamma_{s1} = 0$, namely, $\gamma_{1v} = \gamma_{sv}$ and $\gamma_{1v} = \frac{1}{b}$, but in the region between these values of γ_{1v}, γ_{s1} is negative. This negative region may be above or below $\gamma_{1v} = \gamma_{sv}$ depending upon the value of b. The minimum value γ_{s1} can be calculated by substituting γ_{1v} from equation (25) into equation (24a), i.e.,

$$(\gamma_{s1})_{minimum} = -\frac{1}{4b}(b\gamma_{sv} - 1)^2$$

and the value of this negative quantity is greatest when there is a large difference between 1/b and γ_{sv} and when 1/b is large.

Gray[57] has proposed to call γ_{1v} in equation (25) the surface tension for optimum wetting when it refers to advancing contact angle and "surface tension for optimum adhesion" for receding

SURFACE-CHEMICAL CRITERIA

contact angle. In conclusion, Gray's analysis reveals that for γ_{sl} to be negative, γ_{lv} should be between γ_s and $1/b$.

Condition based upon the penetration of the adhesive (liquid). In order to obtain maximum adhesion, the liquid adhesive should penetrate and fill each capillary. As a first approximation, according to Zisman[41], it may be assumed that a capillary rise equation

$$h = \frac{K\gamma_{lv}\cos\theta}{\rho R} \tag{26}$$

can be used. R is the equivalent radius of the capillary, $K = \frac{2}{981}$ and ρ is the density of the liquid. Again assuming the validity of Zisman's linear relationship, i.e.,

$$\cos\theta = 1 + b(\gamma_c - \gamma_{lv}) \tag{17}$$

and eliminating $\cos\theta$ between equation (17) and (26) one obtains

$$h = \frac{K\gamma_c}{\rho R}(b + \gamma_c + 1) - \frac{(bK)}{\rho R}\gamma_{lv}^2 \tag{27}$$

and for maximum penetration, putting

$$\frac{\partial h}{\partial \gamma_{lv}} = 0, \text{ we obtain } \gamma_{lv} = \frac{\gamma_c}{2} + \frac{1}{2b} \text{ for maximum } h \tag{28}$$

For example, in the case of smooth polyethylene, the maximum capillary rise, i.e., maximum h will occur when (assuming $b = 0.025$) γ_{lv} is 35.5 erg/cm². It should be noted that equation (28) is the same as equations (23) and (25).

Condition based upon Wu's analysis of spreading coefficient[20]. The spreading coefficient $S_{l/s}$ is defined as

$$S_{l/s} = \gamma_s - \gamma_{lv} - \gamma_{sl} \tag{29}$$

Wu has expressed γ_{sl} in three different ways

$$\gamma_{sl} = \gamma_s + \gamma_l - \frac{4\gamma_l^d \gamma_s^d}{\gamma_l^d + \gamma_s^d} - \frac{4\gamma_l^p \gamma_s^p}{\gamma_l^p + \gamma_s^p} \tag{30}$$

$$\gamma_{sl} = \gamma_s + \gamma_l - 2\sqrt{\gamma_l^d \gamma_s^d} - \frac{4\gamma_l^p \gamma_s^p}{\gamma_l^p + \gamma_s^p} \tag{31}$$

$$\gamma_{sl} = \gamma_s + \gamma_1 - \sqrt{2\gamma_1^d \gamma_s^d} - 2\sqrt{\gamma_1^p \gamma_s^p} \qquad (32)$$

where the superscripts d and p represent the dispersion and polar components of various γ's.

For a given substrate γ_s^d and γ_s^p are fixed. Furthermore, on defining

$$K = \frac{\gamma_1}{\gamma_s}, \quad \beta = \frac{\gamma_s^p}{\gamma_s} \text{ and } x = \frac{\gamma_1^p}{\gamma_1},$$

Wu has optimized the conditions for spreading, with the assumption that γ_{sl} is given by equation (30), and he obtained the condition

$$\beta = x \qquad (33)$$

for optimum spreading. In other words, the optimum thermodynamic wettability condition is when the polarities of the adhesive and the adherend are exactly the same. The greater the disparity between the two polarities, the poorer the wettability will tend to be.

Although Wu has not mentioned, yet the same condition $\beta = x$ for optimum wettability is also obtained (as I have done) assuming γ_{sl} is given by equation (32).

So it is obvious from the discussion presented that a variety of conditions should culminate into optimum thermodynamic adhesion. Table II lists the values of optimum γ_{1v} (as dictated by the various conditions discussed) for a variety of polymeric solids. Obviously, there is a liberal latitude in the selection of the right value of γ_{1v} for an adherend in question. The next section summarizes the published experimental results on adhesive strengths, and the applicability of the various conditions discussed here will be tested.

Before leaving this topic of various optimum conditions, it should be pointed out that these have been derived in terms of γ_{1v} (the surface free energy of the liquid, in case of adhesives, it signifies the surface free energy of the liquid adhesive) but Dyckerhoff etal[18] found that it is not the surface free energy of the liquid adhesive which is important, rather the surface free energy of the adhesive in the solid form decides the optimum adhesion. They concluded that for minimum interfacial tension, the surface free energy of the hardened adhesive is equal to that of the adherend.

EXPERIMENTAL RESULTS AND DISCUSSION

In this section, the literature pertaining to the relationship between the practical adhesion and surface-chemical characteristics -- contact angle, γ_{lv}, γ_{sv}, γ_c's, solubility parameter etc. -- of solids and liquids is summarized. Essentially, the results can be divided into two groups: (a) studies in which conditions were such that the adhesive made a non-zero contact angle (i.e., $\theta > 0°$) with the substrates, and (b) cases in which the substrates (or a substrate modified to provide surfaces of different surface free energies) were so selected as to provide both complete wetting ($\theta = 0$) and incomplete wetting ($\theta > 0$). According to equation (6), thermodynamic work of adhesion, W_A^1, can be calculated provided the adhesive makes a definite contact angle with the substrate; in case of complete wetting, W_A should be quantified using equation (3).

Some workers have plotted the adhesive strength and γ_c of substrates and showed a linear relationship. It should be pointed out that a plot of adhesive strength vs. γ_c can be converted into a plot of adhesive strength vs. $\gamma_{lv}(1 + \cos\theta)$ with similar results.

DeBruyne[1] has compared the nominal failing stress for araldite polyethylene system with the contact angle of water on polyethylene. Water being taken as the analog of araldite, the validity of the remains to be proved. The polyethylene surface and therefore the contact angle of water were modified by treating polyethylene with chromic acid for varying times. His data are replotted in Figure 1 so that the abscissa represents the thermodynamic work of adhesion. Except for the initial portion, a linear relationship is apparent. Raraty and Tabor[2] studied the adhesion of ice to various solids testing the interface in shear. The solids studied included stainless steel with various surface treatments, and four organic solids. Contact angles of water were measured against surfaces and the shear stresses required to rupture cylindrical specimens were determined at various temperatures. Their data on the adhesion of ice on four organic solids are plotted in Figure 2; a correlation between the adhesive strength and W_A^1 is apparent. It should be noted that the datum for wetted stainless steel lies on the plot in Figure 2. This is in accord with the derivation of W_A^1 which implies that the magnitude of W_A^1 will be the same, irrespective of the nature of the substrates, if the adhesive makes the same contact angle on these substrates. It will be interesting to compare the behavior of partially wetted stainless steel and organic solids, but, unfortunately, detailed data are not available.

Lurie[6] has measured the velocities of deformation just great enough to cause adhesive, rather than cohesive, failures between a number of metals and a tacky adhesive. His results are plotted in Figure 3 which suggests a correlation between the mechanical performance of joints and the thermodynamic work of adhesion. His

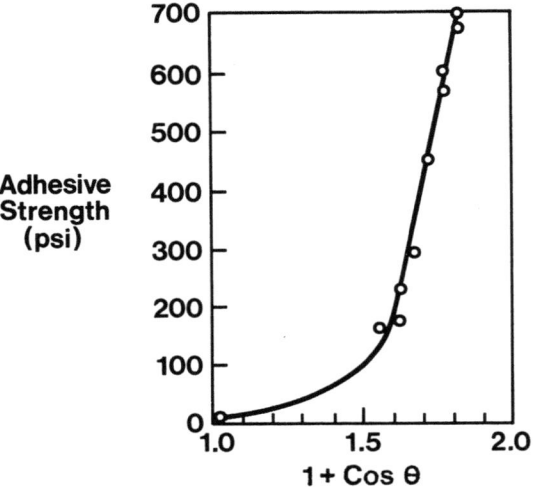

Figure 1 Relationship between adhesive strength and $(1 + \cos \theta)$ using data of reference 1. The contact angles were measured using a liquid different from the adhesive.

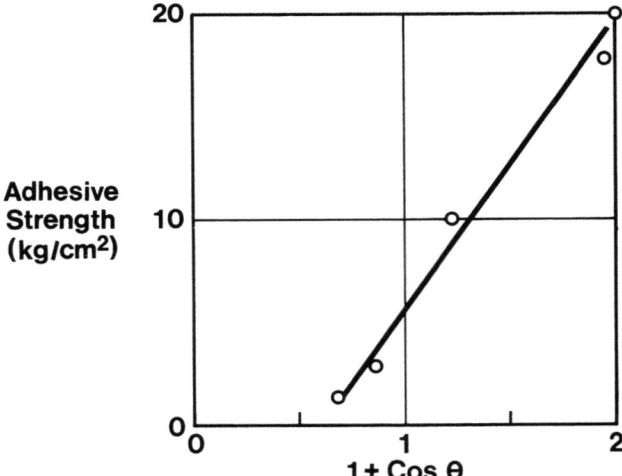

Figure 2 Relationship between adhesive strength and $(1 + \cos \theta)$ using data of reference 2.

contact angle values are based upon the use of a liquid less viscous than, but chemically similar to, the adhesive.

Sprinkle and Taylor[9] in their studies of the tensile strengths of a molding resin on various oxides, obtained a good correlation between the thermodynamic work of adhesion (i.e., $\gamma_{lv}(1+\cos\theta)$) and the measured tensile strengths.

Barbaris[10] studied the adhesion between low density polyethylene and adhesive which was a mixture of bisphenol - type epoxy and a liquid polyamide resin in 70/30 volume ratio. The adhesive had a surface free energy of 41.7 ergs/cm.^{2}, and he measured the contact angles of this adhesive on polyethylene treated in different manners. Figure 4 plots his results, a linear relationship between bond strength and W_A^1, is obvious. Although in Figure 4, the abscissa represents $(1+\cos\theta)$ but similar graph will result if W_A^1, which is equal to the surface free energy of the liquid adhesive multiplied by $(1+\cos\theta)$, were plotted.

According to Barbaris, this increased wettability (i.e., decreasing θ) indicated that γ_c for polyethylene was being increased by the different treatments. But he did not calculate γ_c's to check the relationship between γ_c and bond strength. Using Barbaris' results, I have calculated γ_c's and the results are plotted in Figure 5. γ_c's were calculated in the following manner: Assuming the validity of Zisman's linear relationship

$$\cos\theta = 1 + b(\gamma_c - \gamma_{lv})$$

we obtain

$$\gamma_c = \gamma_{lv} + \left(\frac{\cos\theta - 1}{b}\right)$$

Taking b = 0.026 for polyethylene and γ_{lv} = 41.7 erg/cm^2, the values of γ_c were calculated as a function of contact angle θ. It is clear from Figure 5 that the practical adhesion increases linearly with increasing γ_c. Furthermore, it shows that plots of practical adhesion vs. W_A^1 or γ_c are similar.

Levine et al.[7] studied the adhesion between polymeric films (e.g. PVC, PVDC, PVF etc.) and adhesive composed of Epon 828 cured with diethylenetriamine. The adhesive exhibited a surface free energy of 50 ergs/cm.^{2} The tensile tests were carried out according to the ASTM D-1205. The design of the experiment was such that failure in all cases occurred between the plastic film and adhesive. They found that as the value of the critical surface tension of wetting, γ_c, decreases, the tensile strength also decreases linearly as shown in Figure 6. Furthermore, using their data, Neumann[7] has calculated the values of the interfacial free energy between the

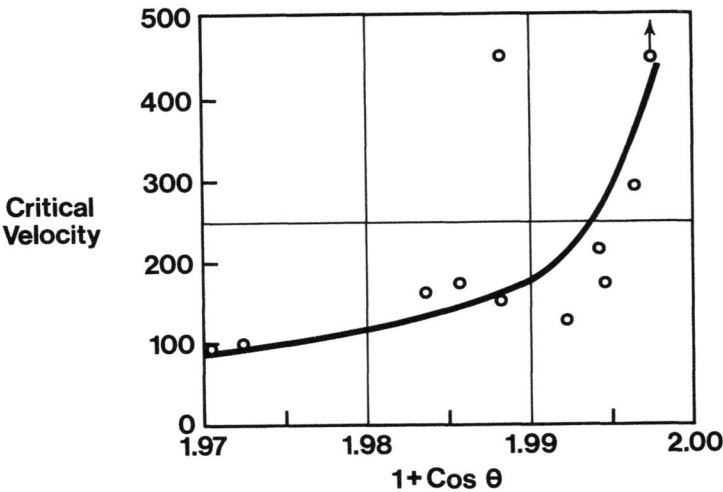

Figure 3 Plot of mechanical performance of adhesive joints and $(1 + \cos \theta)$ (taken from reference 6).

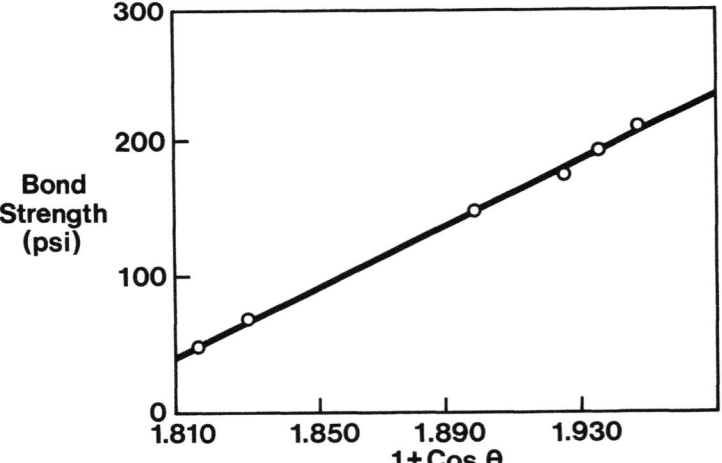

Figure 4 Plot of bond strength against $(1 + \cos \theta)$ on polyethylene using epoxy-polyamide adhesive of surface free energy 41.7 erg/cm^2 (taken from reference 10).

liquid adhesive and the various polymeric films, and found a linear relation between the tensile strength and the reciprocal of the interfacial free energy.

It should be pointed out that although Levine et al. have plotted the tensile strength vs. γ_c's but a linear plot is also obtained when the thermodynamic work of adhesion, W_A^1, is plotted on the x-axis. Using their data, I have calculated cos θ for the polymeric films and a plot of tensile stength vs. (1+cos θ) is shown in Figure 7. Furthermore, it is more appropriate to plot the interfacial free energy rather its inverse, so again using Levine et al.'s results, a graph of tensile strength and γ_{sl} is shown in Figure 7. A direct relationship is apparent.

Dahlquist[8] starts from Kaelble's contention that for a given adhesive at constant peel rate and 180° peel angle, the peel force is to the first approximation proportional to the square of the boundary cleavage stress. He postulated that this boundary cleavage stress should be proportional to the work of adhesion, and that for a given adhesive, with only dispersion forces operating across the interface, the work of adhesion should be proportional to the square root of the critical surface tension of the substrate. Concomitantly, the peel force should be proportional to the critical surface tension. He tested his ideas first on a series of five films (polyester, polystyrene, polychlorotrifluoroethylene, polyethylene, and PTFE) and found a reasonable proportionality. He then re-examined an earlier adhesion study on a series of eight cellulose ester films, from triacetate to tristearate using a rubber resin pressure sensitive adhesive. Again a rough proportionality was observed. More recently, Dahlquist[75] has shown a linear relationship between the boundary cleavage stress and the work of adhesion for a pressure sensitive adhesive and cellulose esters and other surfaces.

Korolev et al.[4] have reported a five to six fold increase in bond strength between polytrifluoroethylene (γ_c=22 dyne/cm) and fluoroelastomers or acrylonitrile rubber by roughening polytrifluoroethylene surfaces. The increase of roughness can lead to an increase in γ_c with the result that the increase in bond strength is attributable to high values of γ_c. According to Kaelble[76]

$$(\gamma_c)_r = \gamma_c + \left[\frac{r-1}{rb}\right]$$

which dictates that greater the value of r, the roughness factor, the larger the $(\gamma_c)_r$. The quantity b is the slope of the Zisman's plot of cos θ vs. γ_{lv}.

DeLollis and Montoya[5] in their adhesion studies on low energy solids such as polytetrafluoroethylene, polychlorotrifluoroethylene, polypropylene, and polyethylene found that the chemical

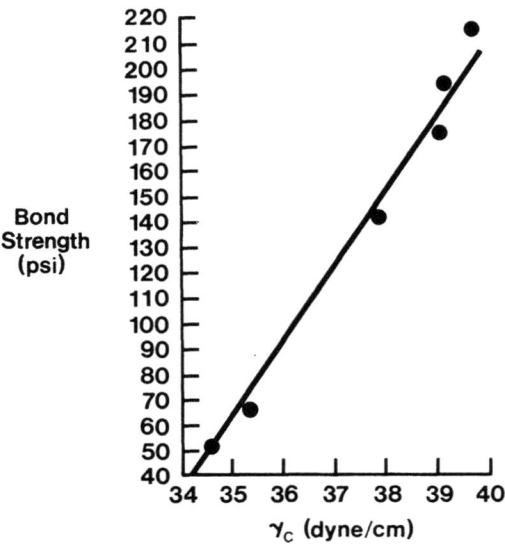

Figure 5 Relationship between bond strength and γ_c using the data of Figure 4.

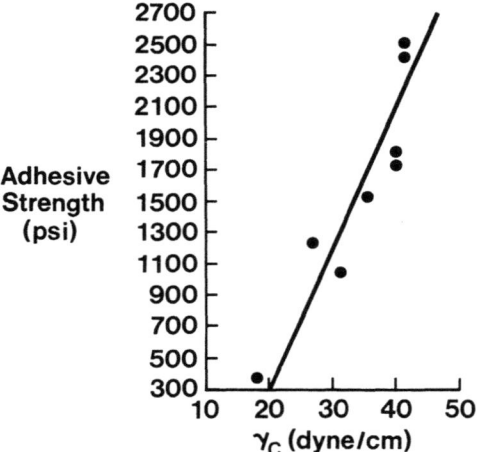

Figure 6 Plot of adhesive tensile strength and γ_c of polymers using epoxy adhesive of surface free energy 50 erg/cm^2 (taken from reference 7).

treatment of the polymer surface to increase γ_c or surface roughening to increase $(\gamma_c)_r$ either singularly or in combination provided substantial improvement in bond strengths with polar epoxy adhesives.

According to Gardon[60], the ideal strength of a butt joint consisting of an adhesive A and a substrate B is given by

$$\sigma_{max} \sim 1.025 \times 10^7 \, \Phi \, \delta_A \delta_B$$

Here Φ is the interaction parameter used to define deviation from the geometric mean rule (i.e., in the expression $\gamma_{sl} = \gamma_s + \gamma_l - 2\Phi\sqrt{\gamma_s\gamma_l}$) and δ's are the solubility parameters. Experimental data show a trend with the trend predicted by the above equation in that the same adhesive often gives higher bond strength with higher cohesive energy density substrates.[77] Apparently, there is some relation of adhesive strength to intermolecular forces (as quantified in Φ) and the product of δ's. However, some other properties -- rheological response functions of the polymers, and probability of interfacial flaw formation -- are the primary properties that control σ_{max}. Furthermore, the cohesive energy density of low energy solids varies with γ_c, this means that Gardon's results suggest higher bond strength with substrates of higher γ_c.

It is obvious from the literature summarized above that wetting plays an important role in determining the final practical adhesion. The lower the contact angle, the higher the thermodynamic work of adhesion which correlates with higher practical adhesion values. One explanation for the increase in adhesive strength with a decrease in contact angle is attributed to the fact that a small contact angle will reduce the stress concentrations at the edge of the flaws, where the adhesive has failed completely to wet the adherend.[41-78]

However, there are reported cases where the wettability, apparently, does not play important, or sometimes any, role in dictating the measured adhesive strengths. Cherry and Muddarris[15] in their studies of adhesion between a series of stainless steel substrates coated with epoxy lacquers and polyethylene as adhesive, showed that the joint strength correlates better with a high wetting constant ($\gamma_{lv}/\eta L$) than with a low contact angle except where this is very low. The wetting constant determines the kinetics of joint formation, exemplified by the rate of wetting, and can be measured by the rate of change of contact angle with time. In the wetting constant expression, γ_{lv} is the surface free energy of the liquid adhesive, η is the viscosity of the spreading adhesive, and 'L' a parameter with dimensions of length which is related to the adhesive/substrate interaction. Their findings are shown in Figures 8 and 9.

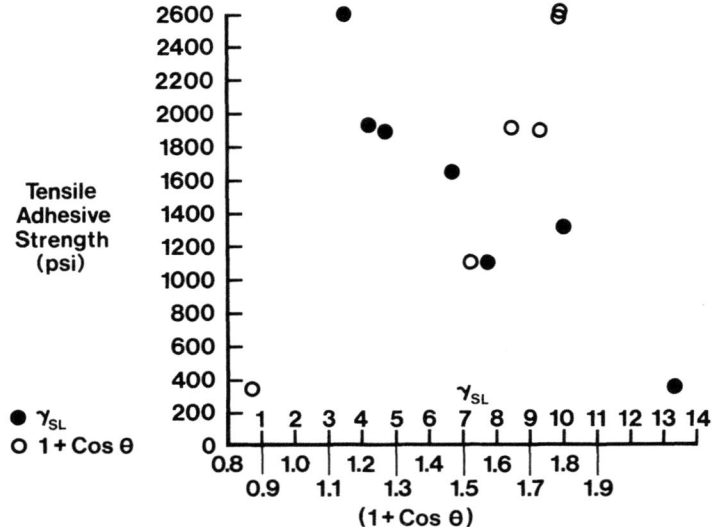

Figure 7   Plots of adhesive tensile strength and $(1 + \cos \theta)$ and γ_{sl} using the data of Figure 6.

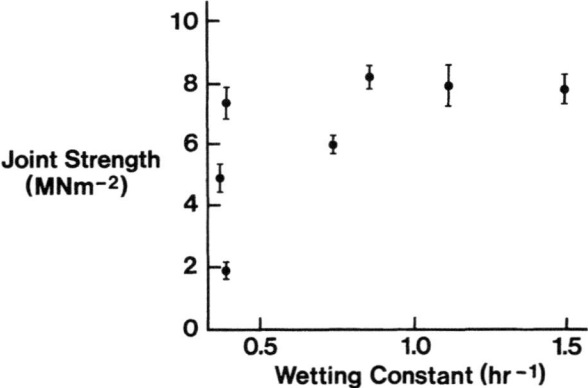

Figure 8   Plot of adhesive strength versus wetting constant (taken from reference 15). x→epoxy urea melamine resins, 0→ other resins.

SURFACE-CHEMICAL CRITERIA

The importance of kinetics of spreading has also been stressed by Bascom and Patrick.[79]

Iyengar and Erickson[17] point out that according to the wetting theory adhesives with $\gamma_{lv} < \gamma_c$ of the substrate should be thermodynamically spreading on the substrate and should show good practical adhesion. Since many of their solvent based adhesives (e.g., polyethylene $\gamma = 31$ erg/cm^2) having surface free energy lower than the γ_c of Mylar (43 dyne/cm) showed poor adhesion with interfacial failure, thermodynamic spreading does not appear to be the sole criterion for good adhesion. On the other hand they found that solubility parameters may be more important.

They measured the force required to peel the sample at 90° angle at 5 cm/min and the mode of failure noted. Their results are shown in Figure 10, where the peel strength is plotted against the solubility parameter, δ, of the adhesives for Mylar - adhesive - Mylar system. Figure 10 suggests that the maximum peel strength is attained when the δ's of the polymeric film and adhesive are equal. Against Nylon the picture was modified considerably by specific interactions, with strongly polar or hydrogen bonding adhesives giving high peel strengths even though the δ's were far from matching.

Two comments regarding the results of Iyengar and Erickson are in order: (i) Instead of using the same adhesive with different substrates, they have used different adhesives with the same substrate. This can result in variable contributions from the inelastic deformation of the adhesives, and as was pointed out earlier, a meaningful investigation should employ the same adhesive; (ii) As will be shown below, for an adhesive whose surface free energy is much lower than the surface free energy of the substrate, poor practical adhesion results. This is attributed to a considerable value of interfacial tension in such systems.

So far we have demonstrated the importance of low contact angle in determining the final practical adhesion. Now let us turn to those studies in which the surface free energy of the substrates were both above and below that of the adhesive, and examine the conditions which dictate the maximum practical adhesion.

Recently Dyckerhoff and Sell[18] have measured practical adhesion by the method of vertical tearing off, between four different adhesives and (i) steel substrates whose surface free energy had been altered by adsorption; and (ii) several polymers having different surface free energy. Their results, plotted as adhesive strength vs. surface free energy of the substrates in Figure 11 show that the practical adhesion has a maximum value when the surface free energy of the hardened adhesive is equal to that of the substrate.

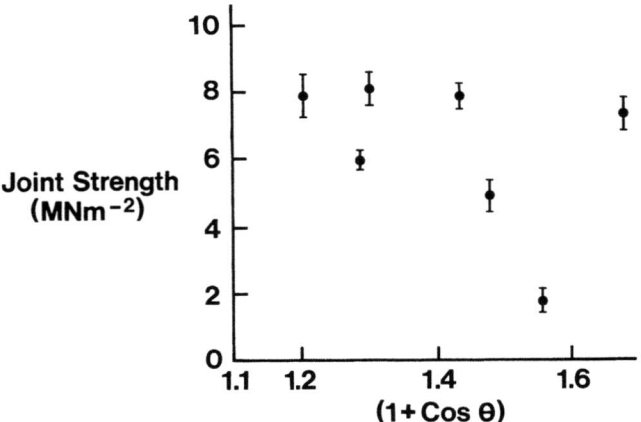

Figure 9 Relationship between adhesive strength and (1 + cos θ) using data of reference 15. x→epoxy urea melamine resins, 0→ other resins.

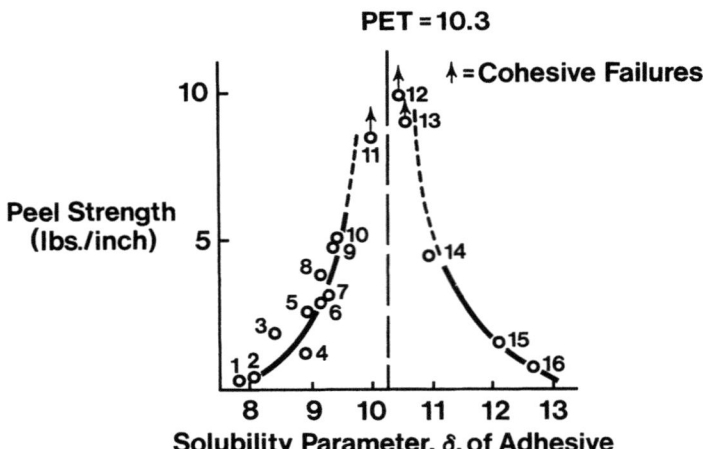

Figure 10 Relationship of compatibility to adhesion for Mylar-adhesive-Mylar system (taken from reference 17). For key to points refer to original source.

Furthermore, under these conditions, interfacial tension between the adhesive and the substrate is minimum. Table III lists values of interfacial free energy for one adhesive, and describes how these are calculated. Figure 12 presents the adhesive strength as a function of the interfacial free energy between the substrate and the hardened adhesive. Unlike Dyckerhoff and Sell, other workers have concerned themselves with the surface free energy of the liquid adhesive. The liquid and solid surface free energies of the adhesives used by Dyckerhoff and Sell are summarized in Table IV. Apparently, the surface free energies of the adhesives in the solid form are higher than the surface tensions of liquid adhesives. In case of polyurethane, there is a very big difference in the values of the two.

Table III. Interfacial Free Energy, γ_{sl}, (erg/cm^2) Between a Variety of Polymers and PVC-PVA† Lacquer as Adhesive. (Taken from reference 18).

Polymeric Solid	γ_s*	γ_{sl}**
Poly(tetrafluoroethylene)	19	1.95
Poly(isopropylene)	26.5	0.19
Poly(ethylene)	28.2	0.05
Poly(vinyl chloride) (red)	33.5	0.18
Poly(vinyl chloride) (grey)	36	0.54
Poly(methacrylic ester)	39	1.21
Poly(formaldehyde)	41	1.80
Poly(amide)	42	2.15

*These are the surface free energy values for polymeric solids.

**These have been calculated by the following equation

$$\gamma_{sl} = \frac{(\sqrt{\gamma_s} - \sqrt{\gamma_1})^2}{1 - 0.015 \sqrt{\gamma_s \gamma_1}}$$

and in this calculation, γ_1 is taken as the surface free energy of adhesive in solid form, i.e., 30 erg/cm^2

† Copolymer of 87% vinyl chloride and 13% vinyl acetate

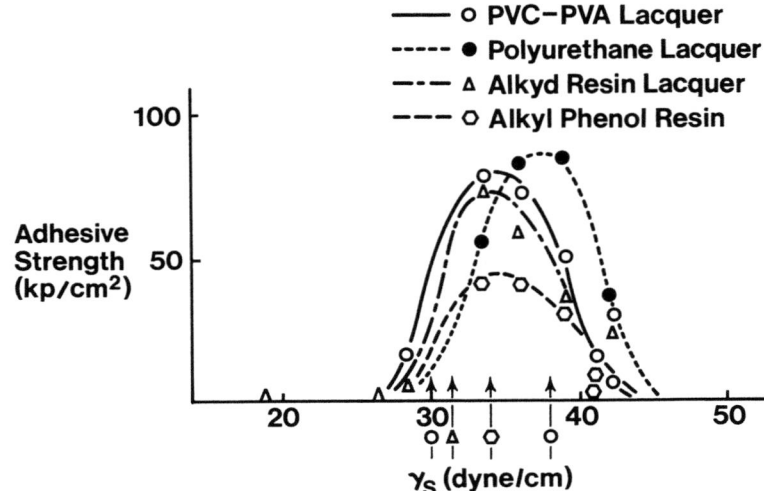

Figure 11 Adhesive strength for various lacquers as a function of the surface free energy γ_S, of plastics. The surface free energy of the adhesives are drawn in on abscissa by means of arrows (taken from reference 18).

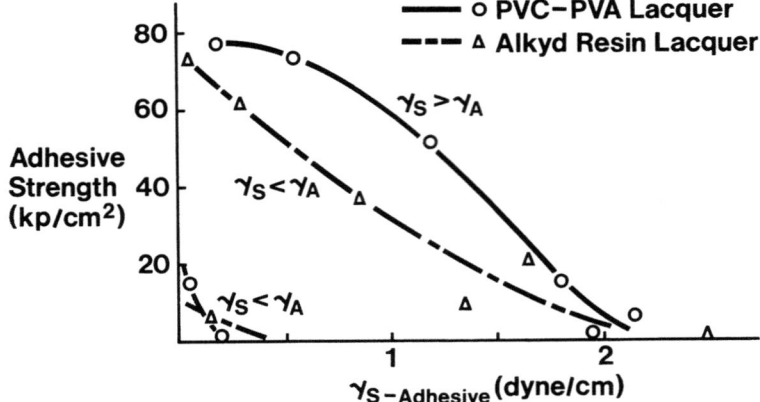

Figure 12 Adhesive strength for various lacquers a function of lacquer-plastic interfacial free energy, $\gamma_{s-adhesive}$ (taken from reference 18).

Table IV. Liquid and Solid Surface Free Energies (in erg/cm^2) of Some Adhesives. (Data of Reference 18)

Adhesive	γ_{Liquid}	γ_{Solid}	γ^+_{Solid}
I. PVC-PVA Lacquer*	28	30	42.6
II. Alkyd Resin	29	31.5	51
III. Alkylphenol Resin	27	34	50
IV. Polyurethane Lacquer	20	38	57.5

*Copolymer of 87% vinyl chloride and 13% vinyl acetate

+Calculated from receding contact angles

Using the data of Dyckerhoff and Sell, I have calculated the thermodynamic work of adhesion, W_A, using equation (3) and the results are plotted in Figure 13. It is obvious from Figure 13 that the practical adhesion increases with W_A up to W_A corresponding to the equality of surface free energy of the substrate and the adhesive, and subsequently decreases. Also, it should be noted that for the same value of W_A, different adhesives yield different adhesive strengths.

In addition, they found that (i) Both the absolute values of adhesive strength and also their relative change with the surface free energy of the substrates vary from adhesive to adhesive, (ii) The increase in adhesive strength with decreasing interfacial energy is less for the case where the surface energy of the substrate is smaller than that of the adhesive, i.e., poor wetting (see Figure 12), (iii) The adhesive strengths of adhesives on the plastics are lower in all cases than the adhesive strengths of the same adhesives on the steel substrates of corresponding surface free energies. This suggests the importance of the specific role of adherends in addition to their surface free energies, (iv) Considering one substrate and various adhesives, the adhesive strength decreases with increasing interfacial free energy in the case of polyethylene and polyamide 6 only.

As mentioned earlier, Kitazaki and Hata[19] have stressed the importance of different γ_c's depending upon the liquid series chosen to measure γ_c, and the proper γ_c should be chosen according to the polarity of the adhesive. To make it clear, if the adhesive is hydrogen bonding then in testing the surface chemical criteria of the substrate for maximum adhesion, γ_c^C of the substrate should be considered.

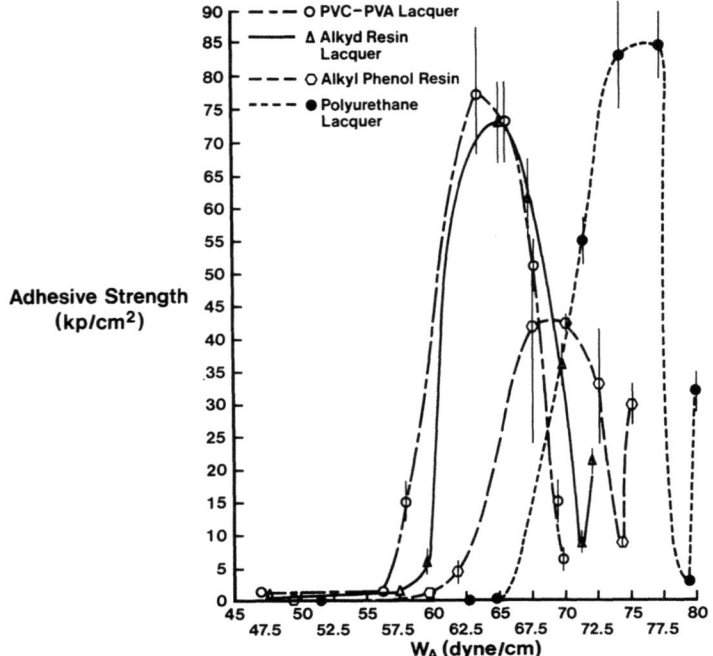

Figure 13 Plot of adhesive strength vs. thermodynamic work of adhesion, W_A, for various lacquers on different plastics. The values of W_A are calculated from equation (3) using data of reference 18.

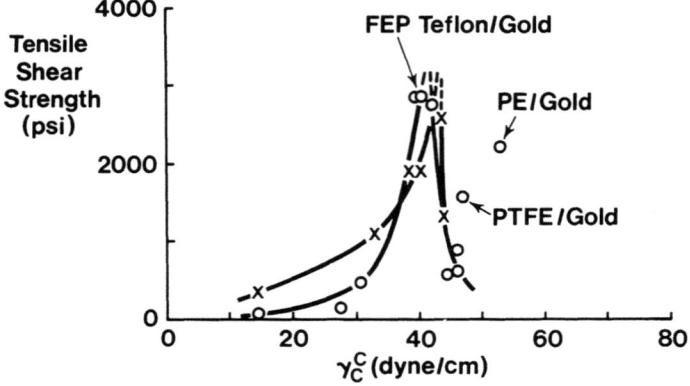

Figure 14 Relationship between tensile shear strength and γ_C^C of adherends (taken from reference 19).

So based upon their proper choice of γ_c, they have shown (see Figure 14) that the tensile strength increases with γ_c^C of up to a value of 40 dyne/cm and then starts declining. Their analysis is based upon the results of Levine et al.[7] and Schonhorn et al.[80,81] who used epoxy adhesive to measure practical adhesion. As the epoxy adhesive is hydrogen bonded, so a choice of γ_c^C for the substrate was made. Figure 14 suggests that the maximum tensile strength is attained when the surface free energy of the adhesive is most near to the proper γ_c of the substrate. As the choice of γ_c is based upon polarity compatibility between the adhesive and the substrate, this assures minimum interfacial free energy.

Recently Toyama et al.[82] studied the adhesion between a pressure sensitive adhesive composed of polybutyl acrylate and various adherends of different surface free energies. Maximum values of the peel force and tack were observed at 33-39 dyne/cm of γ_c; these were a little higher than the surface free energy of the adhesive (γ_{lv} = 28, γ_c = 31 dyne/cm). At higher values of γ_c of adherends, the peel force started declining. Their observations are in agreement with those of Dyckerhoff and Sell, and Kitazaki and Hata, in that the maximum practical adhesion is obtained when the surface free energies of the adhesive and adherend are equal or very close. As pointed out in an earlier section, the equality of surface free energies of adhesive and adherend signifies zero or near zero interfacial free energy. Strictly speaking, the interfacial free energy is minimum when the surface free energy of the liquid adhesive is equal to $1/2b + \gamma_c/2$, and it is not far from the value of the surface free energy of the adherend. Moreover, there is always some uncertainty in the absolute values of various surface free energies values of adherends reported in the literature.

Obviously, so far, we have been concerned with the relationship between adhesive strength and work of adhesion or interfacial free energy. But, Wu[20], based upon his study of adhesion between many pairs of polymers, has pointed out that the spreading coefficient correlates better (see Table V) than the thermodynamic work of adhesion with the adhesive strengths. However, it should be noted that the practical adhesion was rated qualitatively, and he has used different adherends and adhesives. As emphasized earlier, for a meaningful study of the role of surface energetics in controlling adhesive strengths, one should employ one adhesive with different substrates.

SUMMARY AND CONCLUSIONS

Throughout the annals of adhesion science and technology, bipolar views have been expressed relative to the importance of surface energetics in controlling the practical adhesion in adhesively bonded joints involving low energy adherends. According to certain

Table V. Correlation Between Spreading Coefficient and Adhesion of Some Polymer Pairs. (Taken from Reference 20)

Polymer Pairs	Spreading Coefficient[a], S, ergs/cm² at 140°C	Work of Adhesion W_A, ergs/cm² at 140°C	Adhesion[b]
PMMA/L-PE	− 6.5	51.1	poor
PMMA/PS	− 1.6	62.4	poor
PVAc/PS	− 0.2	57.0	fair
PVAc/PnBMA	+ 1.6	49.8	fair
PMMA/PnBMA	+ 6.0	54.2	good
PCP/PDMS	+12.0	40.8	good

PMMA — Poly(methyl methacrylate)
PnBMA — Poly(n-butyl methacrylate)
PVAc — Poly(vinyl acetate)
PS — Polystyrene
PCP — Polychloroprene
PDMS — Polydimethyl siloxane
L-PE — Linear Polyethylene

a The spreading coefficient is larger of the two values.

b The polymer pairs were bonded at 140°C. The adhesion was evaluated at room temperature and rated qualitatively.

SURFACE-CHEMICAL CRITERIA

investigators, it is futile to apply the surface - energetic concepts; but, on the other hand, the import of such concepts has been widely emphasized by the proponents of adsorption theory of adhesion. Furthermore, assuming the role of surface energetics in adhesive joints, there has been a great degree of difference of opinion apropos of what surface - chemical criteria are important and what conditions optimize these criteria.

So in the present paper, I have attempted to juxtapose the various surface - chemical criteria -- thermodynamic work of adhesion, interfacial free energy, contact angle, spreading coefficient, penetration and solubility compatibility -- and have expressed the conditions which optimize these criteria. In the determination of these conditions, the values of γ_{sv}, γ_{sl} and γ_{lv} are required. As the values of γ_{sv} are not easily accessible, so the concept of the critical surface tension of wetting, γ_c, has been introduced. The various forms of γ_c -- γ_c of Zisman, $\gamma_c^A, \gamma_c^B, \gamma_c^C$ -- obtained using different liquid series, and their relationship to the surface free energies of the adherends, γ_{sv} are discussed.

It has been a recurrent hypothesis in the adhesion literature that the adhesive strength should somehow be related to the thermodynamic work of adhesion, and there is a general agreement among the workers that these two quantities are not equal; at the most, a direct relationship between the two can be expected and this is supported by the recent work of Gent and Schultz.[22]

Essentially, the experimental investigations made in the quest for a possible relationship between the practical adhesion (tensile strength, peel strength, shear strength) and surface - chemical criteria can be divided into two groups. (i) The surface free energies of the adherends were such that the adhesive made a definite contact angle ($\theta > 0°$). (ii) The surface free energies of the adherends ran the gamut from lower to higher surface energies than the adhesive itself, i.e., for some adherends there was a non-zero contact angle while others were completely wetted ($\theta = 0°$) by the adhesive.

In the case of (i) above, assuming the equilibrium spreading pressure, π_e, to be negligible, the work of adhesion, W_A^1, can be calculated as $\gamma_{lv}(1 + \cos\theta)$ and in many cases, a direct relationship between the practical adhesion and W_A^1 exists. Furthermore, it should be noted that a plot of adhesive strength vs. γ_c is the same as a plot of adhesive strength vs. $\gamma_{lv}(1 + \cos\theta)$, as shown in Figures 4 and 5, and 6 and 7. Both modes of plotting the results are common in the literature, but in the present paper, these have been represented in a consistent manner. It is very important to note that if the nature of the adhesive is changed, then for the same value of W_A^1, different adhesive strengths are obtained. For the case (ii) above, as soon as $\theta = 0°$, the values of W_A^1 calculated using equation (6) are not meaningful, so in such cases, W_A is calculated using equation (3).

In such cases, it has been observed that the adhesive strength increases with increase in γ_c or γ_{sv} of the substrates up to a certain point, followed by a decrease in adhesive strength. Such plots can easily be transformed into plots of adhesive strength vs. W_A (see Figure 13) and it is patent that the practical adhesion increases with increasing W_A up to a critical value and then starts declining with further increase in W_A. The critical values of W_A are different for different adhesives.

The maximum values of adhesive strengths are attained when the surface free energies of the adherends and adhesives are approximately equal. There is some debate on the choice of the surface free energy for adhesive, i.e. whether it should be for the liquid adhesive or for the hardened adhesive. Ideally, γ_{sl} is not minimum at $\gamma_{sv} = \gamma_{lv}$ (see condition for minimum γ_{sl}, equation (25)) but this equality denotes γ_{sl} values close to minimum. So it is quite reasonable to conclude that the adhesive strength is maximum when γ_{sl} is minimum. Furthermore, the conditions for minimum γ_{sl} are the same as those for the maximum adhesion tension ($\Delta F = \gamma_{lv} \cos \theta$) and the maximum penetration of the adhesive.

In some cases, the maximum adhesive strength has been attained when the solubility parameters of the adhesive and adherend are equal. If it is assumed that the solubility parameters and surface free energies of polymers are related, then, in such cases, it can be construed that the maximum adhesive strength occurs when the surface free energies of the two are equal. Some instances are cited in the literature where the wettability condition ($\gamma_{lv} < \gamma_{sv}$) was satisfied but the adhesive strengths were very low. This can be attributed to the relatively high γ_{sl} in such systems; i.e., such cases lie on the right of Figure 13.

In conclusion, if the nature of the adhesive is the same, the interfacial free energy is the most important surface - chemical parameter in the determination of adhesive strengths. If different adhesives are used, then the situation is somewhat different; i.e., for the same values of γ_{sl} or W_A, different adhesive strengths are possible depending upon the adhesives used.

Also it should be added that there are obviously some other factor or factors, besides those measured by surface thermodynamics, controlling adhesion. On the basis of theories and experiments such as those of Griffith[83], Irwin[84], Gent et al.[21-22], and Good[38], these factors involve dissipative processes. When these factors are, it is to be expected that, adequately taken into account, the correlations between practical adhesion and surface chemical properties will be found to be quantitative.

REFERENCES

1. N. A. DeBruyne, Nature, 180, 202 (Aug. 10, 1957).
2. L.E. Raraty and D. Tabor, Proc. Roy. Soc., 245A, 184 (1958).
3. A.J.G. Allan, J. Polymer Sci., 38, 297 (1959).
4. A.Ya. Korolev et al., Soviet Plastics (English Translation), 5, 33 (1962).
5. N. J. DeLollis and O. Montoya, Adhesives Age, 6, 32 (1963).
6. R. M. Lurie, "Adhesion of a High Molecular Weight Polymer," M.I.T., Ph.D. Thesis, 1955 (quoted by J. E. McNutt, Adhesives Age, p. 24 (Oct. 1964)).
7. M. Levine, G. Ilkka, and P. Weiss, J. Polymer Sci. B-2, 915 (1964).
8. C. A. Dahlquist, in ASTM *Special Technical Publication*, No. 360 ASTM, Philadelphia, 1964.
9. J. K. Sprinkle and H. F. Taylor, "Adhesion of Phenol-Formaldehyde to Various Refractory Oxides." (quoted by J. E. McNutt Adhesive Age, p. 24 (Oct. 1964)).
10. M. J. Barbaris, Nature, 215, 383 (1967).
11. E. A. Boucher, Nature, 215, 1054 (1967).
12. N. L. Bottrell, in *Adhesion Fundamentals and Practice*, Maclaren and Sons Ltd., London, 1969.
13. N. J. DeLollis, Rubber Chem. Tech. 46(2), 549 (1973).
14. W. C. Wake, J. Coated Fabrics, 3(2), 81 (1973).
15. B. W. Cherry and S. Muddarris, J. Adhesion, 2, 42 (1970).
16. L. H. Sharpe, in *Recent Advances in Adhesion*, L. H. Lee, Ed., p. 444, Gordon and Breach, New York, 1973.
17. Y. Iyengar and D. E. Erickson, J. Appl. Polymer Sci., 11, 2311 (1967).
18. G. A. Dyckerhoff and P. J. Sell, Angew. Makromol Chem., 21(312), 169 (1972).
19. Y. Kitazaki and T. Hata, J. Adhesion, 4, 123 (1972), also in *Recent Advances in Adhesion*, L. H. Lee, Ed., Gordon and Breach, New York, 1973.
20. S. Wu, J. Adhesion, 5, 39 (1973), also in *Recent Advances In Adhesion*, L. H. Lee, Ed., p. 45, Gordon and Breach, New York, 1973.
21. A. N. Gent and A. J. Kinloch, J. Polymer Sci. A-2, 9, 659 (1971).
22. A. N. Gent and J. Schultz, J. Adhesion, 3, 281 (1972), also in *Recent Advances in Adhesion*, L. H. Lee, Ed., p. 253, Gordon and Breach, New York, 1973.
23. R. G. Raevskii, J. Adhesion, 5, 203 (1973).
24. K. W. Allen, in *Aspects of Adhesion*, Vol. 5, D. J. Alner, Ed., CRC Press, Cleveland, Ohio, 1969.
25. H. Alter and W. Soller, Ind. Eng. Chem., 50, 922 (1958).

26. W. Brockman, Adhäsion, 335, 448 (1969); 52 (1970).
27. L. H. Sharpe and H. Schonhorn, in *Contact Angle, Wettability, and Adhesion*, Adv. Chem. Ser. No. 43, American Chemical Society, Washington, D. C., 1964.
28. B. V. Derjaguin and V. P. Smilga, J. Appl. Physics, 38, 4609 (1967).
29. B. V. Derjaguin and V. P. Smilga, in *Adhesion Fundamentals and Practice*, Maclaren and Sons Ltd., London, 1969.
30. S. S. Voyutskii and V. L. Vakula, J. Appl. Polymer Sci., 7, 475 (1963).
31. S. S. Voyutskii, *Adhesion and Autohesion of High Polymers*, Interscience, New York, 1963.
32. J. J. Bikerman, Ind. Eng. Chem., 59(9), 40 (1967).
33. J. J. Bikerman, *The Science of Adhesive Joints*, 2nd ed., p. 137, Academic Press, New York, 1968.
34. R. B. Dean, Official Digest, 664 (June, 1964).
35. F. M. Fowkes, private communication, 1973.
36. J. J. Bikerman, J. Paint Tech., 43, 98 (1971).
37. G. J. Crocker, Rubber Chem. Tech., 42, 30 (1968).
38. R. J. Good, J. Adhesion, 4, 133 (1972); see also a paper presented at the International Nonwovens and Disposables Association Symposium on Nonwoven Products Technology, Washington, D. C., Jan. 31, 1973.
39. S. Wu, DuPont Innovation, 2(2), 6 (1972).
40. S. S. Voyutskii, Kolloid-Z. Z. Polym., 214(2), 97 (1966); S. S. Voyutskii and B. V. Derjaguin, Kolloid Z., 27, 624 (1965).
41. W. A. Zisman, in *Contact Angle, Wettability and Adhesion*, Adv. Chem. Ser. No. 43, American Chemical Society, Washington, D. C., 1964.
42. H. Schonhorn and F. W. Ryan, J. Polymer Sci. A-2, 6, 231 (1968).
43. S. S. Voyutskii, Adhäsion, 157 (1960).
44. L. H. Lee, J. Polymer Sci. A-2, 5, 751, 1103 (1967).
45. V. R. Gray, Chem. and Ind., p. 969 (1965).
46. R. J. Good, J. Amer. Chem. Soc., 74, 5041 (1952).
47. G. E. Boyd and H. K. Livingston, J. Amer. Chem. Soc., 64, 2383 (1942).
48. E. H. Loeser, W. D. Harkins and S. B. Twiss, J. Phys. Chem., 57, 251 (1953).
49. J. C. Melrose, in *Contact Angle, Wettability, and Adhesion* Adv. Chem. Ser. No. 43, American Chemical Society, Washington, D. C., 1964.
50. R. J. Good, in *Adsorption at Interfaces*, K. L. Mittal, Ed., ACS Symposium Series No. 8, American Chemical Society, Washington, D. C., 1975.
51. H. W. Fox and W. A. Zisman, J. Colloid Sci., 5, 514 (1950).
52. H. W. Fox and W. A. Zisman, J. Colloid Sci., 7, 109 (1952).
53. H. W. Fox and W. A. Zisman, J. Colloid Sci., 7, 428 (1952).

54. J. R. Dann, J. Colloid Interface Sci., 32, 302 (1970).
55. V. R. Gray, in *Proc. 4th International Cong. Surface Activity* Brussels, 1964. Published Gordon and Breach, New York, 1967.
56. V. R. Gray, in *Aspects of Adhesion*, D. J. Alner, Ed., Vol. 2, CRC Press, Cleveland, Ohio, 1966.
57. V. R. Gray, Forest Product J., 12, 452 (1962).
58. R. J. Good and L. A. Girifalco, J. Phys. Chem., 64, 561 (1960).
59. J. L. Gardon, J. Phys. Chem., 67 1935 (1963).
60. J. L. Gardon, in *Encyclopedia of Polymer Science and Technology*, H. F. Mark, N. G. Gaylord, and N. M. Bikales, Eds., Vol. 3, p. 833, Interscience, New York, 1965.
61. R. J. Good, in *Treatise on Adhesion and Adhesives*, Vol. I, R. L. Patrick, Ed., Marcel Dekker, New York, 1967.
62. R. J. Good and E. Elbing, in *Chemistry and Physics of Interfaces*, II, pp. 72-96, American Chemical Society, Washington, D. C., 1971.
63. R. J. Good and C. J. Hope, J. Colloid Interface Sci., 35, 171 (1971).
64. S. K. Rhee, Mater. Sci., Eng., 11, 311 (1973).
65. R. J. Good, in *Contact Angle, Wettability and Adhesion*, Adv. Chem. Ser. No. 43, American Chemical Society, Washington, D. C., 1964.
66. F. M. Fowkes, Ind. Eng. Chem., 56, 40 (1964).
67. T. Hata, Kobunshi (High Polymers, Japan) 17, 594 (1968); Y. Kitazaki and T. Hata, J. Adhesion Soc., Japan, 7(4), 224 (1971).
68. D. K. Owens and R. C. Wendt, J. Appl. Polymer Sci., 13, 1711 (1969).
69. D. H. Kaelble and K. C. Uy, J. Adhesion, 2, 50 (1970).
70. O. Driedger, A. W. Neumann, and P. J. Sell, Kolloid-Z. Z. Polym. 201, 52 (1965); 204, 101 (1965).
71. P. J. Sell and A. W. Neumann, Angew. Chem., 78, 321 (1966).
72. M. C. Phillips and A. C. Riddiford, J. Colloid Interface Sci., 22, 149 (1966).
73. W. A. Zisman, J. Paint Tech., 44, 42 (1972).
74. W. A. Zisman, Record Chemical Progress, 26, 13 (1965).
75. C. A. Dahlquist, in *Aspects of Adhesion*, D. J. Alner, Ed., Vol. 5, CRC Press, Cleveland, Ohio, 1969.
76. D. H. Kaelble, *Physical Chemistry of Adhesion*, p. 152, Wiley-Interscience, New York, 1971.
77. J. L. Gardon, in *Treatise on Adhesion and Adhesives*, R. L. Patrick, Ed., Vol. I, p. 317, Marcel Dekker, New York, 1967.
78. C. Mylonas, Experimental Stress Analysis, 12, 129 (1955).
79. W. D. Bascom and R. L. Patrick, Adhesives Age, p. 25 (Oct. 1974).
80. H. Schonhorn and R. H. Hansen, Polymer Letters, 4, 203 (1966); J. Appl. Polymer Sci., 11, 1461 (1967).

81. H. Schonhorn and F. W. Ryan, J. Polymer Sci., A-2, $\underline{7}$, 105 (1969); J. Adhesion, $\underline{1}$, 43 (1969).
82. M. Toyama et al., J. Appl. Polymer Sci., $\underline{17}$, 3495 (1973).
83. A. A. Griffith, Phil. Trans. Roy. Soc. London, $\underline{221}$, 163 (1920).
84. G. R. Irwin, in *Treatise on Adhesion and Adhesives*, R. L. Patrick, Ed., Vol. I, p. 233, Marcel Dekker, New York, 1967.

DISCUSSION

On the paper by W. H. Grant, B. W. Morrissey and R. R. Stromberg

D. W. Clayton (*Pulp and Paper Research Institute of Canada*): Did you measure the activation energy for the adsorption of polystyrene on these two surfaces?

W. H. Grant (*National Bureau of Standards*): No. Our objective in this study was to determine the effect of polymer conformation and conformational change on the adsorption process. We have generally, therefore, employed polymer solutions under theta conditions where the polymer conformation in solution is well characterized. Appreciable temperature changes would be necessary to precisely determine an activation energy in these systems.

On the paper by K. Hamann, R. Laible and J. Horn

R. Khanna (*Eastman Kodak Co.*): Your data on the graft vs. the homopolymer are very interesting. 1) What conditions of temperature and concentration were you using? 2) What do you feel are the reasons for the trends and differences in molecular weight?

K. Hamann (*Forschungsinstitut für Pigmente und Lacke*): 1) Normal conditions for polymerization, in case of styrene, for instance, 80° - 130°C in bulk or solution. 2) The lower $\bar{M}$ in the case of graft polymerization depends on steric hindrance on the solid surface by the covalently bound polymers.

D. W. Dwight (*Du Pont Company*): What is the rate of free radical polymerization on silica surfaces? Specifically is the <u>surface</u> polymerization complete in minutes, hours, or days?

K. Hamann: The first part of the graft reaction on the solid surface (40-60°C) only lasts some hours; the reaction then slows down and ends after 20-40 hours.

M. Kendig (*Lehigh University*): Have you tried experiments with iron oxides?

K. Hamann: We have tried TiO, ZnO, SiO_2 but only made preliminary experiments on Fe_2O_3 which I have no data on to date.

A. I. Medalia (*Cabot Corp.*): How do you determine the molecular weight of the grafted polymer?

K. Hamann: By prior removal of the SiO_2 with HF.

On the paper by K. L. Mittal

R. J. Good (*SUNY at Buffalo*): γ_ℓ, γ_s, θ, etc. depend on molecular structure. Numerical predictions of joint strength, from W_a, or γ_ℓ, γ_s, θ, etc., are always orders of magnitude away from observed results; but measured joint strengths can be correlated with molecular structure, particularly with factors that control polymer polarity, chain stiffness, rheological properties, etc. It is logical, then, to conclude that the variation of strength with γ_ℓ or with $\cos \theta$, etc., is the consequence of the fact that both are functions of the molecular properties. And that dependence is complex, and cannot be put in terms of a single independent variable.

K. L. Mittal (*IBM CORP.*): Your comment is very apropos and highly significant. I fully concur with you that both γ_s, γ_ℓ, θ, etc., and the measured adhesive or joint strengths are function of molecular properties. If these molecular properties are better understood and their role in dictating the joint strengths is properly accounted, then hopefully, one should be able to calculate the numerical values of the adhesive strengths. I certainly hope that my paper will galvanize some action in this direction, and in the future we will be able to control joint strengths on the knowledge of the molecular architecture, rheological and other germane properties.

S. C. Sharma (*General Tire & Rubber Co.*): It would be interesting to see if the condition for optimum adhesion, i.e. $\gamma_{s\ell} = 0$ also means that the solubility parameters of the adhering materials are essentially the same.

K. L. Mittal: Actually, the maximum adhesive strength is found for minimum $\gamma_{s\ell}$ and not for $\gamma_{s\ell} = 0$. Ideally, $\gamma_{s\ell}$ is not minimum at $\gamma_{sv} = \gamma_{\ell v}$ but this equality denotes $\gamma_{s\ell}$ values close to minimum. As regards solubility parameters, if it is assumed that δ's and γ's for polymerics are related, then the maximum in adhesive strength should be at the equality of solubility parameters of adherends and adhesives. In fact, Iyengar and Erickson (J. Appl. Polymer Sci., 11, 2311 (1967)) observed the maximum in adhesive strength when $\delta_s = \delta_\ell$. Apropos of relationship between δ's and γ's, it has been pointed out in the literature (J. L. Gardon, J. Phys. Chem. 67, 1935

(1963)) that there is an apparent linear relationship between δ's and γ_c's of polymeric materials.

A. W. Christiansen (*Exxon Chemical Co.*): With regard to minimizing $\gamma_{s\ell}$, what does one do in the case of adhesives composed of polymer blends or of block or graft copolymers?

K. L. Mittal: Your question is of great practical importance, but unfortunately, there is not enough literature available so as to test the criterion for maximum adhesive strength in the case of polymer blends or block or graft copolymers. I remember during the discussion somebody (most probably, Dr. Plueddemann) had cited a case in which copolymer gave good adhesive strength while the single polymer was not effective. This suggests that in the case of copolymers and polymer blends, one polymer may be such as to satisfy the condition for maximum adhesive strength. Obviously, surface characterization of polymer blends or graft and block copolymers, and the conditions for maximum adhesion in these systems need to be investigated.

PART TWO

Synthetic Polymers and Adhesives

Introductory Remarks

F. D. Petke

Research Laboratories

Tennessee Eastman Company

Division of Eastman Kodak Company

Kingsport, Tennessee 37662

    The relationship between the chemical structure of polymeric adhesives and their physical structure, physical properties, and performance characteristics have interested scientists for many years. The very complex nature of these relationships have, however, resulted in a lack of broad generalizations about structure-property relationships; instead, a myriad of technical papers have been published about the work done on specific systems of adhesives and adherends. These papers now form a large part of the adhesives literature; yet many of the questions basic to the design of satisfactory adhesives remain unsolved. The relationships developed for a given series of adhesives seldom apply to another series directly. Thus further experimentation is always needed to optimize formulations and compositions of adhesive products. The objective of the papers in this session is to examine several relationships of molecular structure to properties of several systems of polymeric adhesives which are important in today's technology.

    In the first paper, Petke reviews the current literature for developments in structure-property relationships of adhesives. He illustrates how chemical and physical structure affect both the adhesive and cohesive properties of polymers. In his review, he shows how the balance of adhesive and cohesive properties determine the ultimate bond strength of an adhesive. Relationships are cited for several types of adhesives.

St. Clair and Progar report, in the second paper, the effect of the solvent used in preparation of polyamic acid prepolymers on the bond strength properties of the resulting polyimide adhesives. The effect of chemical structure of the acid and diamine portions of the polyimides are illustrated as well. Compromises in performance of two adhesives are shown to be achievable in these systems by copolymerization or by blending of two polyamic acids.

Next, Kaelble discusses the utility of block copolymers as adhesives. The effect of molecular structure and rheological activity of the interfacial phase between hard and soft segments of the copolymers is examined, using the adsorption-interdiffusion model. The model is applied to styrene-butadiene-styrene block copolymers, polyurethanes, and nitrile-modified epoxide polymers. Kaelble discusses the particular advantages of block copolymers as elastomeric and structural adhesives.

In the next paper, Illinger, Lewis and Barr find that the morphological structure of polyurethane adhesives affects both the optical clarity and the energy-absorbing properties of acrylic-polyurethane adhesive-polycarbonate laminates. The morphological structure of the adhesive is in turn related to the composition of the hard and soft segments of the blocked polyurethane adhesives.

In the final paper, Azrak, Joesten and Hale illustrate the effect of reformulation of acrylic contact adhesives from solvent-based to water-based systems. Requirements on the tackifier and polymer latex are discussed which lead to an adhesive with the proper balance of open time and green strength.

It is certain that these papers will not satisfy all of the needs for knowledge about the relationships between properties and structure of polymers. Ultimately, such knowledge will be helpful in determining with little experimentation what polymer will serve as an adhesive for a specific application. While that state of the art may be some time off, it is perhaps somewhat closer because of this symposium and the papers which follow in this session.

Structure-Property-Performance Relationships in Synthetic Polymeric Adhesives

F. D. Petke

Research Laboratories

Tennessee Eastman Company

Division of Eastman Kodak Company

Kingsport, Tennessee 37662

The objective of this paper is to review the published literature concerning chemical structure-physical property-bond performance relationships in synthetic polymer adhesives. The structure factors that have been identified as affecting the physical properties of the adhesive include molecular weight, composition of the polymer chain, functionality, polarity, chain structure, crystallinity, crosslinking, and additives. The physical properties of the adhesive polymer or formulation which are controlled by the chemical structure include glass transition temperature, melting or softening point, surface energy, bulk and rheological properties, and stability. The relationship of bond strength, locus of failure, bond performance as a function of temperature, and pressure-sensitive tack to those physical properties of the adhesives is reviewed.

INTRODUCTION

Thousands of synthetic polymeric adhesives are in use today that vary from one another in composition or in formulation depending upon the properties necessary for particular applications. The performance of a polymer adhesive is dependent upon the physical properties of the polymeric base from which it is made and the effect on those properties of various formulating agents used to modify the base polymer. Properties of the adhesive are, thus, functions of the chemical composition and structure of the polymers and the modifying additives used in the adhesives. The relationships between the chemical structure of a polymer, its physical properties and its adhesive performance have interested scientists for many years. This has resulted in many excellent technical papers on property-structure relationships and has helped to increase our understanding of the factors controlling adhesive performance. The objective of this paper is to review the published literature concerning chemical structure-physical property-bond performance relationships in synthetic polymeric adhesives as an introduction to the four papers to follow.

Knowledge of basic adhesion theory is a prerequisite for the full understanding of property-structure relationships in adhesives. Over the years, many theories of adhesion have been proposed for different systems. While some have achieved substantial success in explaining adhesion phenomena in certain instances, none has proven to be wholly satisfactory in explaining, let alone predicting, adhesive performance in more than a few selected cases. Adhesion has been described in terms of several mechanisms: mechanical interlocking of adhesive to rough or porous substrates[1], chemical reactions across an interface[2], interdiffusion of macromolecular species between substrate and adhesive[3], electrostatic attraction between adhesive and substrate[3], and physical adsorption of adhesives onto substrate[4,5]. None of these mechanisms is all-encompassing, but since each theory has some experimental support, each may be a potential mechanism in all adhesive systems. Whether or not a particular mechanism or combination of mechanisms operates is dictated by the bonding parameters: the nature and structure of the adhesive and the substrates and the pressure-temperature-time conditions used in bonding.

Adhesive bond strength depends upon both adhesion of an adhesive to a substrate and cohesion of the adhesive layer itself. Adhesion can be examined in terms of the energetic parameters involving wetting of the substrate by the adhesive and kinetic factors which control the rate of spreading of adhesive on the substrate. Cohesive strength of the adhesive layer depends on many morphological, structural, and viscoelastic properties of the adhesive. Each of these will be discussed later.

ADHESION

The foundations of surface chemistry are reviewed by Zisman[5]. The basic relationship at a solid-liquid-vapor interface is

$$\gamma_{sv} - \gamma_{sl} = \gamma_{lv} \cos \theta \qquad (1)$$

where γ_{sv}, γ_{sl}, and γ_{lv} are the interfacial tensions of the solid-vapor, solid-liquid, and liquid-vapor interfaces, respectively; and θ is the contact angle of the liquid on the solid phase. Interfacial tensions cannot be directly measured on solid surfaces, but Zisman's[5] critical surface tension for wetting (γ_c) of a solid serves as an approximation to γ_{sv}, the unmeasurable parameter. γ_c is determined by extrapolating $\cos \theta$ vs. γ_{lv} data to lower γ_{lv} values until $\cos \theta = 1$ at $\theta = 0°$. For many data, the relationship of γ_{lv} and $\cos \theta$ appears to be a straight line, especially for liquids of low surface tension, so that

$$\cos \theta = 1 + b(\gamma_{lv} - \gamma_c) \qquad (2)$$

where b is the slope of the line (b is negative). Values of γ_c have been found to depend on the composition of the surface for many low-energy substrates, such as polymers[5,6], but it is also known that γ_c depends upon the liquids used to determine it, and thus is not a function of the polymer alone[7]. Further work in surface energy showed that surface tensions consist of fractions due to polar (γ_s^p) and nonpolar (γ_s^d) components, such that

$$\gamma_s = \gamma_s^p + \gamma_s^d \qquad (3)$$

Thus, polar sites in the solid and in the liquids used to determine contact angles lead to polar, as well as dispersive, interactions across an interface[7,8,9]. The relative magnitudes of γ_s^p and γ_s^d are functions of composition, functionality, and surface morphology of the substrate and may depend upon the liquids used to determine the contact angles[10].

The utility of surface energies in predicting adhesion improvement has been demonstrated quantitatively many times and thus several criteria have been developed for improved adhesion. It is well known that wetting of a substrate is improved when the surface tension of the adhesive, γ_1, is less than or equal to the critical surface tension, γ_c, of the substrate[5]. The work of adhesion[5], W_A, is

$$W_A = (1 + \cos\theta)\gamma_1 \qquad (4)$$

When measured on a real surface, which is microscopically rough, contact angle θ_r is related to the idealized contact angle by the roughness factor, r,

$$r = \frac{\cos \theta_r}{\cos \theta} \qquad (5)$$

Combining equations 2, 4, and 5, W_A is maximized when

$$\gamma_{lv} = -\frac{1}{br} + \frac{(\gamma_c)_r}{2} \qquad (6)$$

where $(\gamma_c)_r$ is the critical surface tension measured on a real (rough) surface[11]. Dyckerhoff and Sell[12] found that optimum adhesion was obtained when $\gamma_{lv} = \gamma_c$ for a variety of adhesives on several substrates. Kaelble[13,14] has shown that one obtains good adhesion when the wettability envelopes, derived from γ_s^p and γ_s^d for the substrate and the adhesive, overlap. Furthermore, the permanence of a bond in a given environment can be predicted by the wettability envelope method by accounting for the γ_s^d and γ_s^p values of the environment[14]. Thus, optimum adhesion is predicted when γ_l of the adhesive is less than or about equal to γ_c of the substrate to be bonded and when the polar contributions to the surface energies of the adhesive and substrate are similar.

On clean, high-energy surfaces of many important substrates, such as glass, metal, and metal oxides, contact angles cannot be measured. Bolger and coworkers[15,16] have shown that the surfaces of these high-energy substrates are usually covered with multilayers of adsorbed vapors, primarily water, which greatly modify their surface properties and thus influence adhesion. In these systems, polar interactions across the interface dominate the nonpolar dispersion force interactions as the polar adhesive molecules interact by dipole or ionic attraction with the vapors adsorbed on the substrate. The degree of interaction can be described as an acid-base phenomenon and as a function of the difference between the isoelectric point of the surface (IEPS) for the substrate and the pK_a of the adhesive. The concentration of polar groups needed to enhance adhesion to polar substrates is not large. In polyacrylate and polymethacrylate lacquers, 1-2 mole % of carboxylic acid groups was sufficient to improve adhesion to aluminum[17]. Similarly, 1-5 wt % of an acid modifier results in substantially improved adhesion of polypropylene to metals[18]. In a third example, only amines having at least two active hydrogens per molecule were effective adhesion promoters and prevented loss of bond strength in boiling water in epoxy to copper bonds[15]. Use of carboxyl functionality to promote adhesion to a variety of substrates, both high-energy and low-energy, is well known. Thus, adding carboxyl groups by copolymerization of acidic monomers with polymeric bases has been used to improve adhesion of thermoplastic polyesters[19], polyacrylates[17], and polyolefins[18] to many substrates. Other electronegative atoms, such as chlorine[20] and nitrogen[15,16], also promote adhesion by ionic or dipolar attractions across an interface.

Surface energetics of an adhesive are determined primarily by the chemical composition of the components, particularly with regard to polarity and functional groups[5,7]. As shown above, small amounts of certain functional groups can greatly improve adhesion to a given substrate. Thus, matching the solubility parameters of a tire cord and a resorcinol-formaldehyde-latex adhesive was found to give improved adhesion[21], perhaps by allowing interdiffusion at the interface[22]. Since solubility parameter and critical surface tension have been correlated[23], this is equivalent to matching γ_1 of the adhesive with γ_c of the substrate[12].

Favorable surface energetics are a necessary, but not sufficient, condition for bond formation. Proper rheology of the adhesive is required so that the adhesive can spread on the surface of the substrate on the time scale of bond formation[24]. Time-dependent spreading of the adhesive is a major factor in its application, whether or not diffusion processes take place at the interface. This spreading must not be confused with the energetic "wetting" of the surface by the adhesive. The primary factor controlling spreading rate is viscosity, determined mainly by molecular weight of the adhesive base[24,25,26].

Consequently, proper bond formation requires that the adhesive flow over the surface of the substrate and interact with it through intermolecular forces so that it adheres. Proper flow is mainly a function of adhesive viscosity. Adhesion depends on matching, as closely as possible, both the surface energies and the polar characters of the adhesive and the substrate.

COHESION

In addition to having adhesion to a substrate, an adhesive must have sufficient cohesive strength to support a load. Low cohesive strength can be improved by increasing molecular weight[27,28]; but these increases may also decrease adhesion despite improved cohesion[27]. For polymers of equal average molecular weight, the cohesive strength is generally better for one having a narrow molecular weight distribution than for one having a broad distribution[27]. Polymers of moderate to high crystallinity have better cohesive strength than their amorphous counterparts and they retain much of their strength up to their melting points. Thus, chlorinated stereoregular polypropylene gave stronger bonds to stereoregular polypropylene substrates than did chlorinated amorphous polypropylene[20]. However, at some point, increases in crystallinity can harm bond strength, especially peel strength, by embrittling the adhesive[28]. Adhesives based on polyethylene, nylon-12, and poly(ethylene terephthalate) all had better peel strength when their crystallinities were reduced by quenching[29]. Heterogeneous nucleation has been shown to improve the cohesive strength of the surfaces of FEP Teflon, nylon-6, and polyethylene[30,31,32].

Crosslinking is a reliable means of increasing the cohesive strength of polymers, and the properties of these adhesives are known to be functions of the crosslinking agent and the curing conditions. Lin and Bell[33] showed that torsional butt shear strengths of epoxy adhesive/aluminum substrate bonds decreased as M_c, the molecular weight of the adhesive between crosslinks, increased. Similar results were found for a crosslinked, amorphous 60/40 butadiene/styrene rubber bonded to FEP Teflon fluorocarbon when tested at constant test rate and temperature[34]. In a sense, interchain hydrogen bonding can be considered to be a form of crosslinking, although weaker than covalent crosslinking. The excellent cohesive strength of polyamides compared to other common polymers at equivalent molecular weights is attributed to hydrogen bonding[35]. Adhesion of epoxies to aluminum, of surface-treated rubber to glass, and of several polymers to cellulosics has also been attributed to hydrogen bonding across the interface[36].

Mixtures of polymers or blends of polymers with other additives should be compatible for maximum cohesive strength[37]. Grafting of one polymer with the monomer of another is useful in improving adhesion between phases of two incompatible polymers[38]. An epoxy resin cured with a mixture of two hardeners was found to behave like a blend of two compatible polymers[39].

Generally the rigid structural adhesives that display excellent adhesion in lap shear bond tests do not perform well when tested in peel configuration. Peel strength was found to improve as flexibility increased for a series of urethane-modified epoxies, although shear strength and cohesive strength, especially at elevated temperatures, often decrease as flexibility increases[40]. Copolymerization of pyromellitic dianhydride with alicylcic diamines gave adhesives with better flexibility and mechanical properties, but lower decomposition temperatures, than adhesives made from pyromellitic dianhydride and aromatic diamines. Adhesives which had high elasticity had better peel adhesion than those with low elasticity[41]. Epoxy resins have also been made more flexible by chain extension with diisocyanates and acrylates[42]. The flexibility of the long-chain dimer acid moiety in commercial polyamides provides good flexibility for adhesive use[35]. Kendall[43] has shown that cured natural rubber has higher modulus and lower energy dissipation rate than uncured rubber. By choosing joint geometry so modulus effects were eliminated, joint strength depended on loss properties of the adhesive. At high crack speeds, the loss properties of the rubber do not significantly affect the fracture energy, but at low speeds a crack-slowing mechanism occurs which depends on the relaxation properties of the rubber and uncured rubber supports higher stresses. When the modulus affects the bond strength, curing tends to increase strength due to higher modulus, but decrease strength due to reduced dissipation.

The effect of temperature on the performance of adhesive joints is well documented. The strength of an uncrosslinked adhesive at elevated temperatures is controlled by its melting point and glass transition temperature (T_g)[28]. The T_g should be above the upper use temperature of an amorphous adhesive for good bond strength and creep resistance[44]. In semicrystalline polyester adhesives, T_g should be high enough to maintain bond strength at the upper use temperature, but peel strength is low when T_g is appreciably above the upper use temperature and low-temperature properties are limited[28]. Bond strength at elevated temperatures can be increased by raising crystallinity[20,28,29], hydrogen bonding[35], and crosslinking[45]. In situ-toughened epoxies have been prepared in which a much broader use temperature range results from the two-phased structure that develops on curing[46]. Long-term strength of an adhesive at elevated temperature is promoted by minimizing (1) structures subject to free radical decomposition, (2) easily oxidizable sites, and (3) groups susceptible to chemical reaction[45].

BOND STRENGTH

Both adhesive and cohesive forces determine bond strength. Bond failure energy has been shown to be composed of two parts: a reversible work of adhesion and an irreversible work of adhesive deformation[47,48,49,50]. Thus, the strength of styrene-butadiene rubber bonded to several polymeric substrates was found to depend on two components: a viscoelastic energy dissipation term, which is a function of test rate and temperature, and the intrinsic failure energy, θ_o, which agrees closely with the work of adhesion, W_A, when bond failure is apparently interfacial. When bond failure is cohesive in the adhesive layer, θ_o is smaller than W_A[48]. Whether or not true interfacial bond failure occurs is not known. Quite often failure occurs so near to an interface that failure is apparently interfacial, although weak boundary layers and plastic deformation of the adhesive may be the actual causes[1,51]. Good[52] proposes that true interfacial bond failure can occur, and he lists criteria for its occurrence. A particular adhesive may exhibit both cohesive and apparent interfacial failure; transitions from one to the other have been observed in pressure-sensitive adhesives as a function of temperature[53], and in epoxy systems as a function of hardener concentration[54].

Pressure-sensitive adhesives seem to possess a critical balance between adhesion and cohesion. The phenomenon of pressure-sensitive tack has been described as a complex function of flow and diffusion[24]. The role of resins in pressure-sensitive adhesives has been interpreted recently as improving tack by a two-step process: (1) reducing the modulus and viscosity of the adhesive to give faster, more complete wetting of the substrate and (2) raising the T_g of the adhesive. Plasticizers and solvents accomplish the first objective but not the second[55]. This interpretation is consistent with the two-step process of measuring tack according to Bauer[56] in that reduced modulus and viscosity make bond formation easier while increased T_g makes bond failure more difficult. Ionic binding within the molecule has been shown to increase the cohesive strength of acrylic pressure-sensitive adhesives, but to reduce tack simultaneously[29].

SUMMARY

From the preceding information, the complexity of the problems of adhesive design can be realized. For good bond performance, an adhesive must have the proper balance of adhesion and cohesion. Adhesion is determined by the forces acting across the adhesive/substrate interface. These forces are determined by the chemical composition of the adhesive polymer or the additives incorporated into it and the functionality of the polymer. Polymers that match the substrate most closely in polarity and surface energy appear to give the best bonds. The adhesive must flow onto the substrate in such a way that maximum interfacial contact is made between adhesive and substrate. For a polymer to spread spontaneously on the substrate, it must have a surface energy equal to or less than the substrate and it must have a viscosity which allows spreading on a reasonable time scale. For adequate cohesive strength of the adhesive layer, the base polymer requires a balance of toughness and flexibility. Toughness is controlled by tacticity, molecular structures, molecular weight, crosslinking, crystallinity, and hydrogen bonding. Flexibility is a function of molecular composition, side chain structure, and plasticization. Proper design of the polymers to be used in an adhesive will combine the features listed to give a product with the right balance of glass transition temperature, melting or softening point, surface energy, stability and bulk and rheological properties. The task of molecular design is made more difficult by the interrelationship of many of the design factors. Nevertheless, an understanding of the principles will help to enable proper design of an adhesive with the correct balance of properties for a given application.

REFERENCES

1. J. J. Bikerman, *The Science of Adhesive Joints*, Academic Press, Inc., New York, 1961.
2. T. P. Murphy, Ind. Eng. Chem. **58**, 41 (1966).
3. J. R. Huntsberger, The Mechanism of Adhesion, in *Treatise on Adhesion and Adhesives*, Vol. **1**, R. L. Patrick, ed., Marcel Dekker, Inc., New York, p. 119, 1967.
4. R. R. Stromberg, Adsorption of Polymers, in *Treatise on Adhesion and Adhesives*, Vol. **1**, R. L. Patrick, ed., Marcel Dekker, Inc., p. 67, 1967.

5. W. A. Zisman, Adv. Chem. Ser. 43, 1 (1964).
6. W. A. Zisman, ACS Div. Org. Coat. and Plast. Chem. Preprints, 31 (2), 13 (1971).
7. J. R. Dann, J. Colloid and Interface Sci. 32, 321 (1970).
8. S. Wu, J. Adhesion 5, 39 (1973); also, *Recent Advances in Adhesion*, ed. L.H. Lee, Gordon and Breach, New York, p. 45 (1973).
9. F. M. Fowkes, *Recent Advances in Adhesion*, ed. L.H. Lee, Gordon and Breach, New York, p. 39 (1973).
10. H. Schonhorn, Encycl. Polym. Sci. and Tech. 13, 533 (1970).
11. D. H. Kaelble, *Physical Chemistry of Adhesion*, Wiley Interscience, Inc., New York (1971).
12. G. A. Dyckerhoff and P. J. Sell, Angew. Mokromol. Chem. 21, 169 (1972).
13. D. H. Kaelble, Proc. 23rd. Int. Congr. Pure and Appl. Chem. 8, 265 (1971).
14. D. H. Kaelble, J. Appl. Polymer Sci. 18, 1869 (1974).
15. J. C. Bolger, H. E. Molvan, Jr., and Robert W. Hausstein, SPE Technical Papers 18, 408 (1972).
16. J. C. Bolger, SPE Technical Papers 18, 402 (1972).
17. T. R. Bullett, *Recent Advances in Adhesion*, ed. L.H. Lee, Gordon and Breach, New York, p. 201 (1973).
18. R. A. Steinkamp, K. W. Bartz and R. A. Von Brederode, SPE Technical Papers 19, 110 (1973).
19. W. J. Jackson and J. R. Caldwell, Adv. Chem. Ser. 99, 562 (1971).
20. Y. Aoki, J. Polymer Sci., Part C. 23, 855 (1968).
21. Y. Iyengar and D. E. Erickson, J. Appl. Polymer Sci. 11, 2311 (1967).
22. D. H. Kaelble, Trans. Soc. Rheol. 15, 235 (1971).
23. L. H. Lee, Adv. Chem. Ser. 87, 106 (1968).
24. W. C. Wake, *Aspects of Adhesion* 4, 17 (1968).
25. J. W. McDonald and G. L. K. Hoh, TAPPI 51, 46A (1968).
26. A. N. Gent, J. Polymer Sci., Part A2, 9, 283 (1971).
27. G. E. J. Reynolds, *Aspects of Adhesion* 6, 96 (1971).
28. W. J. Jackson, Jr., T. F. Gray, Jr., and J. R. Caldwell, J. Appl. Polymer Sci. 14, 685 (1970).
29. K. Nakao, J. Adhesion 4, 95 (1972); also, *Recent Advances in Adhesion*, ed. L.H. Lee, Gordon and Breach, New York, p. 453 (1973).
30. H. Schonhorn and F. W. Ryan, Adv. Chem. Ser. 87, 140 (1968).
31. H. Schonhorn and F. W. Ryan, J. Polymer Sci. 7, 105 (1969).
32. H. L. Frisch, H. Schonhorn and T. K. Kwei, J. Elastoplastics 3, 214 (1971).
33. C. J. Lin and J. P. Bell, J. Appl. Polymer Sci. 16, 1721 (1972).
34. E. H. Andrews and A. J. Kinloch, J. Polymer Sci., Polymer Physics 11, 269 (1973).
35. R. D. Dexheimer and L. R. Vertnik, Adhesives Age 17, (8), 31 (1974).
36. W. H. Pritchard, *Aspects of Adhesion* 6, 11 (1971).
37. R. D. Bohme, J. Appl. Polymer Sci. 12, 1097 (1968).

38. L. H. Lee, Adv. Chem. Ser. **87**, 85 (1968).
39. R. K. Jenkins, J. Appl. Polymer Sci. **11**, 171 (1967).
40. J. A. Clarke, *Recent Advances in Adhesion*, ed. L.H. Lee, Gordon and Breach, New York, p. 239 (1973).
41. D. Sek, U. Gaik and Z. Jedlinski, Eur. Polymer J. **9**, 593 (1973).
42. Z. Aggias, U.S. Patent 3,488,297 (1970).
43. K. Kendall, J. Polymer Sci., Polymer Physics **12**, 295 (1974).
44. P. E. Cassidy, J. M. Johnson and C. E. Locke, J. Adhesion **4**, 183 (1972).
45. K. W. Humphreys, *Aspects of Adhesion* **1**, 66 (1965).
46. C. D. Weber and M. E. Gross, Adhesives Age **17** (2), 18 (1974).
47. E. H. Andrews and A. J. Kinloch, Proc. Roy. Soc. (London), **332**, 401 (1973).
48. E. H. Andrews and A. J. Kinloch, Proc. Roy. Soc. (London), **332**, 385 (1973).
49. J. Schultz and A. N. Gent, J. Chem. Phys. **70**, 708 (1973); also, *Recent Advances in Adhesion*, ed. L.H. Lee, Gordon and Breach, New York, p. 253 (1973).
50. A. N. Gent and A. J. Kinloch, J. Polymer Sci., Part A2, **9**, 659 (1971).
51. H. E. Bair, S. Matsuoka, R. G. Vadimsky and T. T. Wang, J. Adhesion **3**, 89 (1971).
52. R. J. Good, *Recent Advances in Adhesion*, ed. L.H. Lee, Gordon and Breach, New York, p. 357 (1973).
53. D. Satas and R. Mihalik, J. Appl. Polymer Sci. **12**, 2371 (1968).
54. C. M. Peterson, Koll. Z. & Z. Polym. **222**, 148 (1968).
55. M. Sherriff, R. W. Knibbs and P. G. Langley, J. Appl. Polymer Sci. **17**, 3423 (1973).
56. R. F. Bauer, J. Polymer Sci., Part A2, **10**, 541 (1972).

Solvent and Structure Studies of Novel Polyimide Adhesives

Terry L. St. Clair

Virginia Polytechnic Institute and State University

Blacksburg, Virginia 24061

Donald J. Progar

NASA-Langley Research Center

Hampton, Virginia 23665

Significant improvements in condensation polyimide adhesives have been made by chemical modification of the diamine or dianhydride monomers, by use of copolymer composition and by use of a unique ether solvent as the polymerization medium. The effect on adhesive properties was studied and a family of adhesives developed whose bonding pressures (40-200 psi) and use temperature (up to 300°C) can be varied with composition to cover a wide range of practical applications. Alterations in the polymer backbone have also led to imidized films that have sufficient flow for possible use as film adhesives.

INTRODUCTION

Bonded structures are advantageous to the aerospace industry due to their lightness of weight and excellent resistance to fatigue and corrosion. Supersonic aircraft with Mach 3 capability will experience skin temperatures approaching 300°C. In addition, some future spacecraft structures could require adhesives for use at 300°C and above. Such extreme temperature requirements have necessitated the utilization of titanium and composite materials which will withstand such an environment. The past decade has seen remarkable progress in finding thermally stable polymers which show great promise for use as structural adhesives. However, the actual development of reliable adhesive resins for longtime use with advanced materials at 250-300°C has been less than satisfactory. Among disadvantages to be overcome are the extreme processing conditions required, batch to batch variation in resin properties, and unacceptable variations in initial and aged bond strengths. There is little doubt that additional research and development is in order beyond the products currently being marketed.

Among several classes of polymers which are inherently high in thermal stability, the aromatic polyimides have shown unusual promise as high temperature adhesives with titanium and high performance composites. Since polyimide adhesives in general are applied in the intermediate polyamic acid form from a solvent, it is evident that the solvent stands to play an important part in the bonding of structures. The solvent is beneficial if it causes good wetting of the adherends and is easily removed prior to bonding. On the contrary, if solvent remains in the polymer well into the process, harmful voids can form in the bondline and lead to poor adhesion and aging. Reactivity of the solvent at elevated bonding temperatures can cause degradation of the adhesive. Therefore, the ideal solvent is one which wets the adherends, shows good volatility, and is nonreactive at the elevated temperatures needed for bonding a particular system.

In the development of a polyimide adhesive, the molecular structure of the polymer is just as important as the effect of solvent. The thermal stability of a polyimide is also dependent upon its molecular structure, but such stability does not necessarily relate to good adhesive properties and previous research has emphasized the former. Further, most commercially available polyimide adhesives are based on particular dianhydride and diamine starting materials because of their availability and low cost, which are major determining factors for production. However, the effect of structure variations on adhesive properties should be the prime concern of research to assess the ultimate capacity of a new resin system. Subsequent development should be directed toward economic factors.

This report relates the importance of the polymerization solvent and molecular structure of polyimides to their adhesive properties.

EXPERIMENTAL

Monomers   Pyromellitic dianhydride (PMDA), 4,4'-oxdiphthalic anhydride (ODPA)[1] and 3,3',4,4'-benzophenone tetracarboxylic acid dianhydride (BTDA) were sublimed at 215°C at less than 1 torr. A research dianhydride, bis[p-(3,4-dicarboxyphenoxy)phenyl] sulfone dianhydride (BSDA)[2], was also used. The diamines presented in Table I with literature references were purified by recrystallization before use.

Polymerization   Polymerizations were performed at 20-25°C at a concentration of 15 percent solids in reagent grade bis(2-methoxyethyl) ether (diglyme) as received. First the diamine was completely dissolved or slurried in diglyme. The dianhydride was then added as a solid in four portions allowing sufficient time for reaction to occur between additions. In some cases, the polymer tended to precipitate from solution after an appreciable viscosity had been reached. However, the addition of small amounts of ethanol caused the polymer to redissolve.

Characterization   Inherent viscosities were determined from 0.5 percent solutions in diglyme or N,N-dimethylacetamide (DMAc) at 35°C. Films of the polymers cast on glass plates, dried and imidized for 1 hour each at 100°, 200°, and 300°C in air were studied by infrared spectroscopy. Thermo-mechanical analyses (TMA) were performed on cured films in static air at a heating rate of 5°C/min. on an E. I. duPont Model 990 Thermomechanical Analyzer which was modified to accept film specimens 1 mm in width and 1.5 mm in gage length.[3] Thermogravimetric analyses (TGA) were obtained on 1 mil films at 2°C/min. in static air. Torsional braid analyses (TBA) were determined at 3°C/min. in nitrogen on glass fiber braids coated with polymer resin and treated at 300°C in air for 1 hour.

Adherends   Adherends were either titanium-6Al-4V, glass/P13N polyimide, or unidirectional graphite/P13N composites. The panels measured a nominal 50 mil thickness for titanium, 120 mil for the glass/P13N composites, and 65 mil for the graphite/P13N composites. All adherends were 1 inch in width. Prior to use the titanium panels were cleaned by the Pasa Jell method[4], and composite adherends were lightly grit-blasted with 120 aluminum oxide.

Bonding   The panels to be bonded were coated with the polyamic acid solutions (15 percent solids in diglyme) and allowed to air dry 15 min. in a laminar flow hood and 15 min. in a 60°C oven. A total of five coats were accumulated in this manner. The panels were then bonded with a one half inch overlap in a jig which had been shimmed

Table I

Polymers Prepared in Diglyme

Diamine Structure		Dianhydride	Lap Shear Strength	Diamine Source
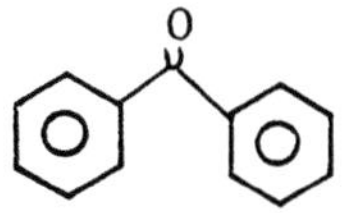 (benzophenone diamine structure)	4,4'-diamino	BTDA	2590 psi	Commercial
	3,3'-diamino	"	6000	Commercial
	3,4'-diamino	"	2465	Commercial
	3,5-diamino	"	2612	Reference 8
	4-methyl 3,3'-diamino	"	2243	Reference 9
	3,3'-diamino	PMDA	0	Commercial
	3,3'-diamino	ODPA	4737	Commercial
	3,3'-diamino	BSDA	2665	Commercial
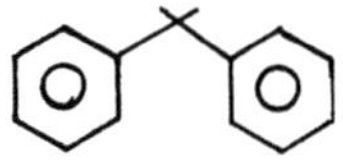 (diphenylmethane diamine structure)	4,4'-diamino	BTDA	1860	Commercial
	3,3'-diamino	"	4229	Reference 8
	3,3'-dimethyl 4,4'-diamino	"	1980	Commercial
	3,3'-dichloro 4,4'-diamino	"	1120	Commercial

Table I (Continued)
Polymers Prepared in Diglyme

Diamine Structure		Dianhydride	Lap Shear Strength	Diamine Source
(bis-phenol with CH-OH bridge)	3,3'-diamino	BTDA	1600 psi	Reference 10
(diphenyl sulfide)	4,4'-diamino	"	4330	Commercial
(diphenyl sulfone)	4,4'-diamino	"	1860	Commercial
(benzophenone-type)	4,4'''-diamino	"	2581	Reference 11
	3,3'''-diamino	"	4200	Reference 9

for a fixed bondline thickness. Bondlines were maintained at 2 to 4 mils depending on the amount of flow that occurred during the bonding cycle. Where no carrier cloth was used, the pressure was held at 50 psi. However, bonding pressures ranged between 50 and 200 psi when fillers or carrier cloths were employed, and bondlines were controlled solely by the cloth or filler. Pressure was applied at the beginning of the bonding operation with a heating rate of 5°C/min. and a heating cycle range of room temperature to 300°C. The panels were held at 300°C for 50 minutes before cooling under pressure to 150°C.

Adhesive Testing The bonded specimens were tested for lap shear strength on an Instron Universal Testing Instrument at a crosshead speed of 0.05"/min. (1200-1400 psi/min.) as detailed in ASTM D-1002. Lap shear strengths in the tables represent a minimum of four specimens. Variations among specimens were no more than ±15 percent of the average values.

RESULTS AND DISCUSSION

Effect of Solvent Polyamic acids were first reported made in ether solvents as insoluble polymers for use as molding powders by Dine-Hart and Wright[5], and Vaughan and Jones.[6] The thrust of their research was to find a solvent in which the reaction between the diamine and dianhydride could afford an insoluble polyamic acid powder.

An evaluation of ether solvents as the polymerization media was undertaken for preparing polyamic acids to be used as adhesives. The systems that exhibited the best adhesive properties were those that seemed to build to high molecular weight in solution before precipitation occurred. In contrast to the systems where the polyamic acids precipitated from ether solvents as powders, the BTDA-diaminobenzophenone (DABP) amic acids tended to precipitate as gummy amorphous materials. For this and other cases where clouding or precipitation occurred, conditions could be reversed and clear solutions obtained by the addition of a small amount of a low molecular weight alcohol (usually ethanol). These solutions were used directly in the bonding process.

The ether solvents were found to enhance the adhesive properties of all polymer systems tested. For example, Table II compares the lap shear strengths for the BTDA-3,3'-DABP when prepared in several different solvents. Diglyme afforded the highest strength (6000 psi--an outstanding value for bonding titanium) and was, therefore, used for all other polyamic acid polymerizations.

Inherent viscosities in DMAc and diglyme were taken for the polyamic acids which had been synthesized in diglyme. The relation-

Table II

Effect of Polymerization Solvent on Titanium-to-Titanium
Lap Shear Strength Using BTDA-3,3'-DABP Adhesives

Solvent	Lap Shear Strength, psi
Diglyme	6000
Dioxane/N,N-dimethylacetamide	2900
N,N-dimethylacetamide	2580
N-methylpyrrolidone	1220
N,N-dimethylformamide	410

ship between values in diglyme and DMAc varied considerably depending upon polymer structure. For example, the polymer BTDA-3,3'-DABP prepared in diglyme had a viscosity of 0.81 in DMAc, and only 0.38 in diglyme. To the contrary, BTDA-3,3'-dichloro-4,4'-diaminodiphenylmethane, also prepared in diglyme, exhibited viscosities of 0.41 in DMAc and 0.38 in diglyme. These examples represent the extremes in comparisons for all polymer systems studied. This effect of solvent on viscosity is not yet fully understood. Although their number and weight average molecular weights have not been determined, all the polymers in Table I, except PMDA-3,3'-DABP, exhibited excellent film forming properties. This implies that molecular weights of polymers made in diglyme are sufficiently high to afford reliable information for adhesives applications.

A relationship between viscosities in the two solvents has been established for the particular system BTDA-3,3'-DABP as shown in Figure 1. The viscosities in diglyme seem to approach a maximum of 0.4. In this region, a small change in the diglyme viscosity reflects a considerably larger change in the DMAc viscosity. The polymers in Figure 1 that had inherent viscosities of 0.28, 0.31 and 0.34 in diglyme and 0.50, 0.61 and 0.81 respectively when run in DMAc had essentially the same lap shear strengths when tested as adhesives.

A current investigation by Wightman[7] shows diglyme to be a better wetting agent for titanium than the more conventional amide solvents. This ability to improve wetting of the adherend coupled with its inertness makes diglyme an excellent solvent for use as a polymerization medium in the preparation of adhesives. An additional advantage to using diglyme is the fact that the commercially available reagent grade solvent needs no additional purification to yield high molecular weight polymer. If small amounts of water are present, it apparently does not hinder reaction and may even be beneficial.[6] However, when amide solvents are used, distillation from such drying agents as sodium hydride is required to obtain the purity necessary for obtaining high molecular weight polymers.

One disadvantage noted in the use of ether solvents is that certain diamines are insoluble and therefore nonreactive. Also certain monomer combinations form insoluble polyamic acids which either require large volumes of an alcohol to bring about redissolution or must be filtered and redissolved in an amide solvent.

Effect of Structure   A compilation of diamine and dianhydride combinations reacted in diglyme and tested for adhesive properties are given in Table I. Generally, the systems based on PMDA had very poor adhesive properties regardless of the diamine, even when the polyamic acid was soluble in diglyme. Polymers prepared from bis-phthalic anhydrides such as BTDA, ODPA, and BSDA performed well as adhesives. The BTDA systems were investigated in greater detail to determine the effect of the diamine structure on adhesive properties. It is apparent from the adhesive data that 3,3'-DABP has a more beneficial effect on the adhesive properties of the polyimides than the other diaminobenzophenone isomers. This modification in structure from the rigid 4,4'-DABP monomer to the 3,3'-DABP resulted in a better than doubled lap shear strength. This same phenomenon was exhibited in the methylenedianiline (MDA) series where the polyimide using 3,3'-MDA was twice as strong as the 4,4'-MDA.

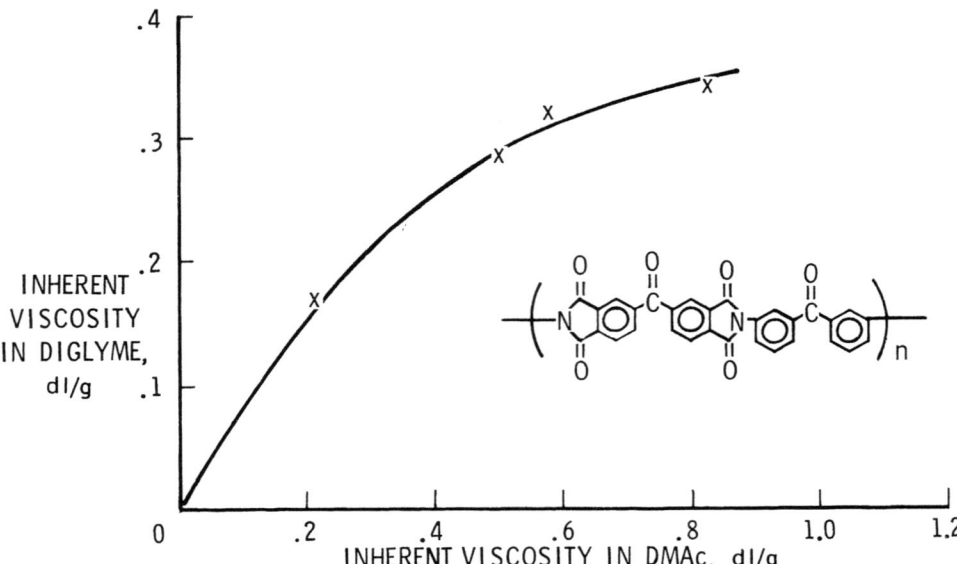

Figure 1. Inherent viscosities of BTDA-3,3'-DABP polymer solutions prepared in diglyme (0.5% concentration at 35°C).

When the imide content was diluted as in the case of the benzyl-benzophenones, the polymer containing the 3,3''-isomer also showed a greatly improved lap shear strength over that with the 4,4''-isomer. Models show that the meta linkage allows the conformations of these polymer systems to be either helical, linear or some intermediate of these extremes, whereas the para systems are strictly linear in conformation. The enhanced flexibility of polymers with the meta linkages may allow better interaction with the adherend surface and/or increase polymer toughness.

The high lap shear strength of 4,4'-diaminodiphenylsulfide (TDA) with BTDA may be explained by the presence of a less rigid system compared to BTDA-4,4'-MDA due to the greater flexibility of the sulfur-carbon bond when sulfur is in this low oxidation state. BTDA-4,4'-diaminodiphenylsulfone had a lap shear strength of 1860 psi which is comparable to that for the 4,4'-MDA polymer. It is possible that the oxidized sulfur atom may restrict the flexibility of the sulfur-carbon bond.

The polymer system with the highest lap shear strength was the BTDA-3,3'-DABP conbination. However, a distinct disadvantage to this system was its low glass transition temperature (T_g) of 256°C and a lap shear strength reduction of one half when tested at 225°C. Two successful approaches to increasing elevated temperature strengths were based on a general rigidization of the polymer backbone. First, the copolymerization of BTDA and PMDA with 3,3'-DABP produced a polymer with a T_g of 290°C when the BTDA to PMDA ratio was 2:1. The second approach involved the blending of equal portions of BTDA-3,3'-DABP and BTDA-4,4'-DABP polymers. Both techniques gave adhesives with improved strengths at elevated temperatures at an acceptable trade-off of room temperature strength as shown in Table III. The addition of aluminum filler further improved both room temperature and high temperature strengths of the 65/35 BTDA/PMDA-3,3'-DABP system as illustrated in Table IV.

Most of the polyimides discussed in this paper were evaluated as adhesives for bonding titanium adherends. Recently, however, a need for polyimide adhesives for use with polyimide composites has developed. Preliminary studies for the BTDA-3,3'-DABP polyimide adhesive system with composites show promising results as reflected in Table V.

Polymer Characterization   The T_g's obtained by TMA and TBA, of polymers prepared in diglyme or other ether solvents were the same as for those polymers prepared in DMAc. Infrared analyses also proved that the polyamic acids and polyimides synthesized in diglyme were essentially identical to those made in the usual aprotic amide solvents. The infrared spectra of polymers made in diglyme and other ethers, and thermally imidized, disclosed a band of medium to low intensity at 1850 cm^{-1} which is characteristic of an anhydride

Table III

Improvement of Elevated Temperature Properties by Copolymerization and Blending

Materials	Lap Shear Strength, psi*			
	RT	225°C	250°C	260°C
BTDA-3,3'-DABP, Control	6180	2600	950	350
90/10 BTDA/PMDA-3,3'-DABP	5050	2580	1365	715
80/20 BTDA/PMDA-3,3'-DABP	4700	2340	1550	980
65/35 BTDA/PMDA-3,3'-DABP	3730	2000	1650	1300
50/50 BTDA/PMDA-3,3'-DABP	2830	1750	-	1330
50:50 BTDA-3,3'-DABP: BTDA-4,4'-DABP Blend	3440	2240	1600	1150

*Titanium adherends

Table IV

Effect of Aluminum Filler on Strengths of Titanium Bonded with Polyimides

65/35 BTDA/PMDA-3,3'-DABP

	25°C	250°C
Unfilled	3105	1150
50% Al*	3425	1285
63% Al*	3625	1450
70% Al*	3855	2360

*Percent aluminum based on total solids

Table V

Composite Bonding with BTDA-3,3'-DABP Polyimide

Adherends	Single Lap Shear Strength, psi
P13N/Glass-P13N/Glass	2700
P13N/Glass-Titanium	2900

Adherends	Double Strap Lap Shear Strength, psi
P13N/Glass-Ti-P13N/Glass	3500
P13N/Graphite-Ti-P13N/Graphite	3200

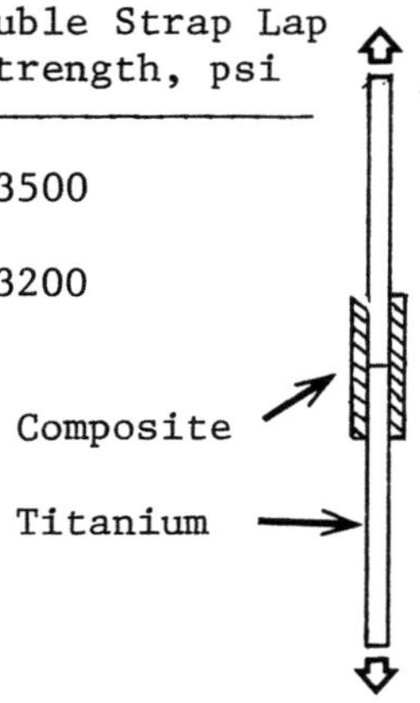

Composite

Titanium

group. The possibility that the presence of residual anhydride in the polymers was responsible for the enhanced adhesion is under investigation. As determined by TGA, the polymers prepared in diglyme occassionally lost weight at slightly lower temperatures than did those made in DMAc. Mass spectrometer analyses showed this loss to be tightly-bound solvent. The amount and effect of residual solvent in lap shear joints has not been determined.

CONCLUSIONS

The use of aliphatic ethers, and diglyme in particular, as solvents for the preparation of aromatic polyimides enhanced the adhesive properties of most of the polymers studied. The ether solvents were better wetting agents for titanium adherends, did not require rigorous purity standards, and were inert at elevated temperatures. The structures of the polyimides were shown to have a substantial effect on their adhesive properties, with the BTDA-3,3'-DABP composition being especially attractive. Slight modifications of the polymer backbone were used to alter the thermo-mechanical characteristics, such as T_g of the adhesives. For example, a copolymer system (BTDA/PMDA-3,3'-DABP) was developed that could be used over a wide range of temperatures. The inclusion of varying amounts of aluminum filler increased the use temperature of the polyimide adhesive system without adversely affecting its strength at room temperature. Initial studies with the unmodified BTDA-3,3'-DABP polyimide indicated great promise for use as an adhesive in bonding titanium and polyimide composite combinations.

REFERENCES

1. G. Kolesnikov, O. Fedotova, E. Hofbauer and V. Shelgayeva, Vysokomol. Soedin., Ser. A., 9: 612 (1967).
2. H. R. Lubowitz, French Pat. 2,030,905 (1970).
3. H. D. Burks, J. Appl. Polym. Sci. 18, 627 (1974).
4. American Cyanamid Co. Inc., Technical Bulletin "FM-34 Adhesive Film", January 25, 1968.
5. R. A. Dine-Hart and W. W. Wright, J. Appl. Polym. Sci. 11, 609 (1967).
6. M. F. Vaughan and J. I. Jones, Brit. Pat. 1,162,203 (1969).
7. J. P. Wightman, Virginia Polytechnic Institute and State University, Blacksburg, VA., unpublished results.
8. V. L. Bell, Org. Coatings and Plastics Chem., Preprints 33 (1), 153 (1973).
9. V. L. Bell, NASA-Langley Research Center, Hampton, Virginia, unpublished results.
10. H. Kuhnis and H. deDiesbach, Helv. Chim. Acta. 41, 894 (1958).
11. L. Gattermann and H. Rudt, Chem. Ber. 27, 2295 (1894).

Block Copolymers as Adhesives

D. H. Kaelble

Science Center, Rockwell International

Thousand Oaks, California 91360

Block copolymers form a new class of molecular composite materials by the phase separation of incompatible hard and soft segments which form their macromolecular structure. Thermoplastic elastomers where the soft segments form the continuous phase have been extensively investigated by means of an adsorption-interdiffusion (A-I) model for the interfacial phase which bonds the hard and soft phases. The molecular structure and rheological activity of the interfacial phase in thermoplastic elastomer block copolymers is shown to play a dominant role in nonlinear viscoelastic response, mechanical hysteresis and energy absorption. Creation of elastomeric microphases in epoxy structural adhesives has been recently identified with in situ block copolymerization between carboxy terminated nitrile (CTBN) rubber and the diepoxide. In this case, the soft phase is discontinuous and acts as a toughening agent for the hard epoxy continuum. The A-I model is shown applicable to describing the rheology and fracture properties of the CTBN modified epoxy adhesives. The specific advantages of developing block copolymers as elastomeric or structural adhesives is discussed in terms of molecular design factors such as composition, block structure and molecular weight. Adhesion properties unique to block copolymer adhesives are discussed and summarized.

INTRODUCTION

Perhaps one of the most stimulating recent developments in adhesive technology is the emergence of block copolymeric elastomers with completely new adhesion performance[1]. The multiphase domain structure resulting from the substantial incompatibility of the block polymer segments produces a new class of molecular composites and viscoelastic response[2,3]. Excellent review papers now document the extensive experimental and theoretical studies of the thermoplastic elastomer class of block copolymer where a distinct glass transition is displayed for each block segment[4-6]. Studies of amorphous and nonpolar block copolymers have until recently been specifically directed toward a better understanding of the structure-property relations which define the morphology and mechanical response, such as the linear triblock copolymer S-B-S where S = polystyrene and B = polybutadiene. A second important class of polar and semicrystalline block copolymer is represented by the repeat segment sequence of the $[-A-B-]_n$ type where A = soft polyether or polyester segments and B = hard semicrystalline polyester or polyurethane segments[7-9]. A third important class of three dimensional block copolymer is represented by the co-reacted epoxy-carboxy terminated nitrile rubber (CTBN) systems which copolymerize and phase separate to form multi-phase crosslinked systems[10-12].

An earlier analysis of the morphology and mechanical response of S-B-S (styrene-butadiene-styrene) block copolymers by Kaelble[3] pointed out that adsorption-interdiffusion theory of bonding predicts that an interfacial (I) phase of mixed (interdiffused) S and B segments intervenes the pure S and B phases. More recently, Kalfoglou and Williams[11] have utilized a similar block copolymer model to analyze the viscoelastic response of the CTBN-epoxy adhesive systems. One objective in the present discussion is to review the special morphology and mechanical response which distinguishes block copolymers as adhesives. A second objective is to discuss the molecular design factors such as composition, block structure, and molecular weight which determine the performance of block copolymers as elastomeric or structural adhesives.

NON-POLAR THERMOPLASTIC ELASTOMERS

The nonpolar thermoplastic elastomers present a simple beginning point for describing the complicated morphology and physical response of block copolymers. A simplified model for two-phase morphology in the symmetrical S-B-S triblock copolymers which follows the description of Holden, Bishop, and Legge[2] is shown schematically in Figure 1 where the polystyrene (S) and blocks have average molecular weight $M_S \simeq 11,000$ and polybutadiene (B) center block a molecular weight of $M_B \simeq 54,000$ as reported by Smith and Dickie[13] for Kraton 101 (Shell Chemical Co.). The two-phase model

Fig. 1. Two-phase model for S-B-S triblock copolymer where S = polystyrene and B = polybutadiene where polystyrene segments form spherical domains in polybutadiene continuum.

of Holden,[2] wherein the hard polystyrene segments form spherical domains of molecular dimensions accounts for the unusually high strength and extensibility shown in the tensile stress-strain curve in the lower view of Figure 2. The two-phase model shows that the chains terminate in a glassy phase and, therefore, act as physical cross-links unless solvolytically or thermally softened. As pointed out by Beecher, Marker, Bradford and Aggarwal[14], the simple two-phase model failed to account for the pronounced mechanical hysteresis in S-B-S copolymer tensile extension-recovery response as shown in the upper view of Figure 2.

The multiphase model for S-B-S block copolymer morphology proprosed by Kaelble[3] presents the three phase morphology for Kraton 101 shown by the schematic of Figure 3. The three phase morphology of Figure 3 has been applied in a detailed analysis of Kaelble and Cirlin[15] of temperature and strain rate effects upon the mechanical hysteresis of Kraton 101. The results of several studies on the hysteresis and autohesion properties of Kraton 101 show that the mechanical hysteresis shown in the upper curve of Figure 2 is accompanied by extensive internal debonding and cavitation whose locus appears to be the interfacial (I) phase which surrounds the polystyrene spherical phase in the schematic of Figure 3[15,16].

It is, of course, now well understood that cavitation and crazing are widely displayed micromechanisms for mechanical hysteresis and high fracture toughness in polymeric materials as detailed in a recent review by Kambour[17]. The extensive cavitation observed in S-B-S triblock copolymer (benzene cast Kraton 101 film) is known to

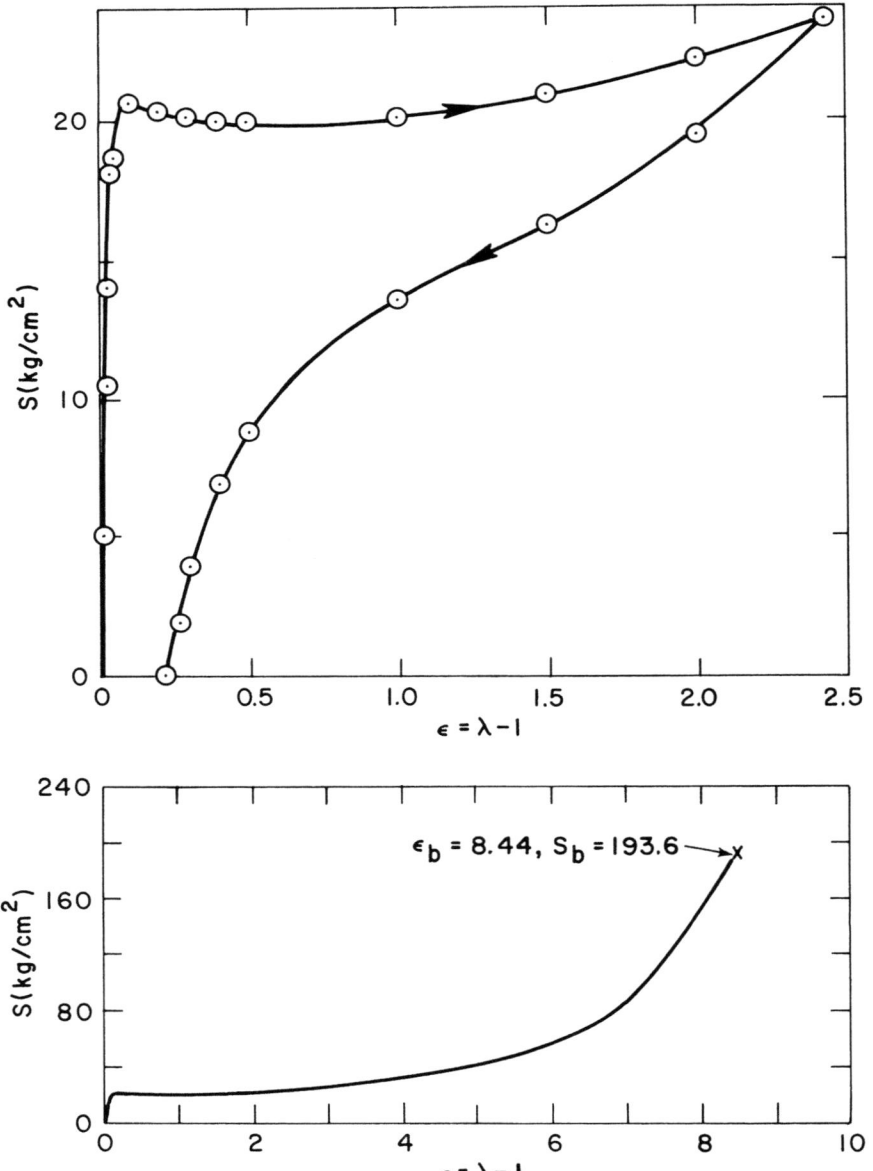

Fig. 2. Nominal stress versus strain curves for S-B-S triblock copolymer (Kraton 101) showing mechanical hysteresis (upper curve) and extension to break (lower curve) at 220°C and constant strain rate $\varepsilon = 0.444$ min^{-1}.

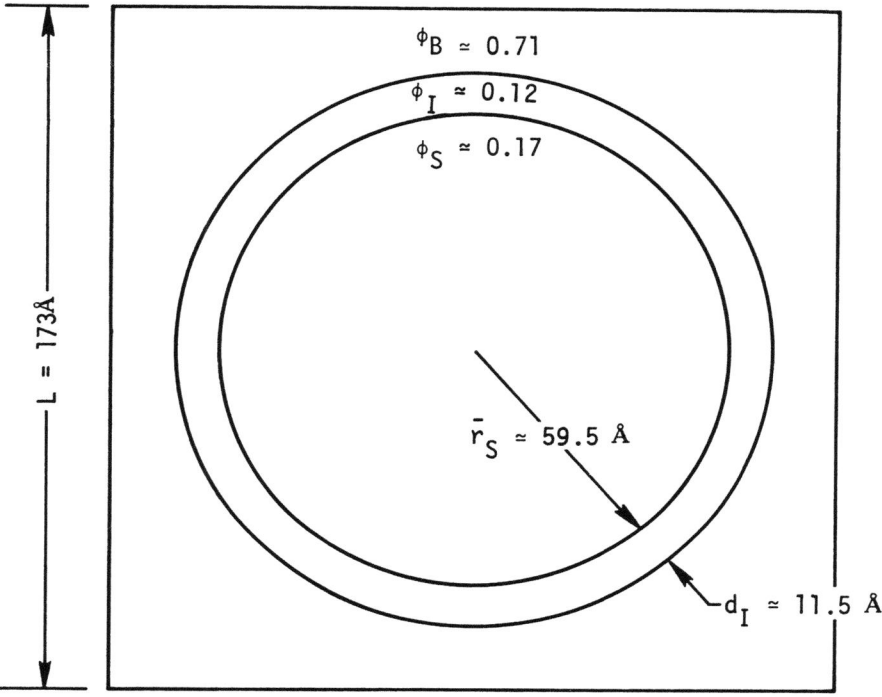

Fig. 3. Cross section schematic of the interfacial morphology of Kraton 101.

occur over a broad range of temperature, from -105° to 90°C, and correlates with mechanical hysteresis, and optical light scattering of the deformed material.[15,16] One of the notable property changes in S-B-S triblock copolymer is the modulus degradation, or stress softening, induced by mechanical deformation. The curves of Figure 4 show the temperature dependence of the tensional modulus E_o prior to cavitation and E_m subsequent to cavitation for Kraton 101 at respective tensile strains $\varepsilon = 0.01$ and $\varepsilon = 2.0$. The curves of Figure 4 show that substantial stress softening occurs in S-B-S triblock copolymer from temperatures above $(T_i)_s = 60°C$ where the polystyrene domains soften to below $(T_i)_B = -96°C$ where the polybutadiene continuum displays glassy response. At low temperatures where $T \simeq (T_i)_B$ the stress softening shown in Figure 4 is accompanied by substantial energy dissipation with a mechanical work loss $W \geq 800$ Kg cm/cm^3. It is important to note that this high mechanical energy dissipation, as measured by the hysteresis between extension and recovery curves in the upper view of Figure 2, does not appear to compromise the high strength and extensibility of the block copolymer. Furthermore, the internal structure degradation, i.e. cavitation and debonding of the interfacial phase, is fully recoverable subsequent

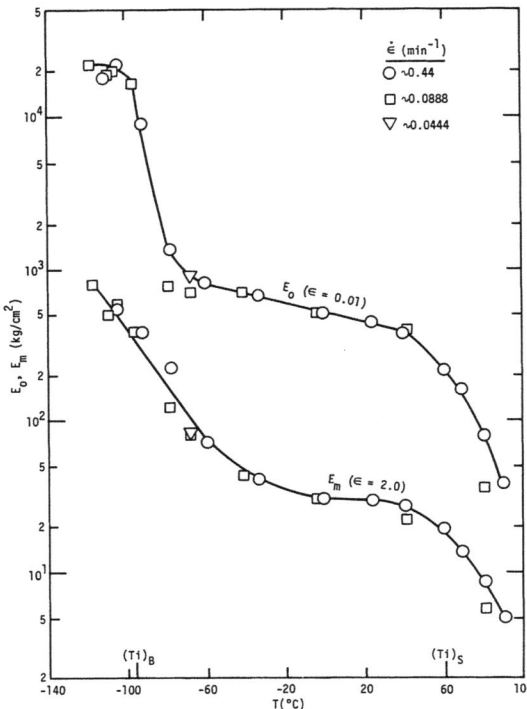

Fig. 4. Temperature dependence of initial modulus E_o and high strain (post cavitation) modulus E_m for Kraton 101 at these strain rates.

to annealing at temperatures where $T \geq (T_i)_S$ for short times.[16] Under long term loading at tensile strains substantially greater than five percent extension, the craze structure of Kraton 101 has been shown to produce porosity and possible susceptibility to environmental attack.[16] This last connection between cavitation characteristics and production of porosity in multiphase block copolymers remains as an important consideration in adhesive applications of these materials. A recent review of pressure sensitive adhesives by Toyama and Ito[18] points out that mixtures of resin tackifier and S-B-S block copolymer have found new applications as hot melt pressure-sensitive adhesives. The reported advantages are no fire hazard, low machinery expense, and no air pollution.

POLAR-SEMICRYSTALLINE BLOCK COPOLYMERS

The phase separation of semicrystalline thermoplastic elastomers is often controlled by the extent of crystallization of the domain forming polar segments. The domain forming segments of polyurethane and polyether-polyester thermoplastic elastomers are generally smaller than the typical molecular weight $M_S \simeq 11,000$ shown in Figure 3 for amorphous S-B-S triblock copolymer[7-9]. Polar semicrystalline block copolymers also show the stress softening and mechanical hysteresis similar to that shown in Figure 2 for amorphous S-B-S block copolymer. One specific technical utilization of a polyurethane block copolymer based upon Adiprene L-100 (polytetramethylene ether glycol of molecular weight about 2000 end capped with toluene diisocyonate) which is stoichiometrically co-reacted with MOCA (4, 4' - methylene - bis - [2-chloroaniline]) is to provide tough adhesives for cryogenic applications.[19] The temperature dependence of the tensile relaxation modulus $E(t)$ at strain $\varepsilon = 0.005$ and relaxation time $t = 15$ sec for a commercial polyurethane adhesive of this general type (Adhesive No. 7450 A+B, Crest Products, Santa Ana) is shown in Figure 5. Also included as a second curve in Figure 5 are manufacturer's data for single lap shear strength (Method MMM-A-132). The tensile modulus data were obtained at strains below the onset of cavitation and, therefore, reflect the continuum properties of the polyurethane adhesive.

Inspection of Figure 5 shows a very broad glass-to-rubber transition range which extends from below -100°C to above 0°C for the polyurethane adhesive. The relaxation modulus $E(t) \simeq 400$ Kg/cm^2 which occurs at the rubbery inflection temperature $T_i \simeq 40°C \simeq 313°K$ describes an effective molecular weight M_x as defined by kinetic theory of rubber elasticity[20,21]:

$$M_x = 3\rho RT/E_i(t) \qquad (1)$$

where M_x is the effective molecular weight between crosslinks or entanglements, ρ is polymer density, R = 84.8 Kg cm/mole deg. is the gas constant and T is the absolute temperature. According to Eq. 1, $M_x \simeq 200$ gm/mole for $\rho = 1.00$ gm/cc, $E_i(t) \simeq 400$ Kgm/cm^2, and $T_i = 313°K$. Based upon an assumed segment molecular weight M = 2000 gm/mole for the tetramethylene oxide segment of Adiprene L-100, it follows that the experimental modulus curve of Figure 5 indicates that a dominant contribution to rubbery plateau modulus derives from morphology rather than molecular structure. The extreme breadth of the glass-to-rubber transition also reflects the restraints of a multiphase morphology on the segment motion of the polyether glycol segments.

The curve for lap shear bond strength in Figure 5 shows a maximum temperature dependence which appears at $T \simeq -100°$ to 0°C and correlates with the glass-to-rubber transition in $E(t)$. The shear bond strength curve of Figure 5 appears to extrapolate to zero strength at temperatures above 100°C where the modulus curve bends downward indicative of thermal softening of the hard polyurethane segments.

Tensile strength σ_b and extensibility ε_b studies are reported by Smith[20] for cured polyurethane elastomers based on Adiprene L-100

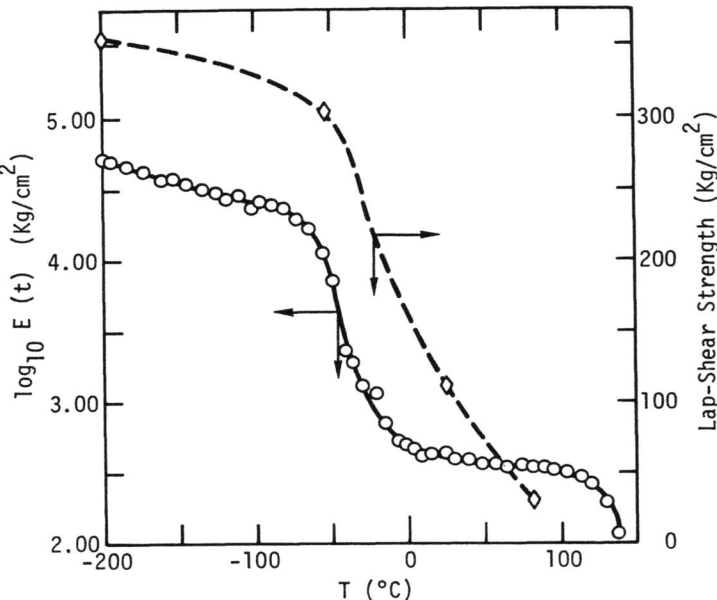

Fig. 5. Temperature dependence of tensile relaxation modulus E(t) and lap shear strength (MMM-A-132) for a polyurethane block copolymer adhesive.

and MOCA for test temperatures from −40° to 150°C. The tensile failure envelope, obtained by cross-plotting σ_b versus ε_b provides a compact and informative representation of tensile failure data. The curve of Figure 6 summarizes the characteristic strength (based on deformed cross section) and extensibility properties of a polyurethane elastomers consisting of 100 parts Adiprene L-100 cured with 12.5 parts by weight MOCA which was cured at 25°C[20]. At test temperatures between −40° and 100°C, the failure envelope of Figure 6 shows a broad upper lobe reflecting an optimized combination of high tensile strength σ_b and extensibility ε_b which appears at $T \simeq 30°C$. Between $T = 100°C$ and 125°C, the failure envelope displays a region of minimum extensibility as strength decreases. The onset of a flow mechanism of deformation between $T = 125°C$ and 150°C produces the beginnings of a lower lobe which is not complete due to lack of network character. The curves of Figure 5 and Figure 6 appear to correlate important new aspects of viscoelastic and fracture response in polar semicrystalline block copolymer systems. Referring to Figure 6, it is evident that the broad region of high tensile strength $\sigma_b \simeq 4422$ Kg/cm^2 (65000 psi) in true stress and extensibility ranging from $\varepsilon_b = 2.0$ to 4.4 (200 to 440% extension) is impressive when displayed at temperatures from −40° to 30°C. The loss of these properties at higher temperatures is, of course, related to the

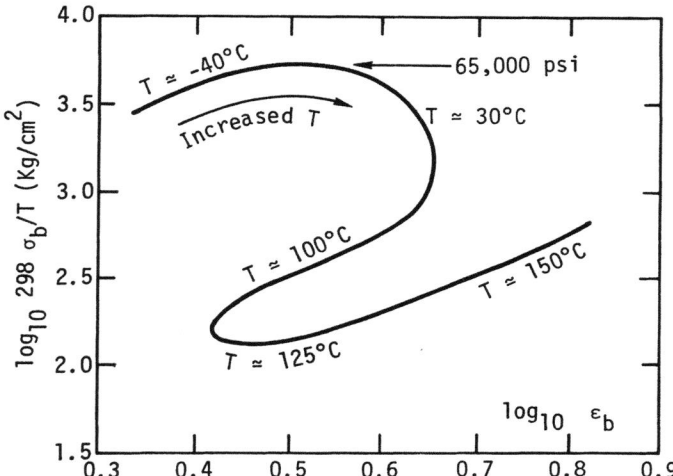

Fig. 6. Tensile failure envelope for Adiprene polyurethane block copolymer elastomer cured at 25°C given by a double logarithmic plot of 298 σ_b/T versus ε_b.

viscoelastic transformations evident in the E(t) versus T curve of Figure 5 which displays thermoplasticity due to melting of the polar urethane segments at temperatures above 100°C. The polyurethane adhesive whose properties are plotted in Figure 5 can be variously utilized as an adhesive, sealant, coating, or casting material thus indicating the versatility of the segmented block polyester-urethane copolymers in industrial applications. The morphology and mechanical response of these systems are complicated by their semicrystalline nature and the complicating effects of prior thermal history upon both crystallinity and crystalline-amorphous phase morphology.[19]

THREE-DIMENSIONAL CROSSLINKED BLOCK COPOLYMERS

In a recent review of the properties of cured epoxy resins, Kaelble[21] has pointed out that even simple formulations of diepoxide monomer and polyfunctional crosslinking agent reflect in the cured state the separate chemistries and structures of the co-reactants. When the molecular weight of both co-reactants in cured resin are high enough to provide more than four chain atoms between junction points, the resultant network begins to display the properties of a three-dimensional block copolymer. Creation of elastomeric microphases in epoxy structural adhesives has been recently identified with in situ block copolymerization between carboxy terminated nitrile (CTBN) rubber and the diepoxide.[11] The adsorption-interdiffusion (A-I) model developed to delineate the morphology of S-B-S triblock copolymers as shown in Figure 3 has also been applied to the CTBN modified epoxy materials.[10] In order to complement other studies of viscoelastic[11] and fracture mechanics[12] response of CTBN modified epoxy resins, Kaelble and co-workers[22] have recently investigated the morphology and thermomechanical response for a series of these block copolymers. Some of the highlights of this study are introduced here.

The co-reactants and curing condition for preparation of the CTBN modified epoxies are summarized in Table I and follow methods reported by Kalfoglou and Williams[11]. Micro-tensile specimens (ASTM Method D1708-66) are cut from the cured epoxy films while they are heated to rubbery state response at T ≃ 120°C and pulled in tension in an Instron TTB-M using the special line clamping previously described for thermomechanical analysis (TMA) testing.[23] Tensile tests were conducted at constant strain rate $\varepsilon = 0.0225$ min^{-1} and test temperatures from T = -200° to 200°C.

The fracture surfaces under tensile deformation of the CTBN modified epoxy polymers at T = -200°C are shown in the scanning electron microscope (SEM) topologies shown in Figure 7. The cured unmodified epoxy (upper left view) shows an essentially smooth fracture failure at T = -200°C. The 17% CBTN modified epoxy (upper right view of Figure 7) shows a roughening of the failure surface and 1.0-2.0 μm spherical "holes" evidently produced by cavitation and crazing during tensile deformation. Bascom and co-workers[12] have previously shown a similar cavitation type failure surface for 15% CTBN modified epoxy tested in adhesive joints at ambient temperature. This tendency of the CTBN modified epoxy to cavitate is not displayed by the 29, 39, and 50% CTBN failure surfaces shown in the lower views of Figure 7 which appear to display undeformed spherical particles suspended in the continuum phase.

Curves of the tensile relaxation modulus $E(t)$, at extension $\varepsilon = 0.005$ and relaxation time t = 15 sec, for temperatures from T = -200° to 200°C are shown in the upper portion of Figure 8. These curves are developed from TMA analysis measurements[23] where the low strain amplitude avoids cavitation and crazing. In general, the curves of $E(t)$ versus T shown in upper Figure 8 display features which parallel low frequency dynamic storage modulus E' as reported by Kalfoglou and Williams.[11] The glass temperature of the CTBN at $T_g \simeq -45°C$ and the epoxy phase at $T_g \simeq 80°C$ accounts for the broad region of viscoelastic response in the upper curves of Figure 8. The mechanics models applied by Kalfoglou and Williams[11] indicate the CTBN is a uniform suspension of rubber particles in an epoxy continuum up to about 20% CTBN and that the CTBN phase becomes co-continuous with the epoxy at higher CTBN content. A parallel analysis of the $E(t)$ versus T functions shown in the upper curves of Figure 8 would appear to produce a similar interpretation of morphol-

Table I

Co-Reactants for Three-Dimensional Epoxy-Nitrile Rubber Block Copolymers

1. Epoxy: DGEBA (Epon 828, Shell Chemical Co.), 100 pbw (parts by weight), $M_n \simeq 380$ gm/mole

$$H_2C \overset{O}{\underset{}{\diagup\!\!\diagdown}} CH - CH_2 - O - \langle O \rangle - \underset{\underset{CH_3}{|}}{\overset{\overset{CH_3}{|}}{C}} - \langle O \rangle - O-CH_2 - CH \overset{O}{\underset{}{\diagup\!\!\diagdown}} CH_2$$

2. Catalyst: Piperidine - 5 pbw

3. Carboxy terminated nitrile rubber (HYCAR CTBN, B. F. Goodrich Chem. Co.) - 0, 17, 29, 39, 50% by weight based on 100 pbw Epoxy + 5 pbw piperidine

$$HOOC - \left[(CH_2 - CH = CH - CH_2)_5 \; (CH_2 - \underset{CN}{\underset{|}{CH}}) \right]_{10} - COOH$$

$\bar{M}_n \simeq 3300 - 3500$ gm/mole

4. Mix items (1), (2), (3) above, degas, and cure for 16 hours at 120°C under dry N_2.

---

ogy as developed by Kalfoglou and Williams. This type of morphology correlates with the "hole" formation in the 17% CTBN failure surface of Figure 7 as opposed to the appearance of intact spherical particles in the failure surfaces for 29, 39 and 50% CTBN where the stress transmission is continuous in both CTBN and epoxy phases.

Above the 100°C, both the CTBN and epoxy phases are in their rubbery state and the presence of a broad rubbery plateau from T = 100° - 200°C confirms the existence of a crosslinked three-dimensional network for the CTBN modified epoxy materials shown in upper Figure 8. The measured values of $E(t) = E_i(t)$ at T = 127°C = 400°K are tabulated in Table II and introduced into Eq. 1 with a nominal density $\rho \simeq 1.0$ gm/cc to provide calculated values of the effective molecular weight M_x between entanglements or crosslinks in CTBN modified epoxy resins. The values of M_x reported in Table II for CTBN modified epoxy are seen to increase with % CTBN and lie intermediate between $M_x \simeq 380$ gm/mole for an ideal network of Epon 828 epoxy or $M_x \simeq 3400$ gm/mole for an ideal CTBN network.

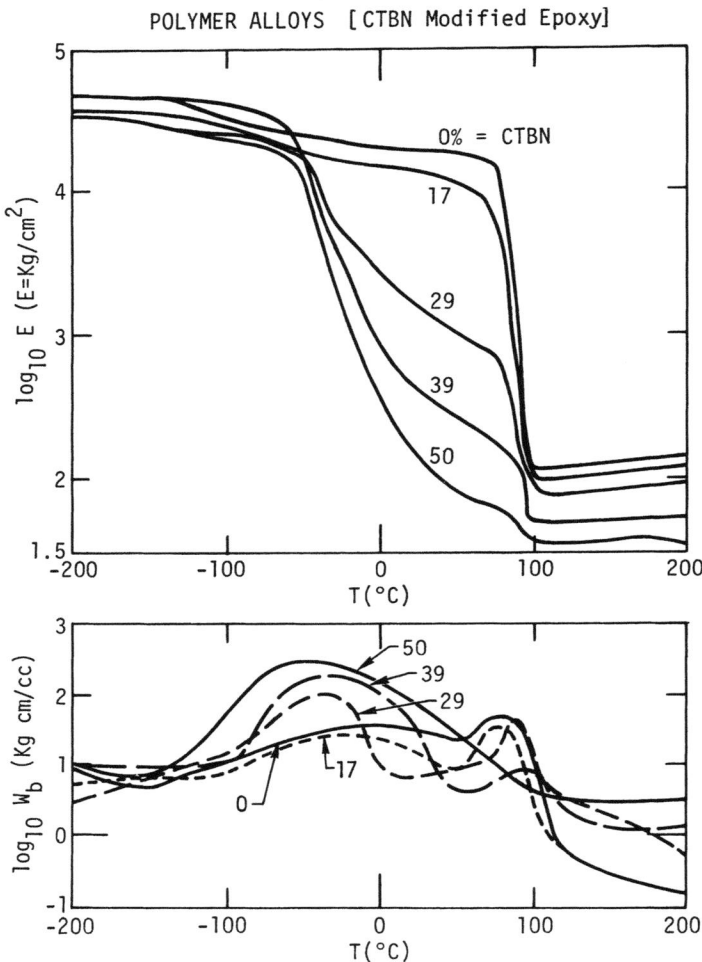

Figure 7. Tensile fracture surfaces for CTBN modified epoxy block copolymers.

In addition to low strain TMA measurements, a detailed study of deformation and failure response of the CTBN modified epoxy systems of Table I has been carried out.[22] Temperatures from $T = -200°$ to $200°C$ and a constant tensile strain rate $\varepsilon = 0.09$ min^{-1} were utilized to define the tensile fracture energy W_b per unit volume of deformed gage section as shown in the lower curves of Figure 8. These measured values of W_b are automatically recorded by an inte-

BLOCK COPOLYMERS

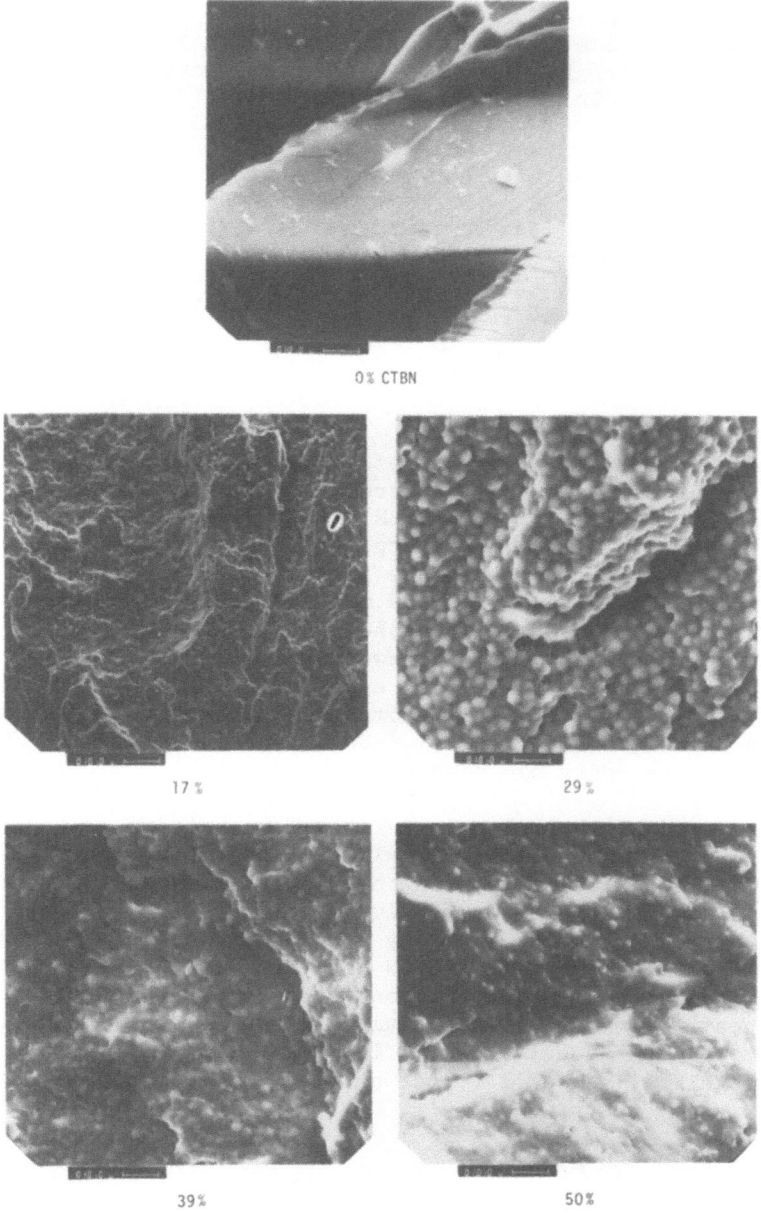

Fig. 8. Temperature dependence of tensile relaxation modulus $E(t)$ (upper curves) and tensile fracture energy W_b (lower curves) for CTBN modified epoxy block copolymers.

Table II

Calculated Values for the Effective Molecular Weight (number ave.) M_x Between Entanglements or Crosslinks at $T = 127°C \simeq 400°K$

% CTBN	$E_i(t)$ (Kg/cm^2)	M_x (gm/mole)
0	119	855
17	112	909
29	85	1197
39	46	2212
50	37	2750

grator accessory to the Instron testing machine and are proportional to the area beneath the tensile stress-strain curve. The curve of W_b versus T for pure epoxy (0% CTBN) shows a low broad maximum centered at $T \simeq 0°C$ and a second sharp maximum at $T \simeq 80°C$ of the epoxy phase. The lower curves of Figure 8 show a closely parallel response of W_s versus T for 0% and 17% CTBN modified epoxy which is consistent with the epoxy continuum morphology discussed earlier. The lower curves of Figure 8 show a new maximum in W_b versus T centered at $T \simeq -50°C$ in 29% CTBN modified epoxy. This low temperature $T \simeq -50°C$ peak in W_s versus T becomes dominant for 50% CTBN modified epoxy and the high temperature $T \simeq 80°C$ peak in W_s versus T characteristic of pure epoxy has disappeared.

It is evident that closely related types of information are conveyed concerning the morphology of CTBN modified epoxy resins by the small deformation data in the upper curves of Figure 8 and the tensile fracture data in the lower curves. Both deformation and fracture data tend to identify a morphological phase change in CTBN modified epoxy above 17% CTBN content wherein the elastomer and epoxy phase become co-continuous. In fact, the curves of Figure 8 appear to imply the essential disappearance of a pure epoxy phase at 50% CTBN as evidenced by loss of the $T = 80°C$ singularities in both $E(t)$ versus T (upper curves) and W_b versus T (lower curves).

In conjunction with the tensile fracture work W_b shown in the lower curves of Figure 8, the tensile strength σ_b (based on deformed cross section) and extensibility ε_b (nominal strain at break) were also measured. The cross plots of σ_b versus ε_b data for the CTBN modified epoxy resins of this study, tested at constant strain rate $\dot{\varepsilon} = 0.09$ min^{-1} from $T = -200°$ to $200°C$ are shown by the curves in Figure 9 as tensile failure envelopes. The upper extremity of the failure envelope curves of Figure 9 reflect a maximum tensile strength

$\sigma_b \simeq 1122$ Kg/cm^2 (16,500 psi) and minimum extensibility $\varepsilon_b \simeq 0.01$ (1.0% tensile strain) characteristic of apparent brittle failure at T = -200°C. The lower extremity of these curves reflect a lower strength limit $\sigma_b \simeq 3.0$ Kg/cm^2 (46.5 psi) and varied extensibilities $\varepsilon_b \simeq 0.02$ to 0.28 (2.0 to 28% strain) at the high test temperature T = 200°C. The failure envelopes for 0, 17, and 50% CTBN display a single outer lobe. The failure envelopes for 29% CTBN displays two well defined outer lobes which appear related to the two maximum in tensile fracture work W_s shown by the lower curves of Figure 8. The failure envelope for 39% CTBN also shows less pronounced evidence of two outer lobes and these also correlate with the W_s maxima shown respectively at T = -50°C and T = 80°C of lower Figure 8.

The lower extremity of the failure envelope of Figure 9 correlate with the presence of a thermoset crosslink network structure. The increasing values of ε_b with % CTBN correlate with the lower value of rubbery network modulus tabulated in Table II. In general, the failure envelopes of Figure 9 are consistent with the failure response of a crosslinked network as contrasted with the previously discussed failure envelope for a polyurethane elastomer shown in Figure 6 whose open lower extremity indicates thermoplasticity.

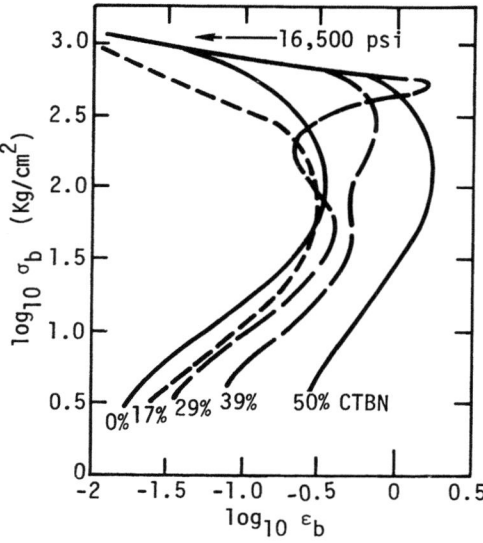

Fig. 9. Tensile failure envelopes for CTBN modified epoxy block copolymers.

SUMMARY AND CONCLUSIONS

This discussion has touched briefly on three types of block copolymers which are currently finding a wide range of new applications as adhesive materials. Thermoplastic elastomers, where a hard discontinuous glassy or semicrystalline phase acts to crosslink and reinforce, provide one class of response. At the opposite extreme, a soft phase of suspended spherical particles may act as a locus of cavitation to toughen an otherwise brittle continuum phase. Intermediate between these morphological extremes is the broad range of co-continuous multiphase systems in which significant segment mixing produces additional interdiffusion phases with thermorheological response characteristic of diffuse transitions. This latter type of interdiffused co-continuous phase morphology for block copolymers may be more the rule than the exceptional case.

The special uniqueness of block copolymers either as thermoplastic elastomers of three-dimensional block copolymers is closely tied to their distinct morphology. A broad range of evidence now supports the concepts of multiphase morphology and possibly substantial interdiffusion to produce broadened thermo-rheological transition response in block copolymer adhesives. The types of morphological models being developed to rationalize the inclusion structure of high impact polystyrene[24], and low profile behavior in unsaturated polyester resins[25] invoke many of the details presently applied to describe block copolymers. The concept of adsorption-interdiffusion (A-I) bonding as applied to define the development of an interfacial phase of mixed segments has an interesting chemical structure parallel in the concept of interpenetrating polymer networks (IPN) as proposed by Frisch[26,27]. In addition to the CTBN modified epoxies discussed here, several new and different types of modified epoxy block copolymers have been reported[28] which indicate further potential in structural adhesive applications requiring both outstanding fracture toughness and high thermal stability.

REFERENCES

1. J. T. Harlan, Jr., "Block Copolymer Adhesive Compositions and Articles Prepared Therefrom", U.S. Patent 3,239,478, March 1965.
2. G. Holden, E. T. Bishop and N. R. Legge, J. Poly. Sci., Symposia 26, 37 (1969).
3. D. H. Kaelble, Trans. Soc. Rheol. 15;2, 235 (1971).
4. J. Moacanin, G. Holden and N. W. Tscheogl, Eds., *Block Copolymers*, Interscience, New York, 1969.
5. S. L. Aggarwal, Ed., *Block Polymers*, Plenum, New York (1970).
6. G. E. Molau, Ed., *Colloidal and Morphological Behavior of Block and Graft Copolymers*, Plenum, New York (1971).
7. G. M. Estes, R. W. Seymour, S. J. Borchert and S. L. Cooper, Ibid., pp. 159-171.

8. S. L. Samuels and G. L. Wilkes, J. Poly. Sci., Symposia 43, 149-178 (1973).
9. W. H. Buck, R. J. Cella, Jr., E. K. Gladding and J. R. Wolfe, Jr., Ibid., 48, 47-60 (1974).
10. J. N. Sultan, R. C. Laible and F. G. McGarry, Appl. Polymer Symposia 16, 127 (1971).
11. N. K. Kalfoglou and H. L. Williams, J. Appl. Poly. Sci. 17, 1377 (1973).
12. W. D. Bascom, R. L. Cottingon, R. L. Jones and P. Peyser, "The Fracture of Epoxy and Elastomer-Epoxy Polymers in Bulk and as Adhesives", A.C.S. Coatings and Plastics Preprints 34 (2), 300-308 (1974).
13. T. L. Smith and R. A. Dickie, J. Poly. Sci., Symposia 26, 163 (1969).
14. J. F. Beecher, L. Marker, R. D. Bradford and S. L. Aggarwal, Ibid., 26, 117 (1969).
15. D. H. Kaelble and E. H. Cirlin, Ibid., 43, 131 (1973).
16. D. H. Kaelble, E. H. Cirlin and M. Shen, in *Colloidal and Morphological Behavior of Block and Graft Copolymers*, Ed., G. E. Molau, Plenum, New York, p. 307 (1971).
17. R. P. Kambour, Macromolecular Rev. 7, 1-154 (1973).
18. M. Toyama and T. Ito, *Polymer Plastic Technology and Engineering*, Ed., L. Naturman, Vol. 2, Dekker, New York, pp. 161-229 (1974).
19. R. E. Keith, *Handbook of Adhesive Bonding*, Ed., C. V. Cagle, McGraw Hill, New York, Chap. 12 (1973).
20. T. L. Smith, J. Poly. Sci., Polymer Phys., Ed., 12, 1825 (1974).
21. D. H. Kaelble, *Epoxy Resins*, Ed., C. A. May and Y. Tanaka, Dekker, New York, Chap. 5 (1973).
22. D. H. Kaelble, L. W. Crane and P. J. Dynes, "Morphology and Thermomechanical Response of CTBN Modified Epoxy Resins", to be published.
23. D. H. Kaelble and E. H. Cirlin, J. Poly. Sci., Part C, 35, 79 (1971).
24. T. O. Craig, J. Poly. Sci., Poly. Chem. 12, 2105 (1974).
25. V. A. Pattison, R. R. Hindersinn and W. T. Schwartz, J. Appl. Poly. Sci. 18, 2763 (1974).
26. H. L. Frisch, D. Klempner and K. C. Frisch, J. Poly, Sci., Part B, 7, 775 (1969).
27. K. C. Frisch, D. Klempner, S. K. Mukherjee and H. L. Frisch, J. Appl. Poly. Sci. 18, 689.
28. A. Nashay and L. M. Robeson, J. Poly. Sci., Poly. Chem. 12, 689 (1974).

Effects of Adhesive Structure on Impact Resistance and Optical Properties of Acrylic/Polycarbonate Laminates

Joyce L. Illinger, Robert W. Lewis and Dennis B. Barr[*]

Army Materials & Mechanics Research Center

Watertown, Ma.

Acrylic/polyurethane/polycarbonate laminates have demonstrated resistance to impact by high velocity projectiles up to 100 percent greater than monolithic polycarbonate, currently considered the best commercially available transparent impact-resistant material. The energy absorbed during impact was found to be dependent on the chemical composition and microstructure of the adhesive interlayer and varied by a factor of two over the range of variables investigated. Polyurethanes based on MDI-butanediol "hard" and block copolyether "soft" segments, used as the laminate interlayer, affect both impact resistance and optical clarity as chemical composition of the urethane is varied. Microstructure as seen in both the optical and transmission electron microscope changes from no visible structure in transparent materials, through few widely scattered isolated spherulitic structures in translucent materials, to densely packed interpenetrating spherulitic structures in opaque materials. Changes in microstructure accompany increased "hard" segment and are not affected by "soft" segment composition. Urethanes with increasing

[*]*Tennessee Eastman Company*
 Kingsport, Tennessee 37662

spherulitic content show decreasing tackiness accompanied by decreasing laminate impact resistance. Dynamic mechanical spectra show sharp low temperature loss peaks (240°K) ascribable to T_g of soft segments and an additional broad loss shoulder around 140°K which sharpens as impact resistance decreases and urethane content increases.

INTRODUCTION

Polymeric materials are widely used as transparent enclosures for windshield and canopies in aircraft and for eye protection in military and riot control applications. For these uses, the transparency must not have only high impact resistance, but, in addition, must not "spall" upon impact (see Figures 1 and 2). Currently, polymethylmethacrylate (PMMA) is the transparent polymeric impact-resistant material most frequently encountered in aircraft applications. PMMA, however, while much tougher and more impact resistant than glass, is relatively brittle and will spall upon high velocity projectile impact, leading to possible injury to personnel behind the transparency (Figure 1). Rubber modification improves the impact resistance substantially, but causes the acrylic to lose transparency at low temperatures[1]. Polycarbonate (PC) eliminates the spall problem (Figure 2) but has only slightly higher resistance to impact by high velocity projectiles than PMMA (62 vs 56 joules absorbed by impact with a 1.1 gram projectile). In addition, PC has poor abrasion and poor solvent resistance.

A search for transparencies with improved impact resistance coupled with non-spalling characteristics has led to the evaluation of the potential for plastic/plastic laminates[2]. In the course of the evaluation, it was found that PMMA/PC laminates markedly enhanced the energy absorbing capacity (Figure 3). At the extreme left in Figure 3 is the energy absorbance of 3/8" thick 100% PC with a value of 62 joules. With PMMA facing the impact (upper curve), the optimum energy absorbing capacity of the combinations tested occurred at a 2:1 weight ratio of total thickness 3/8" PMMA to PC (Energy = 100 joules). With the ductile PC facing the impact (the lower curve), however, there was a reduction in energy absorbing capacity compared to that of the homogeneous materials over the range of compositions investigated. The enhanced resistance of the clamped laminate was attributed[2] to the brittle PMMA spreading out the impact over a wider area than a ductile material, while the ductile PC serves both to absorb the impact and to prevent the brittle material from spalling. Thus, an improvement in energy absorbing capacity of about 67 percent was obtained over either material alone.

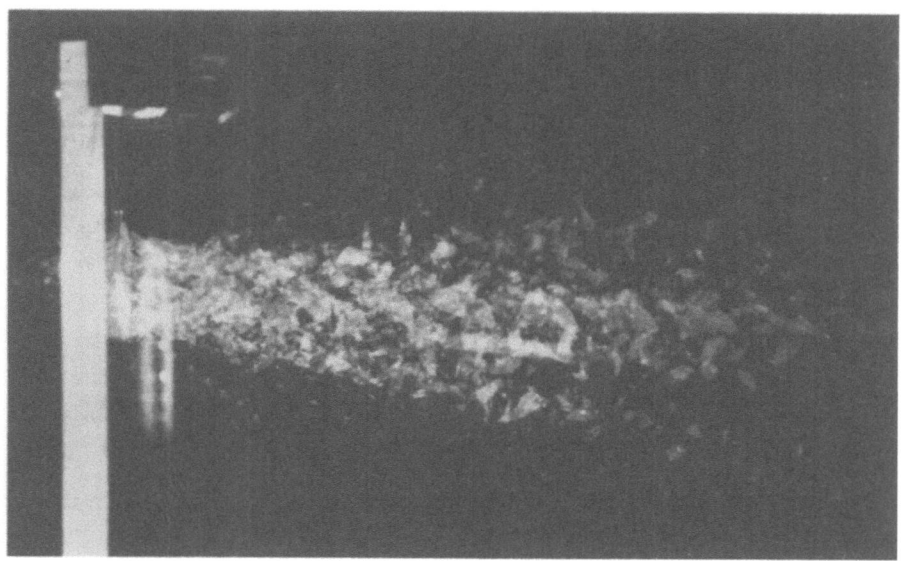

Fig. 1. Multiflash photograph of 1/4" PMMA under high velocity impact. Edge of sample is shown, missile is moving from left to right. Projectile was completely stopped by sample.

Fig. 2. Multiflash photograph of 1/4" PC under high velocity impact. Edge of sample is shown, missile is moving from left to right. Note absence of spall.

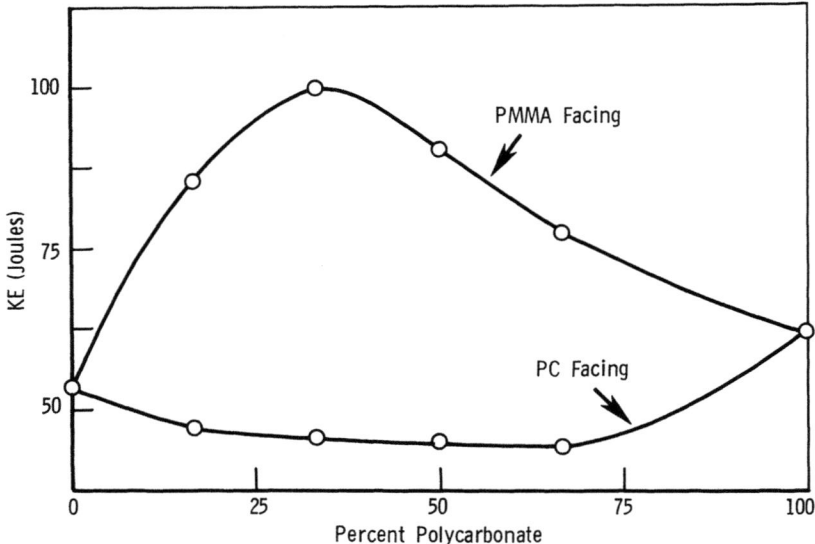

Fig. 3. Impact Resistance of PMMA/PC laminates without interlayer.

In order for a laminated system to be useful for aircraft applications, however, it is necessary to bond the materials together, either thermally or by the incorporation of an adhesive interlayer. Earlier work demonstrated, though, that both brittle adhesives (conventional polyvinylbutyral as well as transparent epoxies) and thermal bonds resulted in poor impact performance accompanied by spalling of the PC, caused by failure of the interlayer to cushion the impact loads transmitted from the facing PMMA. Consequently, it was found that a flexible interlayer is required to maintain the enhanced impact resistance.

This paper describes an investigation of the relationships between microstructure of transparent segmented polyurethane (PU) adhesives, their optical properties and laminate impact resistance that has led to a one hundred percent advance in the state-of-the art of impact resistance of transparent plastics.

EXPERIMENTAL

Synthesis and Composition

A standard two-step synthesis was used to prepare three series of adhesive polymers. The general structure may be represented by

M ～ [(MB)$_x$M～]$_y$ M

where   M (MDI) is O=C=N-⬡-CH$_2$-⬡-N=C=O;

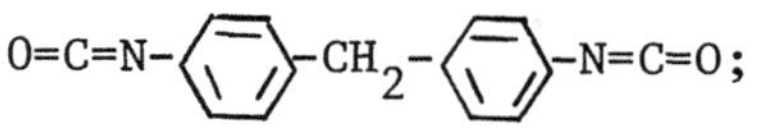

B (Bd) is 1,4 butanediol, HO CH$_2$CH$_2$CH$_2$CH$_2$OH; and

～ is a polymeric glycol, HO-(CH$_2$CH$_2$O)$_a$-(CH$_2$CHO)$_b$-(CH$_2$CH$_2$O)$_a$-H
                                                    |
                                                   CH$_3$

Table 1 shows the variation in composition and structure of the available polymeric glycols, which are made from polypropylene glycol (PPG) and polyethylene glycol (PEG).

Table I

Glycol Composition and Structure

Composition	a	b
100% PEG	35	0
50/50 PPG/PEG	11	17
60/40 PPG/PEG	10	23
70/30 PPG/PEG	6	23
80/20 PPG/PEG	3	22
90/10 PPG/PEG	2	33
100% PPG	0	34

All three series had the degree of polymerization, y, held constant. Series A was made using 50/50 PPG/PEG as the glycol and varying x from 0 to 1.0, plus two additional polymers at x = 2.0 and 5.0. Series B was made holding x constant at 0.4 and making polymers using each of the seven polymeric glycols for the soft segment. Series C was similar to series B except that x = 0.5.

Sample Preparation for Electron Microscopy (Ultracryotomy)

Due to the nature of the polymers, room temperature sectioning of samples for microscopy proved impossible. Hence cryogenic techniques had to be developed. The equipment used was an LKB Ultratome III with Cryo Kit attachment (LKB Instruments, Inc.) and Diamond Knife (DuPont). Small pieces were mounted in the sample holder on the microtome. The Diamond knife reservoir was filled with a mixture of DMSO and water (60/40). Sample temperatures were set between -70°C and -167°C, although -100°C was generally the best temperature for all samples. The knife temperature was set at -50°C. Nominal section thickness was 1000A. Sections wrinkled slightly as they were cut, but gentle heat supplied by resistance-heated platinum caused the sections to relax sufficiently. No morphological changes due to this heating were observed in the electron microscope. Sections were picked up on copper grids, washed in distilled water, and dried in a desiccator.

Impact Testing

The energy absorbing capacities of the transparent laminates were obtained by measuring the resistance to impact by a chisel-nosed steel projectile (Figure 4) fired from an air gun at known velocity using standard techniques[4]. High speed flash photography verified that the projectile consistently impacted the laminate at 0° obliquity, chisel-nose forward. Enough impacts were obtained to measure the average velocity at which complete penetration of the laminate occurred (usually about eight impacts were required). From the impact velocities, the energy absorbing capacities were calculated using the projectile mass of 1.1 g. The impact velocities were accurate to approximately 20 fps and energies accurate to 5 joules.

RESULTS AND DISCUSSION

Description and Characterization

The first series of polymers ranged from extremely sticky with very little elastomeric behavior to slightly tacky with very little

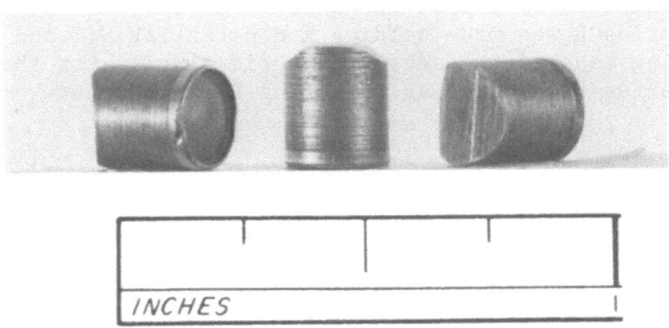

Fig. 4. Chisel-nosed projectile for high velocity impact test.

permanent set (Table II). From x = 0.0 to x = 0.6 the polymers were transparent becoming translucent at x = 0.8 and 1.0. The molecular weights of all but x = 0 are 40,000 so that the gradual change in tackiness, elastomeric properties, and transparency must be due to change in structure due to increasing butanediol content.

Table III summarizes the properties of series B and C.

Whereas Table II showed that at 50% PPG translucency does not occur until the Bd content reaches 0.8, series B and C show that translucency can occur at Bd values of 0.4 and 0.5 when the PPG content is increased to 80 or 90%. Apparently, then, the transparency of the adhesive depends both on the diol content and composition of the soft segment glycol.

Microscopy

Figure 5 portrays optical micrographs of four of these materials showing changes in microstructure as a function of butanediol ("hard segment") content. Figure 5a is a formulation with less than 1 mole

Table II

Variation of Hard Segment

MDI	Bd	Glycol	Description	$\bar{M}_n$	Optical Properties
1.05	0	1	sticky, gradual flow set	30,000	transparent
1.261	0.2	1	sticky, set, slightly elastomeric	39,000	transparent
1.471	0.4	1	tacky, less set, elastomeric	40,000	transparent
1.576	0.5	1	less tacky, still some set, elastomeric	40,000	transparent
1.681	0.6	1	slightly tacky, very little set, elastomer	41,000	transparent
1.891	0.8	1	little tack, very little set, elastomer	42,000	translucent
2.1	1.0	1	nontacky, no rapid set, elastomer	45,000	translucent
3.15	2.0	1	nontacky, elastomer	40,000	opaque
6.3	5.0	1	nontacky, elastomer	40,000	opaque

of butanediol, and the polymer shows no structure. This material is transparent. Figure 5b shows a polymer which was translucent (1 mole Bd), and here isolated scattered spherulitic structure is apparent. At increased hard segment content more spherulites appear until finally (Figure 5d) at 5 moles hard segment the spherulitic structure is predominant. These latter materials are completely opaque.

Table III

Properties of Series B and C

Glycol	Series B 1.471 MDI:0.4 Bd:1 glycol	Series C 1.576 MDI:0.5 Bd:1 glycol
100% PEG	tacky, opaque in bulk (transparent in laminate)	tacky, opaque in bulk (transparent in laminate)
50/50 PPG/PEG	tacky, transparent	tacky, transparent
60/40 PPG/PEG	tacky, transparent	tacky, translucent
70/30 PPG/PEG	sticky/tacky, transparent	tacky, transparent
80/20 PPG/PEG	sticky, transparent	sticky/tacky, transparent
90/10 PPG/PEG	sticky, translucent	sticky, translucent
100% PPG	very sticky, translucent	sticky, translucent

Figure 6 shows transmission electron micrographs of the same polymers. At these much higher magnifications the lack of any structure in the transparent materials is obvious (Figure 6a). Structure in the isolated spherulites of the translucent materials is not highly regular but still enough to scatter light leading to loss of transparency. In the very high hard segment composition material, Figure 6d shows how highly intertwined these spherulitic structures have become. This accounts for the total opacity of the material. The dimensions of these spherulites are sufficiently large that they must include both soft and hard segments.

Impact Resistance of Laminates

Laminates from 1/4" PMMA, 10 mil adhesive, and 1/8" PC were fabricated at press conditions of 100 psi at 95°C for 40 minutes residence time cooled under pressure. Two laminates were made for each test. Samples were subjected to impact testing from three to twenty days following lamination. In one case where delay between fabrication and testing was five days for one laminate and fourteen days for the other no difference in results was seen. The results of the impact testing are shown in Tables IV and V.

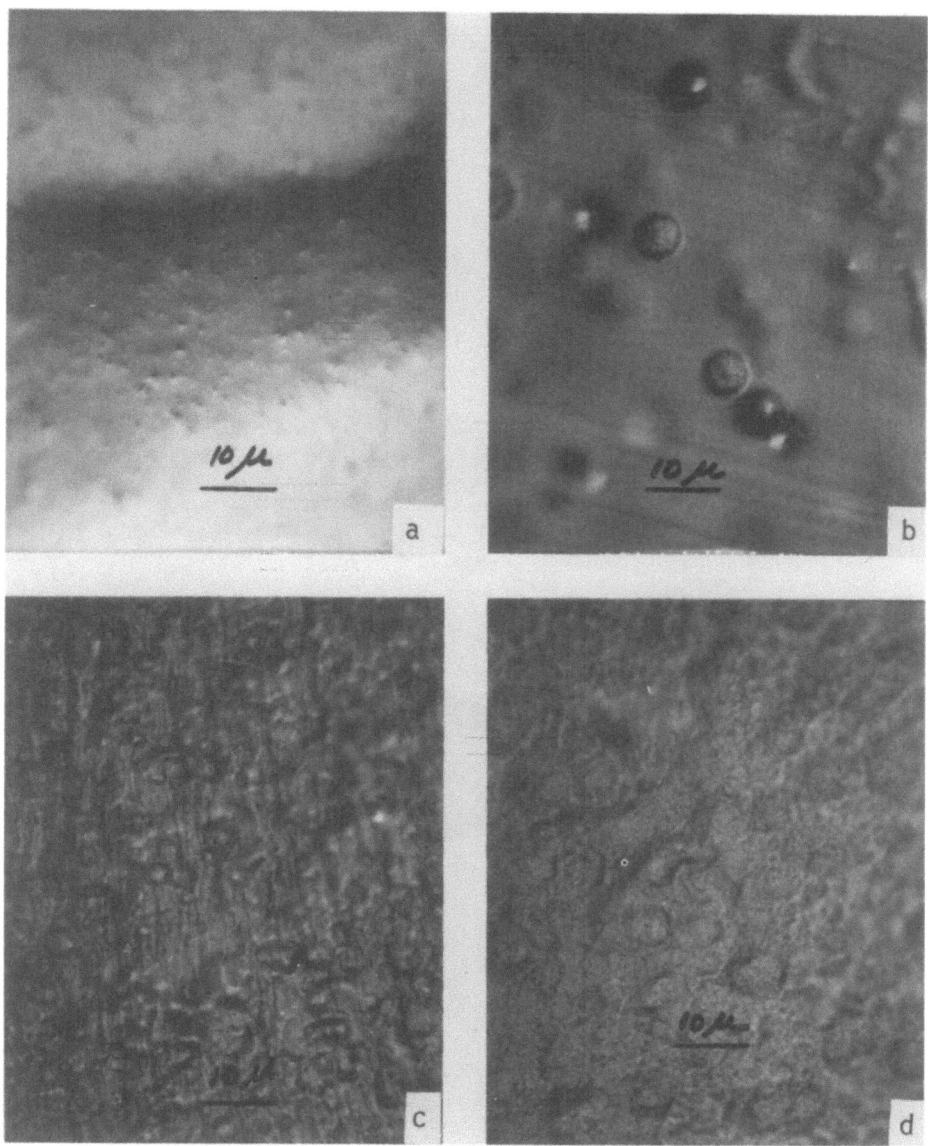

Fig. 5. Optical micrographs development of spherulitic structure
   a. 1.57 MDI: 0.5 Bd: 1.0 Polyol
   b. 2.10 MDI: 1.0 Bd: 1.0 Polyol
   c. 3.15 MDI: 2.0 Bd: 1.0 Polyol
   d. 6.30 MDI: 5.0 Bd: 1.0 Polyol

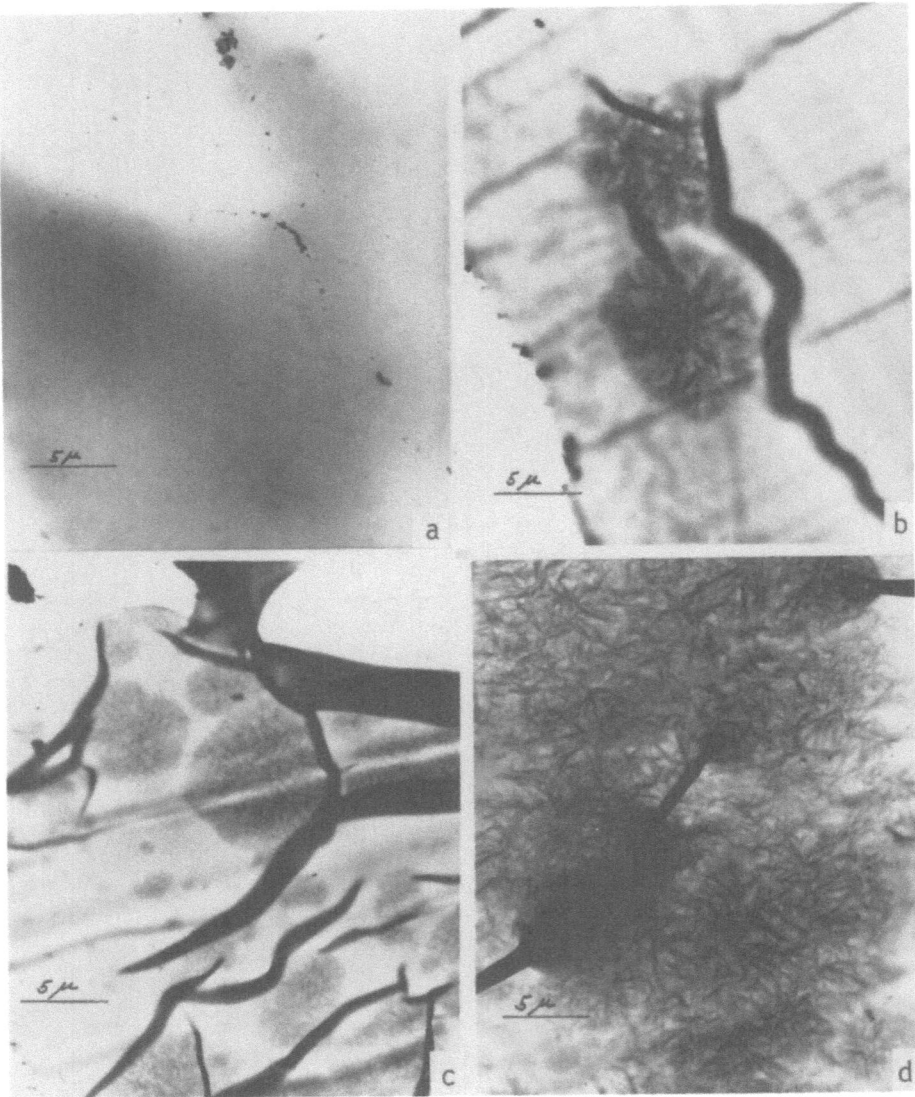

Fig. 6. Transmission electron micrographs showing details of spherulitic structure
  a. 1.57 MDI: 0.5 Bd: 1.0 Polyol
  b. 2.10 MDI: 1.0 Bd: 1.0 Polyol
  c. 3.15 MDI: 2.0 Bd: 1.0 Polyol
  d. 6.30 MDI: 5.0 Bd: 1.0 Polyol

Table IV

Effect on Butanediol Content on Impact Resistance

Polymer Series A Butanediol Content	KE (Joules)	Optical Properties
0.0	125	clear
.2	126, 131+	clear
.4	117	clear
.5	123, 127+	clear
.6	125	clear
1.0	101	translucent

0° obliquity, PMMA facing,
+ Determination of KE for duplicate polymer syntheses

Table V

Effect of Variation of Polymeric Glycol, Series B and C
KE (joules)

Glycol	Series B 1.47 MDI:0.4 Bd: 1 glycol	Series C 1.576 MDI:0.5 Bd: glycol
100% PEG	127	128
50/50 PPG/PEG	117	123, 127+
60/40 PPG/PEG	---	129
70/30 PPG/PEG	128	129
80/20 PPG/PEG	136	128
90/10 PPG/PEG	130	131
100% PPG	---	127

0° obliquity, PMMA facing,
+ Determination of KE for duplicate polymers

As long as the adhesives used were transparent (no spherulitic structure present as seen by microscopy), the energy absorbed by the laminates was essentially the same. The 1.0 butanediol content polymer was nontacky, indicating some embrittlement. It was also translucent, attributed to isolated spherulites seen in Figures 5 and 6. The lowered impact resistance is in agreement qualitatively with earlier observations on the effect of brittle adhesives.

As seen in Table V, the structure of the polymeric glycol (molecular weight 2000) at constant MDI and butanediol composition had no significant effect on the ballistic performance of the laminates. These polymers were all transparent at the 10-mil thickness of the adhesive layer used for laminate evaluation.

The range of energy absorbed from 125 to 130 joules represents a 100 percent improvement over either material alone (KE = 56 to 62 joules) and a 30 joules increase over the laminate without interlayer. In addition, the laminate did not spall, unlike the situation where a brittle adhesive was used (Figures 7 and 8).
Thus, the transparent flexible PU adhesives satisfactorily meet the criteria for enhanced impact resistance while eliminating spallation.

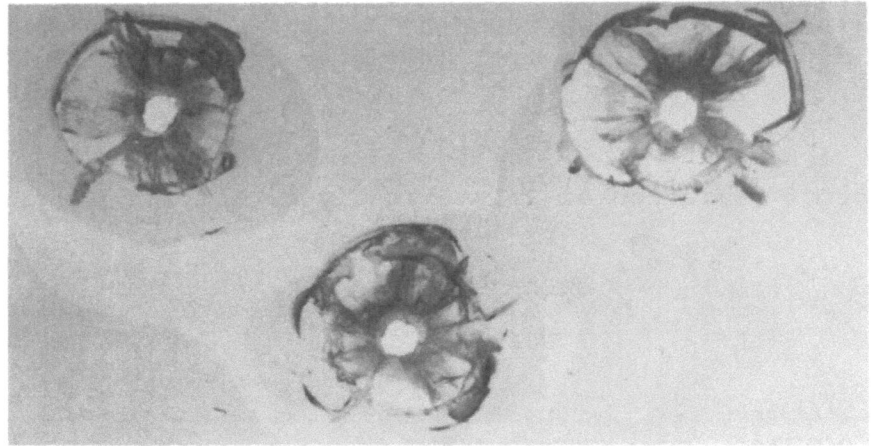

Fig. 7. Brittle failure of thermally bonded PMMA/PC laminate.

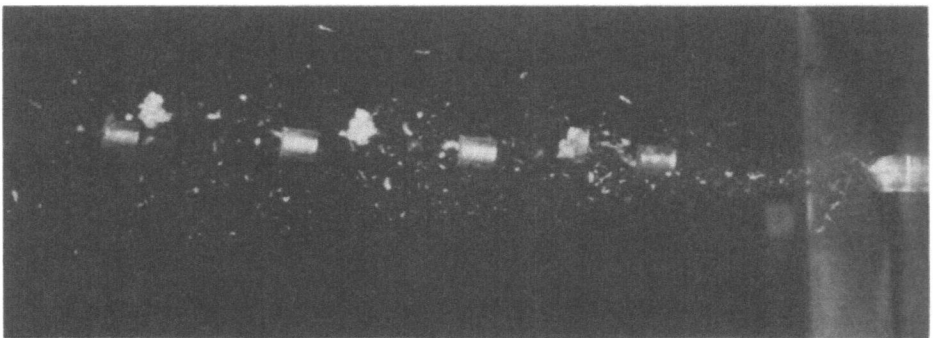

Fig. 8. Multiflash photograph of 1/4" PMMA/Pu Adhesive/1/8" PC under impact. Note absence of spallation.

Dynamic Mechanical Spectra

Previous investigators have qualitatively related macromolecular structural changes, as determined by dynamic mechanical spectroscopy, to the ability to absorb energy under a variety of impact loading rate conditions. The dynamic mechanical spectra of several compositions of "hard" segment in polymers made using 50/50 PPG/PEG are shown in Figure 9. The spectra were determined on a Chemical Instruments Torsional Braid Analyzer which operates in a frequency range around 1 Hz. (The spectrum for the 3.15:2.0:1 polymer is from reference 3 taken on a Rheo vibron DDV II.) At lower Butanediol content there is a broad loss shoulder around -150°C which gradually sharpens into a peak with increasing hard segment composition. The glass transition temperature gradually moves to higher temperature with increasing hard segment content. Spectra for series C (1.576 MDI:0.5 Bd:1 glycol) are essentially the same for all glycols used.

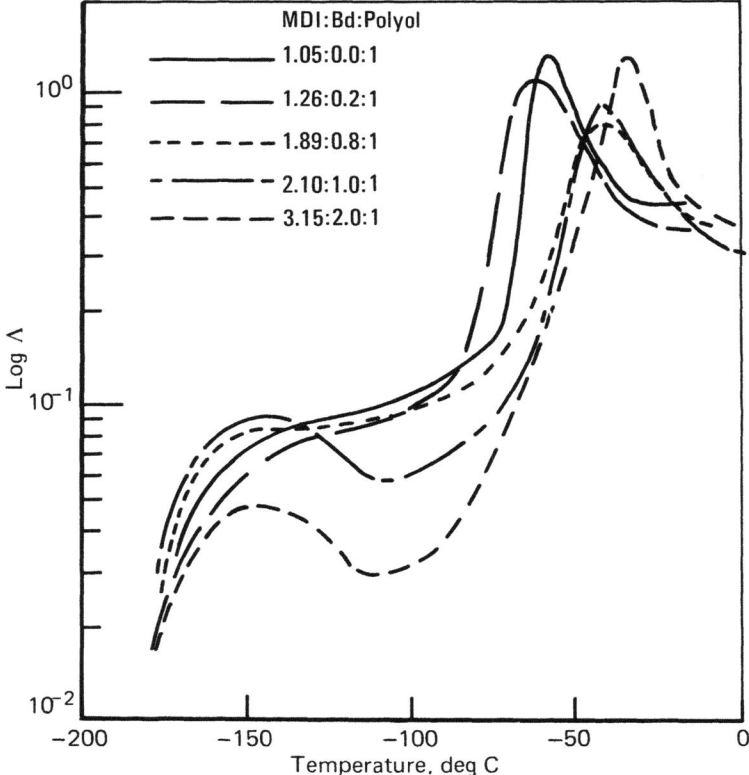

Fig. 9. Dynamic mechanical spectra of polyurethane adhesives (logarithmic decrement).

The composition at which the low temperature shoulder sharpens into a loss peak coincides with the first evidence of spherulitic structure, which also correlates with lowered impact resistance of laminates. Thus, the appearance of a low temperature loss peak is indicative of structural changes in the adhesive that is accompanied by reduced impact resistance.

It is apparent from Figure 10 that as long as there is no evidence of spherulitic structure and the material stays transparent, Bd content has little effect on the stiffness (up to Bd = 0.6). As the Bd content is increased to 0.8, spherulitic structures appear. Since the modulus also decreases, it is possible that the spherulites concentrate the hard segments leaving the matrix softer than it was at lower Bd content. The softer, or less stiff, matrix would account for the significant overall lowering of the modulus as seen in Figure 10. As the Bd content is increased still further, more spherulites would appear causing an increased stiffness (analogous to behavior observed on filling a rubber with rigid particles).

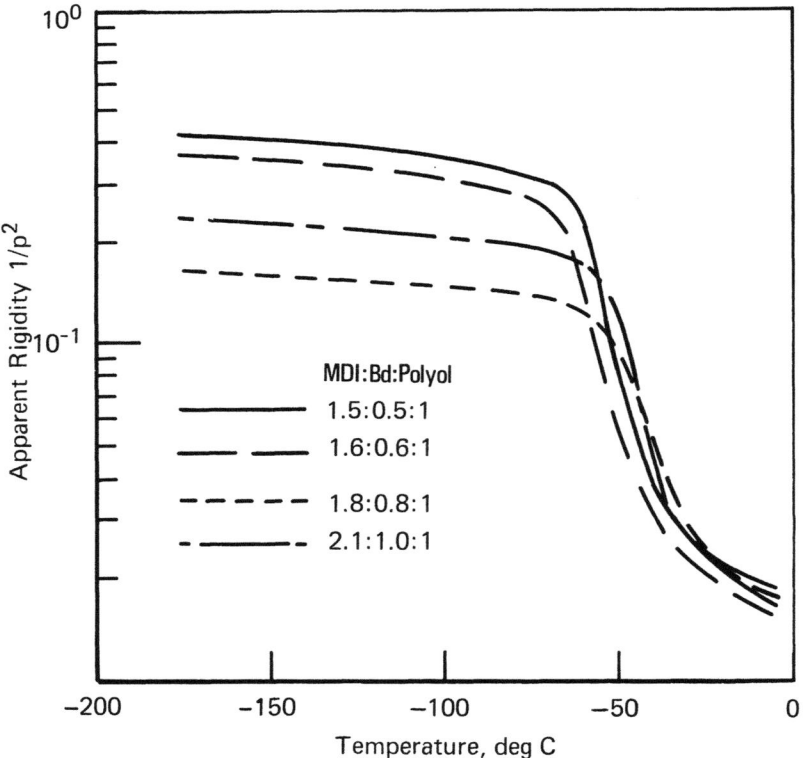

Fig. 10. Dynamic mechanical spectra of polyurethane adhesives (apparent rigidity).

CONCLUSIONS

Acrylic/polycarbonate laminates for windshield applications fabricated using transparent flexible polyurethane adhesive interlayers have demonstrated resistance to impact up to 100% greater than polycarbonate, currently considered the best commercially available transparent impact-resistant material.

Optical and transmission electron microscopy show that the translucent and opaque materials have spherulitic structures which are few in numbers at low "hard" segment concentration and densely packed at high concentration. Dynamic mechanical spectra revealed a glass transition temperature gradually increasing with hard segment content and a broad low temperature loss shoulder. As the hard segment content increases, the broad shoulder sharpens into a loss peak, coinciding with the formation of spherulitic structure accompanied by reduced impact resistance.

So long as the materials showed no spherulitic structure, no real variation in impact performance was found with increasing "hard" segment. Variation in soft segment chemical composition also had no effect on the impact resistance. The mode of failure of these laminates was ductile contrary to thermally bonded laminates or laminates using a brittle adhesive, thereby eliminating the spallation problem inherent in the use of monolithic acrylic windshields. Thus, as long as spherulitic formation can be prevented, the high velocity projectile impact resistance of polyurethane interlayered plastic laminates is superior to all other transparent plastic systems.

ACKNOWLEDGEMENTS

The authors wish to thank Joseph M. Rogers for carrying out the ballistic tests, Richard W. Matton for preparing the polyurethane polymers and Robert Coulehan of CIBA-GEIGY Corporation for obtaining the dynamic mechanical spectra.

REFERENCES

1. R. W. Lewis, M. E. Roylance and G. R. Thomas, "Rubber Toughened Acrylic Polymers for Armor Applications", Army Science Conference, West Point, New York, June 1970.
2. M. E. Roylance and R. W. Lewis, "Development of Transparent Polymers for Armor", Army Materials and Mechanics Research Center, AMMRC TR 72-23, July 1972.
3. J. L. Illinger, N. S. Schneider and F. E. Karasz, "Low Temperature Dynamic Mechanical Properties of Polyurethane-Polyether Block Copolymers", Poly. Eng. & Sci. $\underline{12}$, 25 (1972).

4. A. F. Wilde, D. K. Roylance and J. M. Rogers, Tex. Res. Jour. **43**, 753 (1973).
5. C. R. Desper, R. W. Lewis, S. L. Lopata and M. E. Roylance, "Structural Characterization of XP Films as Related to Mechanical Properties and Ballistic Performance", Army Science Conference, West Point, New York, June 1972.
6. R. F. Boyer, Poly. Eng. and Sci. **8**, 161 (1968).

Physical Property-Performance Correlations in Contact Adhesive Systems

R. G. Azrak, B. L. Joesten and W. F. Hale

Union Carbide Corporation

1 River Road

Bound Brook, New Jersey   08805

Toxicity, flammability, and materials shortages associated with organic solvents have resulted in the development of several new water-borne adhesive systems. Many of the previously effective adhesive formulations are not readily converted to water-borne analogs. Many of the previously understood solution property-performance correlations remain valid in water-borne systems; some do not. In this work an effort has been made to improve our understanding of the adhesive mechanisms and property-performance correlations in solvent and water-borne contact adhesives. These adhesive systems include phenolic/neoprene and phenolic/acrylic compositions. Some of the physical properties of the individual components and of the phenolic/elastomer blends have been related to contact adhesive performance. These properties include molecular weight, polarity, glass transition temperature, particulate nature, and component compatibility. The effect of these parameters on open time tack are discussed. Also discussed is the ability of some of these compositions to exhibit wet tack in the complete absence of organic solvent.

INTRODUCTION

Toxicity, flammability, and material shortages associated with organic solvents have resulted in a trend towards water-borne adhesive systems. The transition from solvent to water-borne systems involves more than simple reformulation; additional variables are encountered and a properly designed system should control each of these variables to optimize adhesive performance.

One adhesives area which is expected to move heavily towards water-borne systems is the contact adhesives segment. Both water-borne neoprene and acrylic based systems are expected to contribute to this growth.[1] A contact adhesive is a material which, when coated and allowed to partially dry on two surfaces, may have little residual surface tack (to the touch) but forms a strong joint when the two coated substrates are brought together under low to moderate pressure.

The characterizing performance parameter for a contact adhesive is the "open time-green strength" relationship. <u>Open time</u> is the time elapsed between the spreading of the last coat of adhesive and the assembling of the adhesive joint. <u>Green strength</u> may be defined as the strength of the adhesive joint shortly after assembly (e.g., less than 1 minute); since the adhesive is normally only partially dry at the time of joint assembly, green strength is usually measured when the joint contains residual carrier. The development of good green strength requires that the adhesive build cohesive strength without its surface losing the ability to deform or exhibit autotack (i.e., its ability to bond to itself) under moderate pressure.

In addition to exhibiting this critical property of contact tack, a good contact adhesive should also have high specific adhesion (adhesion to a variety of substrates) and good elevated temperature performance.

In this paper we discuss how some of the physical properties of adhesive compositions affect their performance in the areas of contact tack, specific adhesion, and elevated temperature strengths. We concentrate on water-borne acrylic/phenolic systems, but, also draw upon our experience in solvent and water-borne neoprene/phenolic systems.

The physical parameters investigated in this study include the glass transition temperature (T_g) of the acrylic resin, the phenolic particulate nature, the phenolic molecular weight, the effect of water on the phenolic T_g, the phenolic and polymer compatibility, the nature of the surfactants, and the reactivities of phenolic and polymer.

EXPERIMENTAL

With a few exceptions indicated below each of the Tables and related text describe the materials, sample preparation procedures and test procedures employed to obtain the data discussed.

Stable blends of UCAR Acrylic Latex 152 and phenolic dispersion BKUA-2260 were prepared by preneutralizing the acrylic latex with ammonia and adding the phenolic. Acetone-water solutions of acrylic-CK-1834 phenolic were prepared by adding the acrylic latex to an acetone solution of CK-1834 at the level necessary to yield a 15% solids solution. Indicated surfactants were then added to this solution. Metal adhesion samples were assembled wet and peel tested at 2"/min. after 1 week of drying at room temperature.

Glass transition temperatures were measured with a Perkin-Elmer DSC-2 differential scanning calorimeter. Samples were cooled at 10°C/min. from above the T_g prior to measuring the T_g at a heating rate of 10°C/min. The reported T_g values represent the extrapolated onset of the increase in specific heat which occurs at the glass transition. The relationship between the toluene content and the glass transition of CK-1834 was determined by adding known amounts of toluene to pulverized resin. The relationship between the water content and the glass transition of BKUA-2260 dispersion resin was determined from measurements of T_g after various drying cycles.

RESULTS AND DISCUSSION

A. Performance Overview

Consider the data of Tables 1-3, which briefly illustrate the adhesive performance of a few contact adhesive systems. Table 1 illustrates that the proper phenolic can significantly enhance the contact tack of an acrylic or neoprene latex. Table 2 indicates the extent to which specific adhesion can vary between contact adhesion formulations. Table 3 shows an apparent conflict in elevated temperature performance between two different test procedures. In subsequent sections we examine these performance variations in greater detail with the most in-depth analysis being given to the phenomenon of contact tack.

B. Contact Tack

Microscopic studies of acrylic/phenolic dispersion blends show the phenolic particles dispersed in the acrylic matrix. This and the data of Table 1 suggest at least three possible mechanisms for green strength enhancement. 1) The phenolic particles may be serving merely as fillers to improve the wet strength of the system.

Table 1

Contact Tack Properties

Green Strength[1] (lbs./inch width) as Function of Open Time and Assembly Pressure

Open Time-Pressure	Acrylic Latex[2]	Acrylic Latex/ Phenolic Dispersion A[3]	Neoprene Latex/ Phenolic Dispersion B[4]
1 hr. open time low pressure assembly	4 C	5 A/C	5 C
2-1/2 hr. open time low pressure assembly	4 C	12 C/A	7 C
high pressure assembly	4 C	19 C/A	15 C
5 hr. open time high pressure assembly	6 C	17 C/A	16 C

1. Canvas-canvas T-peel bonds. 2"/min. peel rate. A = adhesive failure at canvas; C = cohesive failure; A/C = mixed failure. Two coats of adhesive applied to each piece of canvas with #52 wire wound rod. 1/2 hour between coats. Green strength tested within 1 minute of joint assembly under indicated pressure.
2. UCAR Latex 152 (Union Carbide).
3. UCAR Latex 152/BKUA-2260 dispersion 2/1 on solids (Union Carbide).
4. Neoprene Latex 750/CK-1834 dispersion, 2/1 on solids; (DuPont/Union Carbide).

Table 2

Specific Adhesion

Canvas Web to Substrate Bonds[1] (lbs./inch width)

Substrate	Acrylic Latex/ Phenolic Dispersion A[2]	Acrylic Latex/ Phenolic Dispersion B[3]	Neoprene Latex/ Phenolic Dispersion B[4]
Cold Rolled Steel	20 AW	2 AS	2 AS
Aluminum	16 AS	1 AS	2 AS

1. Canvas-substrate peel; 2"/minute; 15' open time; AW = adhesive failure at canvas web; AS = adhesive failure at substrate.

2. UCAR Latex 152/BKUA-2260 dispersion 2/1 on solids.

3. UCAR Latex 152/CK-1834 dispersion 2/1 on solids.

4. Neoprene Latex 750/CK-1834 dispersion 2/1 on solids.

Table 3

Elevated Temperature Performance

Adhesive	Dead Load Shear[1] (°F)	T-Peel[2] at 160°F (lbs./inch width)
Acrylic/Phenolic Dispersion A[3]	350	6 C
Water-Borne Neoprene/Phenolic	200	14 C
Solvent Neoprene/Phenolic	160	14 C

1. Shear strength: Canvas-steel; 15 mil glue thickness; 1 in.² contact; dry 1 week at room temperature; mount vertically in oven with 5# weight suspended from canvas; program heat from 130°F at 10°F/15 minutes; note temperature (°F) at which canvas slides off.

2. T-peel: Canvas-canvas; 20 mil glue thickness; dry 1 week at room temperature; peel rate 2"/minute at 160°F after 30' at 160°F.

3. UCAR Latex 152/BKUA-2260 dispersion 2/1 on solids.

2) The phenolic particles concentrate at the surface yielding some surface tack. 3) Some intermixing or solubility of the two phases occurs and affects particle coalescence and surface tack. The next four subsections discuss these mechanisms.

Acrylic-Latex Properties. The choice of acrylic latex is critical to obtaining true contact adhesive performance. The first criterion is that the latex and phenolic must yield a stable blend when combined directly, with pH modification, or with the aid of additional surfactants. The properties of the acrylic resin particles then become important: A latex resin which is too hard and/or does not properly coalesce may never develop contact tack in the course of drying; a latex resin which is too soft may exhibit tack even when fully dry, but may not have adequate cohesive strength for most applications.

Figure 1 illustrates this for several UCAR acrylic latexes. With room temperature as the reference point, the value of the glass transition temperature (T_g) relative to room temperature can be used as a measure of acrylic resin hardness. For example, the further the T_g occurs below room temperature, the softer the resin will be at room temperature. In Figure 1, the T_g of the latexes investigated is plotted vs. the green strength of a canvas-canvas bond after a fixed open time (2-1/2 hours). Clearly, for these acrylic latexes, there is an optimum T_g range (near 0°C) for green strength development. Those resins with a much lower T_g have tack but poor strength, and those with a much higher T_g have insufficient tack to form a bond.

Figure 1 shows further that the addition of a phenolic dispersion, BKUA-2260, does serve as a tackifier to increase the green strength for most of these resins and that the most dramatic results are obtained in the 0°C T_g range. For the three resins tested in this range (UCAR Latex 152, 153, 154), green strengths were in excess of 15 lbs./inch width and failure occurred at the surface of the canvas and not between the surfaces being mated. From the fact that at least some of the phenolic particles remain distributed throughout the continuous acrylic phase, it is not too surprising that the performance of a formulated system should follow the general dependence of the unformulated latex. The properties of the phenolic dispersion are described below.

Phenolic Dispersion: Particulate Nature and Molecular Weight. Phenolic dispersions of the BKUA-2260 type are solvent free, gum arabic stabilized dispersions of heat reactive resins prepared from the condensation of formaldehyde with variously substituted phenolic monomers. A typical set of properties is shown in Table 4.

How important is the particulate nature of the phenoic dispersion to its role of enhancing green strength in water-borne systems? Is it simply serving as an inert filler to enhance wet strength or

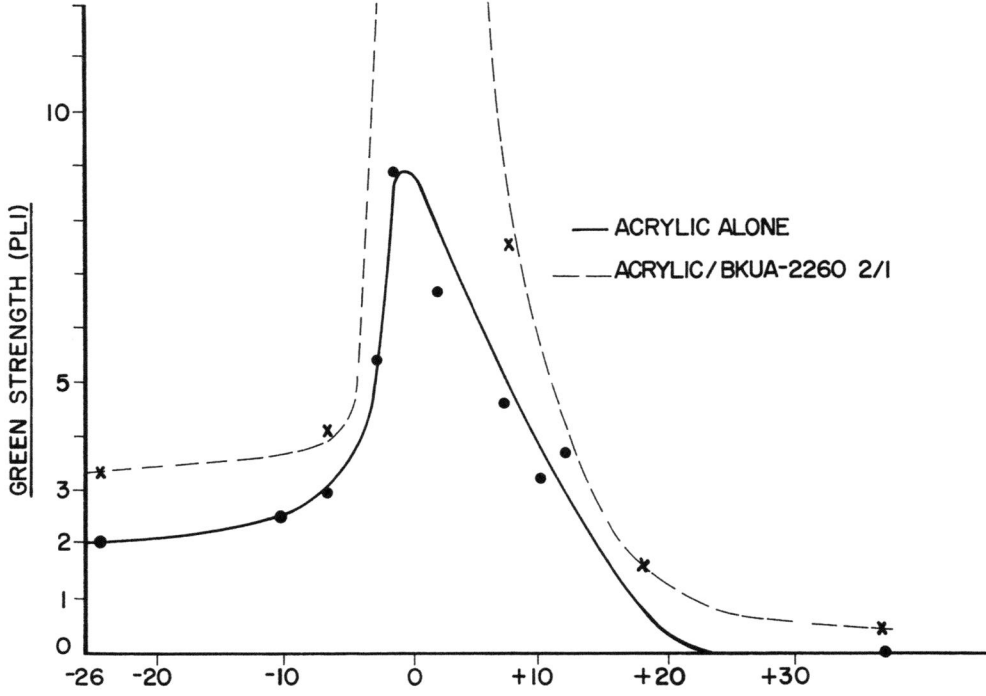

Fig. 1. Green Strength Versus Glass Transition Temperature of UCAR Acrylic Latex

is it, as in conventional solvent systems, actually enhancing surface tack? Several experiments were performed to answer these questions and they demonstrated that the phenolic dispersion is serving as a tackifier rather than merely as an inert filler. One of the experiments involved replacing the phenolic with a comparably sized inert filler--$CaCO_3$. The contact properties of the $CaCO_3$-loaded acrylic vs. the phenolic-loaded acrylic were compared and showed that by the time the $CaCO_3$-loaded system had developed cohesive strength, it had lost tack. The comparably phenolic filled acrylic had substantial green strength.

In another set of experiments the water-borne system was made to approach a solvent system by adding the adhesive blend to excess quantities of acetone. Films of adhesive cast from this acetone "solution" showed no signs of phenolic particulate nature. The green strength of an acetone "solution" of UCAR Latex 154 was compared to that of an acetone "solution" of UCAR Latex 154/BKUA 2260. The presence of the phenolic, even in its non-particulate form, enhanced the green strength of the acrylic; this suggests again that the BKUA-2260 is a true tackifier.

Table 4

Typical Properties of BKUA-2260 Phenolic Dispersion

Solvent Content	None
Solids Content (1.5 gm/3 hrs./135°C)	48%
Gel Time (150°C, 1 gm sample)	180 sec.
pH	6.7
Viscosity (#3 spindle/10 rpm)	2000 cps
Mean Particle Size	8 μ
Dispersion Stability	Storage life is ≥ 1 yr. at 40°F Gel Time decreases $\approx$ 20 sec. after 2 months at 75°F Some stratification on standing for prolonged periods

    Also examined has been the effect of phenolic molecular weight as it affects green strength and final bond strength. Table 5 lists the strengths of canvas-canvas bonds prepared using UCAR Latex 154/ phenolic dispersion (2/1) adhesive where the phenolic molecular weight was varied. The green strength data shows that for a fixed phenolic monomer composition, increasing phenolic molecular weight (decreasing gel time*) decreases system tack. Dry bond strengths remain high.

    However, there are lower limits on phenolic molecular weight. If molecular weights are reduced to the point where the phenolic is water soluble, dramatic decreases in green and dry bond strength are obtained at high phenolic concentrations; the phenolic now forms the continuous phase and prevents the latex from fusing. Thus to

---

*Gel time is the length of time required for a heated resin to crosslink or advance in molecular weight to the gel point, where it will no longer flow readily. Therefore, for a given monomer composition, under a given set of conditions, a resin with a shorter gel time will, in general, have a higher initial molecular weight.

Table 5

Phenolic Molecular Weight

UCAR Latex 154/Phenolic (3/1 Solids)

Phenolic Resin Dispersion	Green Strength[1] 2-1/2 Hours Open Time	Wet Assemble- Dry Bond Strength[2]
Gel Time[3]: 210 sec.	10 A/C	20 C/A
160 sec.	9 A/C	22 C/A
110 sec.	4 C	18 C
60 sec.	4 C	17 C
No Phenolic	3 C	17 C

1. Low pressure assembly – 2 drawdown coats; strengths in lbs./inch/width; A = adhesive failure at canvas; C = cohesive failure; C/A = mixed failure.

2. Samples assembled at 0 open time and allowed to dry for 1 week at RT.

3. Gel Time: is roughly inversely related to molecular weight (see text); 150°C stroke cure hotplate gel time on 1 gm sample.

the extent that the phenolic resin does not form the continuous phase, its particulate nature is important. However, from the other experiments described above--including the contact adhesive dependence on phenolic molecular weight--it is obvious that the phenolic is more than an inert filler.

Phenolic Glass Transition Temperature and Its Dependence on Water Content. A t-butyl phenolic resin such as CK-1834 (Union Carbide) or thoroughly dried material from a BKUA-2260 phenolic disperion (gel time = 140 sec.) has a glass transition temperature, T_g, in the range of 30-35°C. Such hard resins might not be expected to tackify an already soft acrylic. However, except for heat activated samples, contact tack is measured when there is residual carrier. Therefore, the effect of organic solvents or water on the T_g must be considered.

Residual toluene can be removed from CK-1834 only with great difficulty and can easily depress the T_g below room temperature. For example, 5% toluene decreases the T_g to 10°C. This degree of T_g depression is consistent with results calculated from the well-known Fox equation[2] which predicts a T_g of 11°C for [CK-1834] containing 5% toluene (assume T_g[toluene] = -150°C[3]; T_g [CK-1834] = 32°C). This suggests that solvent plasticization is important in the tackifying process. In fact, aqueous dispersions of CK-1834 are ineffective in tackifying neoprene latex unless organic solvents are present!

The acrylic/phenolic contact adhesive systems discussed above do not contain an organic solvent. However, the T_g of the phenolic BKUA-2260 in these systems was found to depend on residual water. Table 6 shows how the T_g changes during a drying program for two phenolic dispersions which differ only in gel time. Figure 2 shows the relationship between water and T_g for the same two dispersions. For the dispersion with 140 sec. gel time, the T_g is 32°C when dry and near 10°C with 3% water. (Fox equation predicts T_g = 18°C for 3% water, assuming T_g [water] = -146°C[4] and T_g [BKUA-2260] = 32°C).

Table 6

Glass Transition Temperature of
BKUA-2260 - Effect of Drying Time

Drying Cycle	T_g, °C	
	140 Sec. Gel Time[1]	50 Sec. Gel Time[1]
1 hour, ambient	Not Measurable	24
2 hour, ambient	Not Measurable	36
4 hour, ambient	-5	35
2 hour, vacuum, 40°C	8	62
4 hour, vacuum, 40°C	9	64
6 hour, vacuum, 40°C	19	64
8 hour, vacuum, 40°C	24	65
10 hour, vacuum, 40°C	27	64
16 hour, vacuum, 40°C	32	

1. 150°C Stroke Cure hotplate gel time on 1 gm sample.

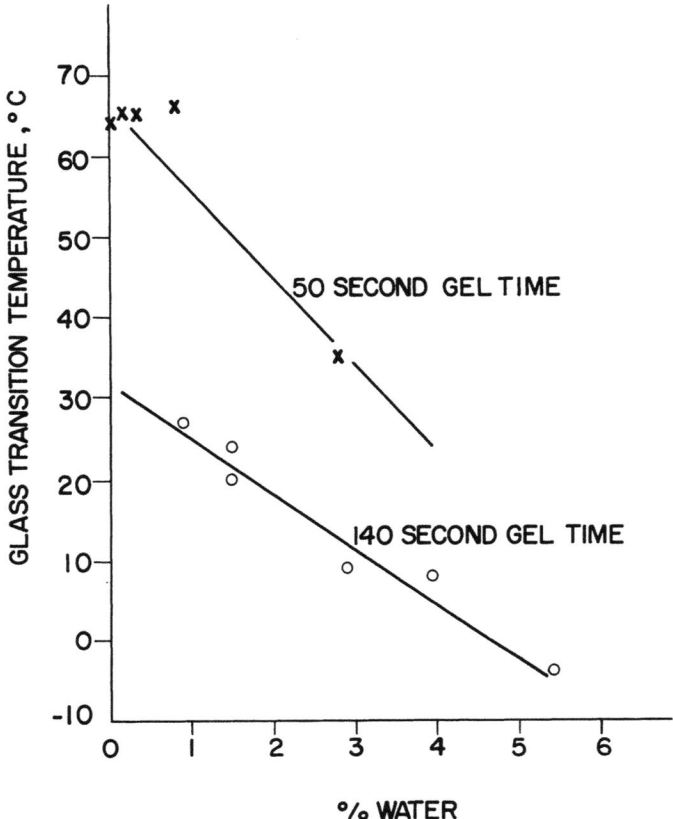

Fig. 2. Glass Transition Temperature of BKUA-2260 Versus % Water

This dispersion requires long drying times before its T_g increases above room temperature. Therefore, a long gel time dispersion of this type can yield a system with tack for extended open times--even in the absence of organic solvents. Once the system is dry, it can become hard enough to not substantially detract from room temperature cohesive strength.

The T_g is 64°C for dry resin from the 50 second gel time dispersion. The high dry T_g results from a higher molecular weight relative to the dispersion with 140 second gel time. Very short drying times are required for this dispersion to reach a T_g greater than room temperature. At 3% water the T_g is well above room temperature.

In general, as we have indicated, for one-step resins with a given phenolic monomer composition and a fixed set of reaction conditions, long gel times correlate with low molecular weight and low crosslink density. Furthermore, for phenolic resin dispersions of the type discussed in this paper, longer gel time also indicates that the resin has high polarity because it has a large number of methylol groups and methylene ether bridges. Therefore, long gel time resins should be somewhat softer when dry, should adsorb water more readily and should bind water more tightly than short gel time resins. These observations are consistent with the previously described dependence of contact tack on phenolic molecular weight.

Compatibility of Components. While plasticization of the phenolic component (by solvent or water) appears to give a partial explanation of its tackifying mechanisms, it certainly cannot be the entire explanation for green strength development. Neither solutions of CK-1834 nor dispersions of BKUA-2260 exhibit contact properties by themselves. Their interaction with the other components in the formulation (neoprene or acrylic) must be considered.

A prevailing theory on tackification suggests the necessity of controlled compatibility between tackifier and matrix polymer. Low polarity t-butyl phenol resins are most effective with neoprene rubber, somewhat higher polarity phenolic resins are recommended for nitrile rubber adhesives, and we have found that relatively polar resins are preferred as tackifiers for acrylic latexes. However, limited actual data is available to demonstrate the existence of phenolic-matrix compatibility in any of these systems. Is a nonyl phenol resin, which is not an effective tackifier, really less compatible with neoprene than is a t-butyl phenol resin? Is there an intermixing of components in the dispersed aqueous systems? These questions are answered affirmatively by our initial investigations into component compatibilities via modulus-temperature studies (E-T) and differential scanning calorimetry (DSC) on dried samples of components and blends. Compatible systems have a single T_g which is intermediate to the T_g's of the individual components. Incompatible systems reveal the T_g's of the unblended components[2]. E-T curves show neoprene to be far more compatible with a t-butyl phenol resin than with a nonyl phenol resin [T_g neoprene = $-40°C$; T_g neoprene/nonyl (2/1) = $-40°C$; T_g neoprene/t-butyl (2/1) = $-14°C$]. They also show partial compatibility in a neoprene latex/CK-1834 dispersion containing some toluene, and in acrylic/BKUA-2260 systems. The DSC results agree with the E-T results. Additional work in this area is in progress.

These results do indicate then that effective adhesion compositions exhibit some component compatibilities in dry form and they suggest that the phenolic resin and matrix polymer do function in concert to product contact properties.

C. Specific Adhesion

The data of Table 2 indicates that water-borne acrylic/BKUA-2260 adhesives have excellent metal adhesion whereas water-borne acrylic/CK-1834 dispersion adhesives and water-borne neoprene adhesives have poor metal adhesion. The dispersing aids employed in the last two systems were investigated as possible causes of this lower metal adhesion. The surfactant system employed to disperse CK-1834 and similar phenolics is listed at the bottom of Table 7. In order to separate out the effects of the dispersing agents and at the same time maintain a stable system, the CK-1834 phenolic and the acrylic resin were put into acetone solution. The components of the dispersant system were individually added and their effect on the metal adhesion of the system was determined. The data of Table 5 clearly shows that the oleic acid and its salts are the principle cause of the reduced metal adhesion: The peel strengths fall off dramatically when these components are added. Furthermore, the dark violet color, characteristic of the phenolic-iron complex, disappears, indicating that the oleic acid interferes with the phenolic substrate interaction and forms a weak boundary layer.

An additional point on specific adhesion is that acrylic based contact adhesive systems showed markedly better bond strength retention to plasticized vinyl substrates than do neoprene based adhesives. This results from the inherently better plasticizer migration resistance of acrylic resins.

D. Elevated Temperature Dependence

As shown in Table 3 in an elevated temperature shear test, acrylic/BKUA-2260 systems perform extremely well, while elevated temperature peel shows some fall off in performance. Since the BKUA-2260 is self-crosslinking at elevated temperature and has the ability to co-cure with certain acrylic and vinyl-acrylic latexes, the elevated temperature performance of bonds depends, to some extent, on the previous thermal history of the bond. The shear test involves a slow heat up to test temperature and allows the bond to cure partially. Mechanical properties of thermally cured films of UCAR Latex 154/BKUA-2260 show 10 fold modulus increases over the uncured films. Vinyl-acrylic latexes that do not have the ability to co-cure with the phenolic do not exhibit the same high elevated temperature shear values.

Table 7

Component Effects on Metal Adhesion
UCAR Latex 154/CK-1834 (3/1)
15% Solids in 85/15 Acetone/H₂O

Additive (On pph Phenolic Solids)	Canvas-Steel Peel Value (PLI)	Interface at Which Fails	Color at Steel
None	15	Canvas	Dark Violet
6.0 Gums	16	Canvas	Dark Violet
1.5 NaOH	14	Canvas	Dark Violet
6.0 Oleic Acid	2	Steel	Mottled Brown-Violet
1.5 NaOH } 6.0 Oleic Acid	1	Steel	Yellow-Brown
UCAR Latex 154/1834 Dispersion	2	Steel	Brown
CK-1834 Dispersion Contains: (pph on Phenolic)	6 Gums, 1.5 NaOH, 6 Oleic Acid		

CONCLUSIONS

As in conventional solvent systems, phenolics can serve as effective tackifiers in properly designed water-borne systems. Most notable is the fact that water, as well as solvents, can serve to plasticize the phenolic. Also care should be taken to insure that the phenolic is the proper molecular weight, that the components are at least partially compatible, that the phenolic does not form the bulk of the continuous phase, that the surfactants employed do not form weak boundary layers at substrates, and that elevated temperature tests approximate use conditions.

ACKNOWLEDGEMENTS

We wish to acknowledge discussions with and the assistance of Drs. N. J. McCarthy and O. Olabisi in respectively supplying the samples of phenolic dispersion and the modulus-temperature data. The technical assistance of Messrs. M. F. Patrylow and J. E. Simborski is greatly appreciated, as is the permission of Dr. J. Koleske to quote his unpublished data.

REFERENCES

1. R. G. Azrak and B. P. Barth, J. of Adhesives and Sealant Council, Spring, 1974.
2. L. E. Nielson, *Mechanical Properties of Polymers*, New York, Reinhold Publishing Company, 1962.
3. J. Koleske, private communication.
4. I. Yannes, Science $\underline{160}$, 298 (1968).

DISCUSSION

On the Paper by F. D. Petke

A. J. Kinloch (*Ministry of Defense, U. K.*): On the subject of boundary layers, Wake et al. (Faraday Special Discussions, No. 2, 1972, 1) have used phase contrast interference microscopy to examine the fracture surfaces of a titanium alloy joint bonded with a filled structural adhesive. They found a boundary region of the adhesive, about 1 μm in depth from the substrate surface, to be free from filler particles which had been homogeneously dispersed in the adhesive prior to joint preparation and curing. They further identified the locus of joint failure as being in, or close to, this boundary layer.

A. A. Gertzman (*Ethicon, Inc.*): What is the relationship between surface energy and bond strength in adhesive bonds?

F. D. Petke (*Tennessee Eastman Co.*): According to a discussion I had with Dr. A. J. Kinlock this morning, bond energy is composed of two parts - adhesive strength, θ_o, and adhesive deformation energy, ψ. For a case of actual interfacial bond failure, the θ_o factor is equal to the work of adhesion, W_A,

$$W_A = \gamma_{\ell v}^o (1 + \cos \sigma)$$

where $\gamma_{\ell v}^o$ is the surface tension of the adhesive and σ is the contact angle. In addition, however, ψ also depends on $\gamma_{\ell v}$ and σ, but in a very complex manner which has not been determined.

J. Vullo (*Mohasco*): What are specific examples of tacifiers raising glass transition temperature of polymer?

F. D. Petke: The work I refer to by Sherriff, et al. [J. Appl. Polym. Sci., 17, 3423 (1973)] reports results of blends of natural rubber with each of four resins: Pentalyn H (pentaerythritol ester of hydrogenated rosin; Piccolyte S115 (poly[β-pinene]), Piccolyte S70 (poly[β-pinene]); and Arkon P125 (reported to be polymerized dicyclopentadiene).

R. Schure (*Swift & Company*): What effect will moisture play on the overall effect of the final bond strength on your adhesive design?

F. D. Petke: The problem is probably due to surface preparation rather than the adhesive design. Care should be taken to remove any moisture from the surface of substrates. Silane adhesive promoting agents can be used to improve adhesive joint strength and resistance to environmental moisture for many adhesives.

W. Newby (*Alcox International Ltd.*): What effect on bond strength is there due to the fact that the surface on which the polymer is cast can affect the morphology of the polymer?

F. D. Petke: By combining the principles discussed by Dyckerhoff and Sell and by Schonhorn and Ryan, one might be able to control the surface energy of a substrate to match that of an adhesive and thus maximize bond strength. This, however, seems to me to be an undesirable approach. It is normally more desirable to model the adhesive to match the surface properties of the substrate.

On the Paper by T. J. St. Clair and D. J. Progar

A. J. Kinloch (*Ministry of Defense, U. K.*): What surface pre-treatment did you employ for the titanium alloy substrate prior to bonding?

T. J. St. Clair (*NASA Langley Research Center*): The PASA-JELL 107 method was used for cleaning all of adherends used in the presented work. The phosphate-fluoride treatment was studied in our lab and was found to give the same lap shear results at room temperature on the systems that were tested.

A. J. Kinloch: Have you conducted any experiments to determine the durability of your titanium bonded joint to water?

T. J. St. Clair: The optimized (copolymerized 65/35 BTDA/PMDA; $3,3^1$ DABP) system mentioned in the talk has been exposed to cyclic 95% relative humidity for 250 hours (at present) and the lap shear samples are essentially the same as before exposure. Testing is continuing.

R. Schure (*Swift & Company*): Do you have any information regarding adhesive strength as a function of film thickness?

T. J. St. Clair: The adhesive was generally applied to that bond-line thickness was from 3-5 mils (and certainly 1-5 mils). It was observed that better adhesive strength was obtained from thinner bond-lines.

J. S. Noland (*American Cyanamide*): What is responsible for the diglyme effect? It looks like a plasticization effect.

DISCUSSION

T. J. St. Clair: There is evidence that diglyme is remaining in the adhesive and the possibility of plasticization is feasible.

K. D. Nisbet (*Glidden-Durkee*): Did you analyze the films specifically for solvent retention? It seems to me that amide retention would be quite detrimental to lap shear testing. This in itself would explain the lower force failures.

T. L. St. Clair: Yes, GC-mass spectral data indicate that both solvents are still present after bonding when they are used as the polymerization media. A greater amount of diglyme was retained. I feel that the amide-type solvents would be more detrimental when left in the curing adhesive.

A. Forschirm (*Celanese Research Co.*): Could you comment on the elongation characteristics of these adhesives and their adhesion to the glass plates used for infrared analysis?

T. L. St. Clair: We have not studied the elongation characteristics of these adhesives. The films for infrared analyses which were cast on glass and cured to 300°C were extremely difficult to remove and in some cases actually pulled pieces of glass from the surface of the plate on which it was cast.

On the Paper by J. L. Illinger, R. W. Lewis and D. B. Barr

E. C. Chenevey (*Celanese Research Co.*): What is the effect of projectile velocity?

J. L. Illinger (*Army Material & Mechanics Research Center*): All of the impact tests were performed using 1.1g projectiles so that the kinetic energies quoted are computed from velocities. Obviously, then, as the projectile velocity increased, a thicker laminate would be required to prevent penetration. It would be instructive, however, to vary the mass of projectile at constant total kinetic energy.

E. H. Mottus (*Monsanto Co.*): Did you look at and what was effect of using lower molecular weight soft segment diols?

J. L. Illinger: We only looked at one polymer with half the soft segment length. In that case the polymer showed reduced ductility and the laminate showed lowered resistance to impact.

B. Dimick (*Institute of Paper Chemistry*): Would these laminates be feasible for applications such as school windows?

J. L. Illinger: Yes. The material cost would be expensive initially, but labor costs for replacement of windows are high and increasing, such that use of laminates could pay off.

Although, for school windows, where low mass high velocity projectiles are not generally the major problem (rather, the problem is high mass-low velocity projectiles - rocks) a more efficient solution would be a homogeneous plastic such as polycarbonate.

On the Paper by R. G. Azrak, D. L. Joesten, and W. F. Hale

S. C. Sharma (*General Tire and Rubber Co.*): Regarding the effect of molecular weight of the phenolic resin, isn't the curability of these resins different with different molecular weights? How does this affect the green strength?

R. G. Azrak (*Union Carbide Corp.*): Yes, the time required for the phenolic resin to thermoset is molecular weight dependent: the gel times and methylol contents will, of course, vary with degree of resin advancement. The rate of chemical cure is not directly related to green strength development; however, the differences in residual methylol content would be expected to affect green strength development by affecting water plasticization and component compatibilities.

P. K. Chatterjee (*Personal Products Co.*): What is the definition of tack? Is ther any difference between tack and adhesion? If so, what?

R. G. Azrak: Tack has almost as many definitions as there are adhesive chemists, but what we mean by contact tack is the ability of the adhesive material to set and bond to itself under minimal load. Although it may not be rigorously true, I tend to visualize tack in these systems predominantly as a property of the adhesive surface or surface layers; these should deform readily and wet each other under this minimal pressure. To many people contact tack and adhesion are synonomous. For the purposes of our discussion today we have attempted to distinguish two components of green strength (adhesion) - tack and cohesive strength. In these terms the development of good green strength requires that the surfaces wet each other (i.e. exhibit contact tack) but also that the adhesive have adequate cohesive strength - from the surface down to the substrate - to resist being pulled apart. The tack resists having the locus of failure occur at the interface of the two mated surfaces; the cohesive strength resists failure elsewhere in the adhesive. Together they contribute to green strength adhesion.

J. Vandesaer (*Alcolac*): Did you study surfactant variation in the acrylic latex?

R. G. Azrak: We examined the effect of acrylic latex surfactants on the latex - dispersion blend compatibility and found that several surfactants markedly reduced blend stability. In the area of specific adhesion we did evaluate latex systems with varying surfactants; none of these exhibited poor specific adhesion properties that we could directly relate to latex surfactant effects. I might point out, however, that excess surfactants could have a detrimental effect on tack development.

F. D. Petke (TENNESSEE EASTMAN CO.): I would have thought that an adhesive which resists plasticizers for PVC, such as the acrylic/phenolic would allow a build-up of plasticizer at the adhesive-adherend interface. How does this behavior agree with your statement that the acrylic/phenolic adhesive has better aging properties on PVC than the neoprene/phenolic adhesive?

R. G. Azrak: A build-up of plasticizer at the adhesive-adherend interface can occur only if the driving force for plasticizer exudation at the interface is high enough to displace the adhesive; thus a strong adhesive-adherend interaction should retard plasticizer buildup.

PART THREE

Rubber Adhesives and Sealants

Introductory Remarks

M. E. Gross

The B.F. Goodrich Company

9921 Brecksville Road

Brecksville, Ohio   44141

As anyone experienced in the technology of elastomers well knows, the problems of adhesion are omnipresent whether one is constructing a tire, a belt, a hose, or the multitude of composites combining rubber parts with metals. The problems can be as varied as adhesion between dissimilar elastomers, adhesion between plies of rubber having dissimilar filler loadings and cure rates, adhesion between rubbers and reinforcing cords or fibers, adhesion between the rubber and reinforcing fillers, or the adhesion problems associated with bonding elastomers to high modulus substrates such as metals and plastics. Having successfully achieved the solution to any of these bonding problems, the engineer is still confronted with the question of bond durability. How well it withstands the aggressive environment of heat and dynamic stress encountered in the operation of a tire or a power transmission belt? Will the plies in the tire separate? Will the bond between the rubber and reinforcing cords withstand the punishment encountered in actual service? Is the rubber to metal bonding stable to heat, moisture, salt, stress cycling, or mechanical shock loading? If an organic adhesive is used, will be be chemically stable during a service life which might extend to several years? Obviously, such questions cannot be safely answered without extensive as well as expensive laboratory and real life testing.

In the first paper, Dr. Sexsmith briefly reviews elastomer adhesion mechanisms involving vulcanization bonding to metals and textile substrates. He relates these same mechanisms of chemical interaction and diffusion to the practical applications of bonding elastomers to textiles. Dr. Gent's paper deals with the relationship between strength of adhesion and density of chemical bonding at the interface between an elastomer and a glass substrate.

In line with the old adage that a picture is worth a thousand words, Mr. Smith will show that when elastomeric materials are joined together as in the plies of a tire or belt construction, useful information can be gained by examining the resultant bondline with optical and/or electron microscopes. A skilled observer can relate gross morphological characteristics such as bondline thickness, surface roughness, and flow of materials during cure, to adhesion performance. It is also possible to characterize the bonding problems arising from juxtaposed dissimilar elastomers, fillers, and curatives at an interface. Such observations made on a microscale can be useful in interpreting bonding problems and can frequently lead to a successful resolution. The key to the solution is arrived at through proper interpretation of the microscope observations.

Mr. Given's paper points up the fact that the conventional, slow speed, essentially static type measurement of rubber to metal bond strength is inadequate for many real life applications where high speed mechanical shock stresses are encountered.

Over the past 25 years, a considerable amount of effort has been expended to develop and study high efficiency, reinforcing mineral fillers as partial replacement for the carbon blacks. Even though it has been but moderately successful the impetus to continue this work for economic considerations is obvious. In his paper on silane treated clays as reinforcing fillers for rubber, Dr. Plueddemann will discuss the nature and extent of adhesion between matrix and filler. This is a subject which has been and still is highly controversial.

As a change of pace for the final paper, Dr. Greenlee will delve into a specialty area of polymer science by presenting the chemistry of how one part RTV silicone polymers are crosslinked to elastomers while providing good adhesion to substrates. This paper deals primarily with the mechanisms of improved adhesion.

Mechanisms of Adhesion in Elastomer-to-Textile Bonding

F. H. Sexsmith and E. L. Polaski

Hughson Chemicals, Division of Lord Corporation

Erie, Pennsylvania   16509

Elastomer adhesion mechanisms are discussed first from the viewpoint of vulcanization bonding to metals. The approaches of brass plate bonding, bonding with ebonite, with butadiene-methacrylic acid copolymers, and with proprietary adhesives are cursorily reviewed to invoke theories of chemisorption, diffusion, and directed chemical reaction. The time dependency of peel strength as an adhesion response through the vulcanization cycle is cited as a possible indication of mechanism.

Examples and data are then prepared to show that the same mechanisms of migration and chemical interaction are operative in various approaches for vulcanization bonding of elastomers to textiles. The systems discussed are the work-horse RFL (resorcinol-formaldehyde-latex) dips, self-bonding elastomers (via two different approaches), and a versatile proprietary bonding agent.

Kinetic effects and the effects of varying the proportions of key ingredients are related for both the proprietary adhesive and a typical RFL formulation.

This paper will review briefly a number of adhesion mechanisms that have been put forth to describe the joining of vulcanizing elastomers to metals. It will then be argued that some of these same mechanisms are equally useful to explain the adherence of vulcanizing elastomers to textiles. Considerable emphasis will be placed on directed chemical reaction, since there appears to be much evidence for this in most practical instances of excellent bonding performance.

I. ELASTOMER-TO-METAL ADHESION REVIEW

Over the years, a fairly large number of methods have been used to bond vulcanizates to metals.[1,2] These approaches, in their variety, have pointed up several features that are significant from the mechanistic viewpoint. One of the earliest, but also of great commercial importance, was the approach of plating metal members with brass so that ordinary sulfur-vulcanizing compounds of natural rubber might adhere to them directly. There is considerable literature on this particular technique of bonding,[3,4] but many adhesion technologists attribute it to chemisorption or direct bond formation between the sulfur of the stock and copper atoms in the brass (Figure 1).

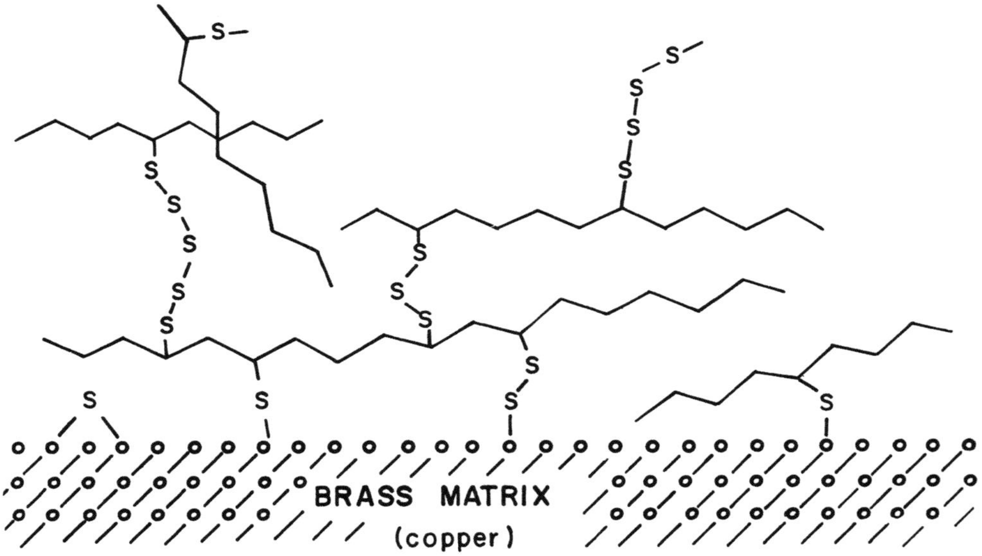

Figure 1. Chemisorption on Brass via Direct Copper-Sulfur-Rubber Linkages.

ELASTOMER-TO-TEXTILE BONDING

Another of the very early practical techniques was to use one or more layers of ebonite as the adhesive entity. Ebonite is by definition a sulfur-rubber mixture with an extremely high sulfur ratio and which sets to a hard but thermoplastic mass. We are dealing here with semi-reversible sulfur bindings and interdiffusion, certainly of the sulfur but perhaps also of the rubber chains, is a likely consideration for the adhesion mechanism. Sulfur migration or diffusion could occur from the adhesive layer to the rubber being bonded. This would provide a smooth gradient of hardness through the interface from the ebonite to the softer vulcanized rubber compound. Figure 2 shows the approximate variation in modulus and sulfur concentration that one might visualize in moving from the metal interface through the adhesive and on into the rubber.

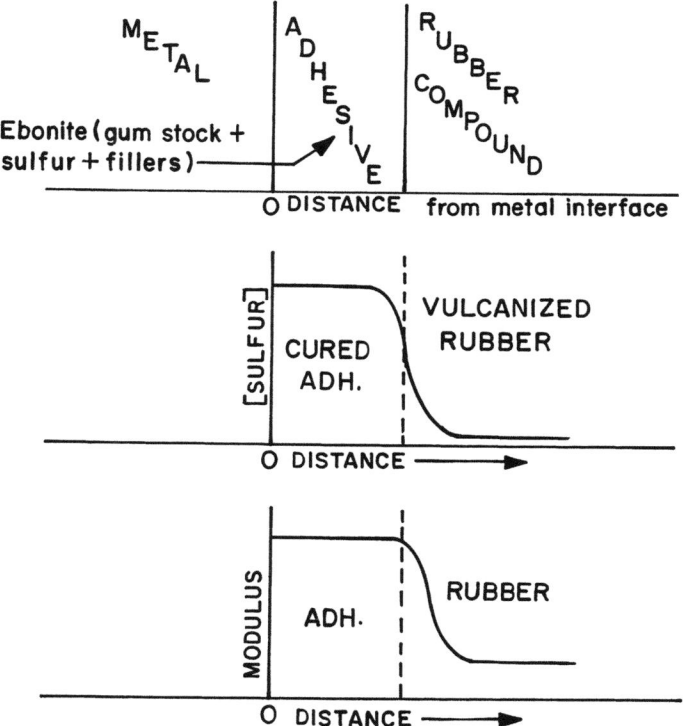

Figure 2. Hypothetical Concentration and Modulus Gradients.

The desirability of balanced polarity is often mentioned in the adhesion context, and it is usually true that materials of very different polarity will be reluctant to bond to one another. Perhaps, a better way of describing this balance in simple rubber-to-metal bonding compositions would be by reference to simultaneous metal-seeking and elastomer-seeking factors. One of the best and simplest demonstrations of this was some work by Kraus et al[5] over 20 years ago. Butadiene-methacrylic acid copolymers were prepared at various ratios of the two monomers and were used to bond vulcanizing NR to steel. The results are shown in Figure 3.

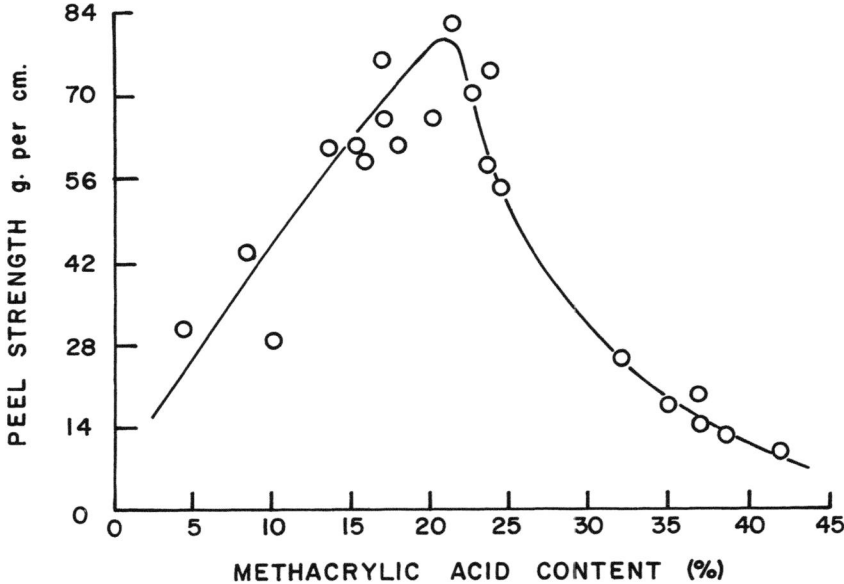

Figure 3. R-to-M Bonding with Butadiene - Methacrylic Acid Copolymers.

The points in the neighborhood of the left descending portion of the curve were characterized by predominantly cement-to-metal interfacial failure. Those to the right of the peak showed cement-to-rubber failure. The points on the graph certainly reflected progressive variations in polarity, but one might note too that the hardness of the adhesive layer was also increasing with the higher methacrylic acid contents. Probably more important than either of these in the mechanistic vein were the different chemical capabilities of the respective mer-units.

The union of polybutadiene segments with the rubber is accountable if the two polymers intermingled or if sulfur from the NR member entered into crossbridging reactions. A likely role for the methacrylic acid moiety would be in adsorbing on the steel substrate.

In fact, the tendency for certain chemical groups or entities to
chemisorb on hydrated metal oxide surfaces is inherent in most practical adhesive systems that enable the metal bond to prevail through
aggressive service environments. Contributing in this sense are
organofunctional silanes, certain polyisocyanates and heat-reactive
phenol-formaldehyde condensates.

In addition to this metal-seeking feature, proprietary adhesive
systems for rubber-to-metal often have either of two capabilities
for chemical activity at the elastomer end. In one type, the bonding
agent will contain a mobile vulcanizing agent which is capable of
reacting across this interface.[6] In the other type, the adhesive
composition will embody polymers that are themselves attachable
through a migrating moiety from the elastomer member. Usually, this
is soluble sulfur or sulfur plus accelerators, but it can also be
a peroxide or free radical type crosslinker. The principal adhesive
polymers in these cases are often themselves unsaturated or capable
of generating double bonds at the vulcanization temperature.

Some evidence for two types of reactivity at the elastomer end
comes from rate measurements. The experimental method for following
peel strength development as a rate process was described in a 1970
paper.[6] In Figure 4 a sulfur-cured EPDM of the commercial Nordel

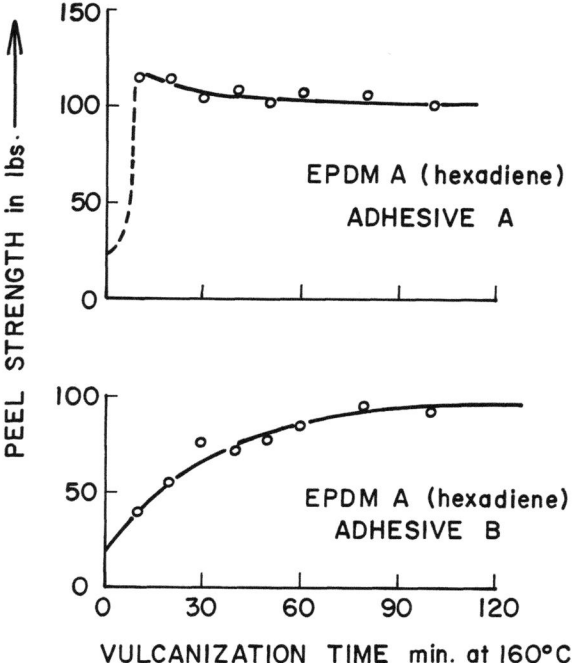

Figure 4. EPDM to Metal Bonding.

family was bonded first with a proprietary adhesive that contained a strong crossbridging agent, and then with a different formulation that was based in part on a butadiene polymer. The very fast bond formation in the first instance was characteristic of this particular crossbridging species; the relatively slow development of peel value in the second case we attributed to rather slow diffusion of sulfur across the interface for attachment to the adhesive's diene polymer.

The curves in Figure 5 are a bit of a digression and are included only to show that the vulcanization time dependency of peel strength values is not necessarily diagnostic of adhesion mechanism. Both EPDM's were bonded with the fast adhesive system referred to previously. But the dicyclopentadiene terpolymer recipe was so very slow to vulcanize itself that it imposed a profile of sluggishness in the curve for peel value development.

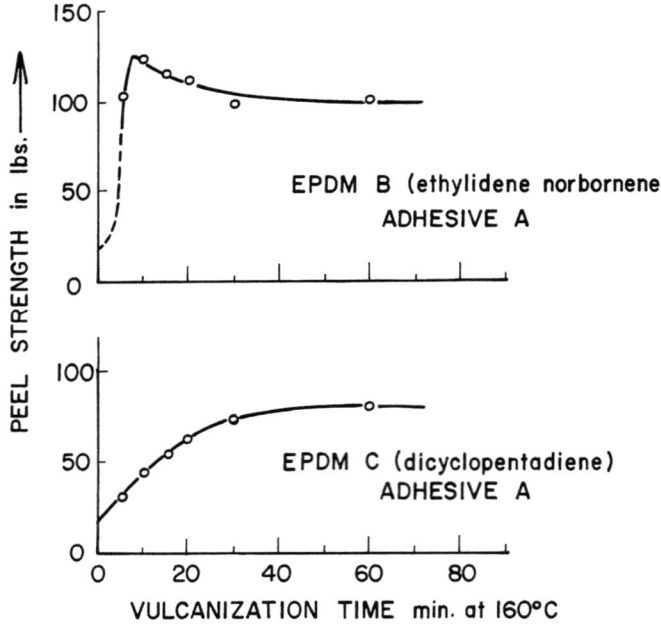

Figure 5. EPDM to Metal Bonding.

There are other examples of material approaches that have contributed to our interpretation of elastomer-to-metal bonding mechanisms, but all can be rationalized under one or more of the processes in the following schematic (Figure 6). Here, we have tried to identify all of the phenomena that may occur in practical industrial bonding systems of the primer-covercoat type.

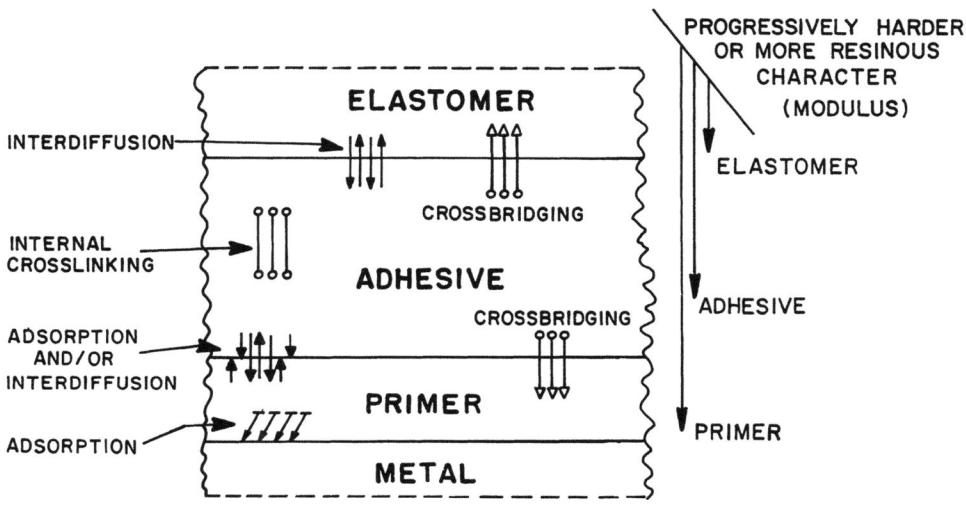

Figure 6. Elastomer-to-Metal Bonding Processes

II. ELASTOMER-TO-TEXTILE ADHESION

We can now shift our attention to various processes for adhering vulcanizing elastomers to textiles. Generally, these fall into the same categories and sub-categories as for rubber-to-metal assemblies -- self-adhering stocks, and one- or two-coat adhesive systems. Within the latter category are familiar polymers that have the capability to react with the mobile sulfur of the vulcanizing elastomer, but also reactive adhesive ingredients that contribute powerfully by crossbridging to the elastomer.

The most common textile bonding compositions are, of course, the RFL (resorcinol-formaldehyde-latex) dips used to adhere tire carcass rubber to cord reinforcing members. A typical RFL composition calls for a latex terpolymer of styrene-butadiene-vinyl pyridine as well as the resorcinol and hexamethylene tetramine. Some natural rubber latex or regular SBR latex may also be added. The R-F resin which forms in situ is able to react chemically with rayon by methylol etherification of the cellulose's hydroxyls. Similar reaction is likely with the amide groups of nylon fibers as shown in Figure 7.

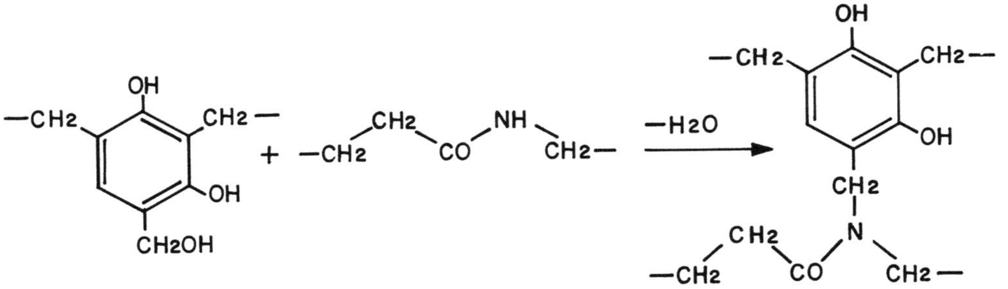

Figure 7. Resorcinol - Formaldehyde Reaction with Amide Linkage.

Resorcinol monomer is a good swelling agent for oriented nylon 6 and 66 polymers. This could enhance the accessibility of the amide linkages for R-F reaction.

Rayon and nylon are the main fibers with which RFL dips function well directly. For polyester and glass, the textiles require a special treatment, usually involving a polyisocyanate and a reactive silane respectively, to afford appropriate levels of RFL adhesion. In the glass-silane case, the mechanistic contribution is almost surely one of strong adsorption and attachment through the silane's organofunctional end with reactive R-F species.

The mechanism with polyester is more open to conjecture; many investigators believe effective pretreatments involve specific penetration or swelling of the polyester's amorphous regions as well as chemical networking capability.

At the elastomer end, it is reasonable to draw a closer analogy with rubber-to-metal bonding agents. The RLF's latex terpolymer is a diene rubber, and is likely to be vulcanized by any sulfur or accelerators that diffuse over from the carcass rubber. It has further been conjectured that the pendant pyridine rings of the same terpolymer should be attacked by the reactive R-F species. Thus, a total chemical network, but involving two or more kinds of crosslinking or covalent bonding reactions, is plausible.[7,8]

Some indirect evidence in support of this is shown in Figure 8 where some typical RFL dips were prepared at constant solids but with varying ratios of resorcinol (plus formaldehyde donor) to latex terpolymer. Boundary conditions were no latex at all at the left extreme and no R-F at the right limit. In keeping with the purported mechanism, maximum vulcanization adhesion between nylon fabric strips and SBR was achieved at a critically balanced ratio. (The fabric bonding and testing procedures are described in an appendix to this paper.)

Additional evidence for sulfur's role was provided by a paper two years ago by K-D Albrecht[9] of Bayer. With both nylon and rayon, RFL dips furnished markedly improved adhesion as the level of sulfur was increased in the carcass compound over the range from one part to three parts. This is shown in Figure 9.

More indirect support comes from the observation that at least one non-sulfur vulcanization system for NR does not result in adequate adhesion to RFL treated textiles. When rubber was crosslinked with a diisocyanate-nitrosophenol adduct (or "diurethane" system),

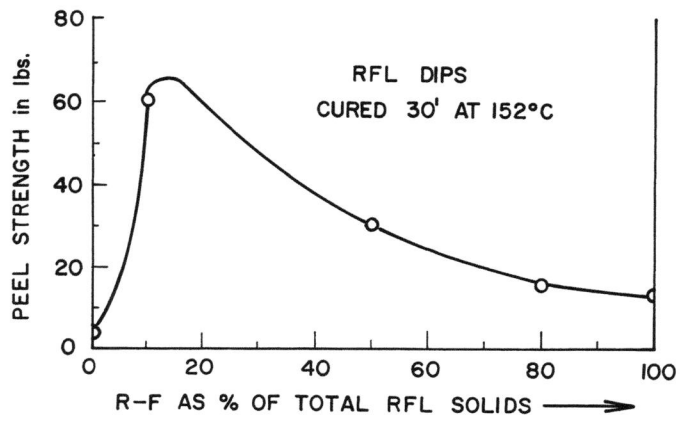

Figure 8. Nylon to SBR.

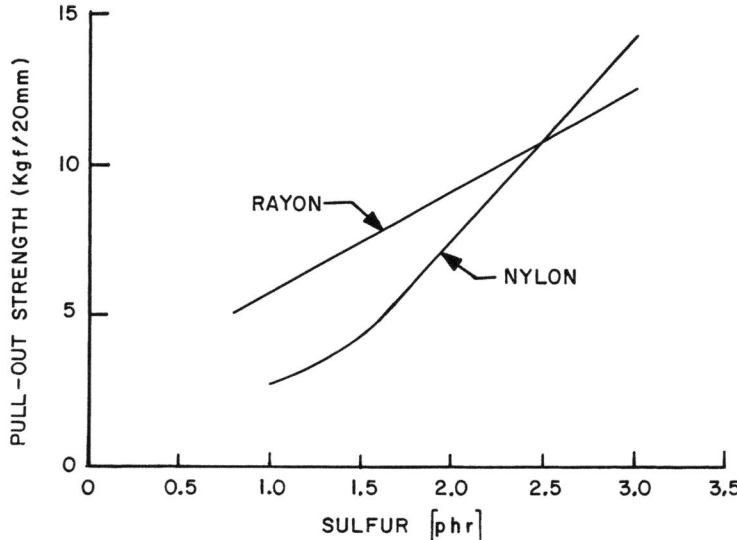

Figure 9. Rubber Adhesion to RFL-Treated Cords (after Albrecht).

the attendant poor adhesion was attributed to nondiffusion, this, in turn, being due to large molecular size and/or too rapid reaction of the diurethane agent with the carcass stock. This being the case, it might be expected that addition or sulfur to the diurethane vulcanizing compound would enhance its adhesion to RFL treated cord. This, in fact, is what we observed (Figure 10). Only the point corresponding to the totally sulfur-free vulcanizate showed interfacial failure at rupture. The other points were all carcass-tearing bonds.

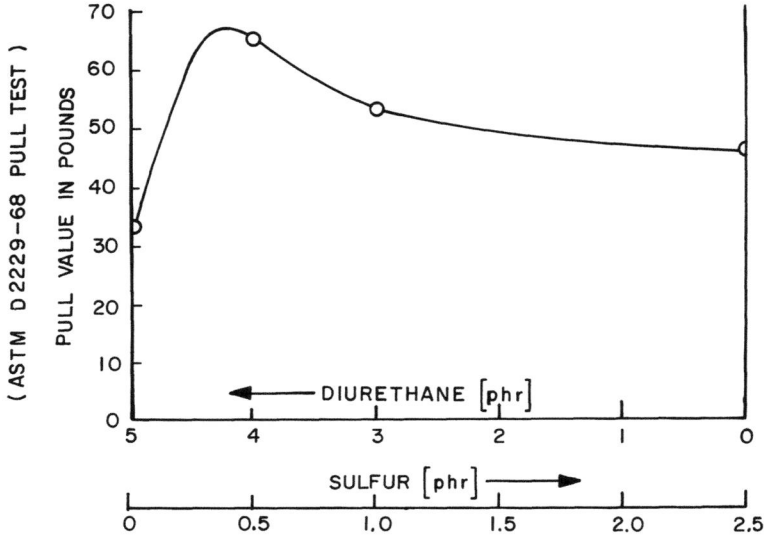

Figure 10. Bonding of RFL-Treated Glass to Diurethane Vulcanizates of N.R.

The diurethane vulcanization chemistry is even more interesting in the context of bonding mechanisms since certain of its rubber formulations exhibit self-adhering properties. For this reason, it would be instructive to review the diurethane chemistry briefly.[10]

It is based primarily on the ability of aromatic C-nitroso compounds to attach themselves to diene polymers (Figure 11). When diisocyanates are also present, the pendant secondary aromatic amine is attacked with formation of a ureido bridge (Figure 12). Since isocyanates are both toxic and abortively reactive with moisture, it is convenient to prereact them with the o-nitrosophenol so that the compounding agent, which breaks up with heat, is actually a type of diurethane species (Figure 13).

Early in the investigations of the diurethane systems it was noted that improved cure efficiency could be achieved through the

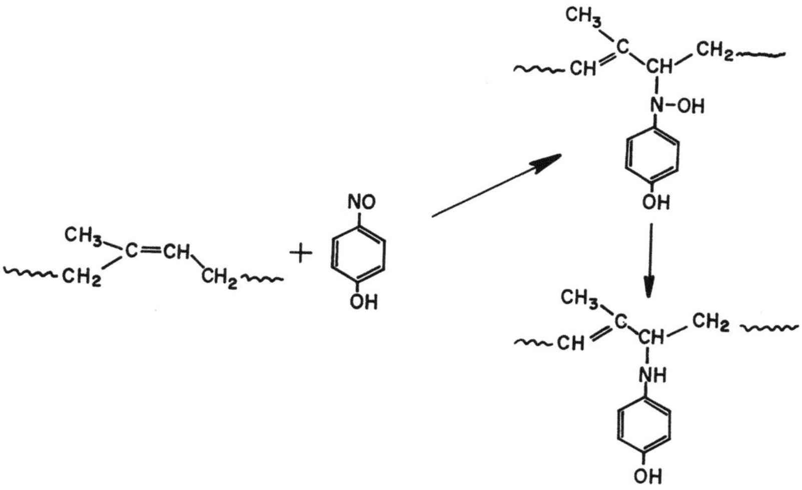

Figure 11. Reaction of Nitrosophenol with Rubber.

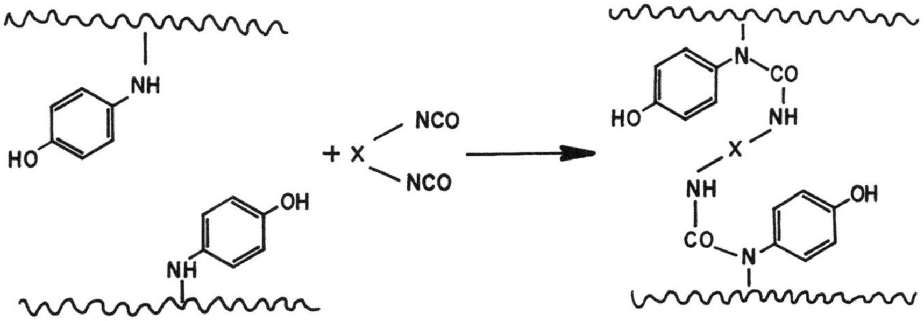

Figure 12. Diisocyanate Vulcanization of Nitrosophenol -- Modified Rubber.

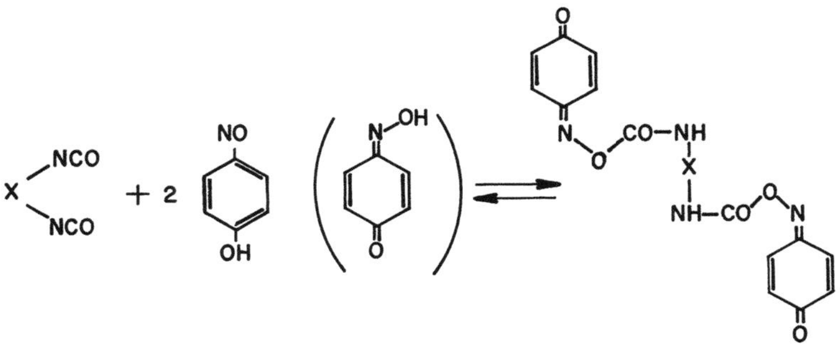

Figure 13. Preparation of Diurethane Crosslinker.

use of excess polyisocyanate, for example, the diurethane of nitrosophenol and toluene diisocyanate with additional free TDI. Rubber compounded and vulcanized in this way was found to have direct adhesional affinity for metals and textile yarns, a capability not shown by ordinary sulfur vulcanizates except in the case of brass.

In quantifying and elucidating this sort of adhesion with polyester fabric, the data of the next graph were obtained (Figure 14). DUT is an abbreviation for the nitrosophenol-TDI diurethane. The polyisocyanate was one commercially available from phosgenation of low molecular weight aniline-formaldehyde condensates. The numbers on each curve correspond to the DUT level in each compound in parts per hundred of rubber (p.h.r.). All of the polyester samples were pretreated with the same concentration of polyisocyanate solution.

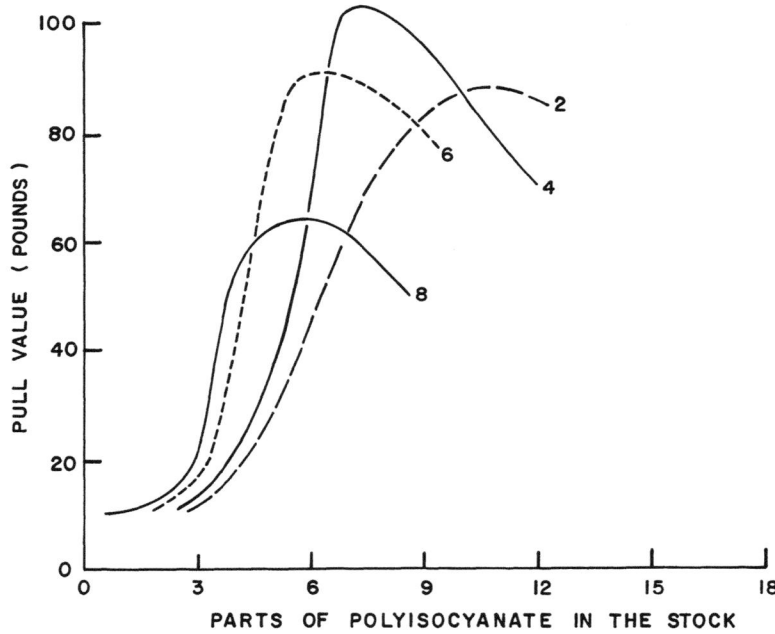

Figure 14. Natural Rubber Vulcanized with DUT-Polyisocyanate; Direct Bonding to Polyisocyanate-Treated Polyester Fabric.

Most of the experimental points for the upper regions of the curves represented excellent adhesion with rubber tearing bonds. A control compound with an ordinary sulfur vulcanization system plus free polyisocyanate developed no adhesion at all to the polyisocyanate treated fabric.

Figure 15 shows how similar adhesion effects were demonstrated to unprimed steel coupons. With the metal, it was possible to promote good adhesion by increasing the level of free polyisocyanate in the rubber stock beyond five parts per hundred. Again, without the presence of the key diurethane, no substantial adhesion could be promoted regardless of the free polyisocyanate level.

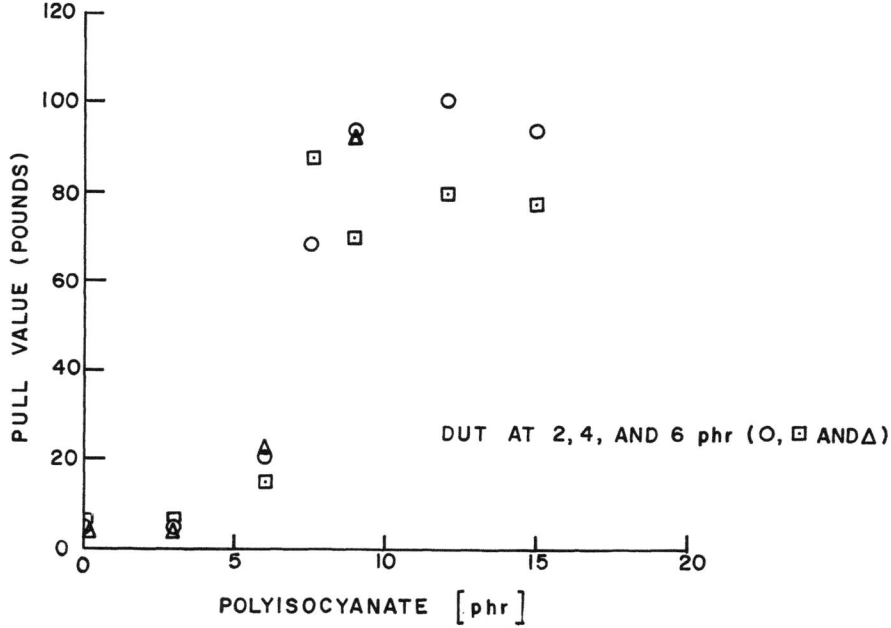

Figure 15. Diurethane Vulcanizates to Unprimed Steel.

It is reasonable to attribute this type of adhesion to thermally induced adsorption of the aromatic polyisocyanate on the metal surface, with linkages from the same isocyanate condensate simultaneously attaching to the elastomer through the diurethane vulcanization network.

No discussion of self-bonding elastomers and their possible adhesion mechanisms would be complete without reference to the most widely used system in this category. This is the approach of elastomer additives comprised of silica, resorcinol and formaldehyde donor (or "SRH" if hexamethylene tetramine is the formaldehyde carrier). In many practical cases, such additives are able to promote enough adhesion to textile fibers or even metal substrates that use of a primer or tie-coat proper can be obviated.

In the applications of SRH additives for typical industrial molded or extruded elastomers, there has been considerable study of formulating variables which has thrown some light on the mechanism. Chemisorption or methylol reaction with the non-elastomer substrate is, of course, probably similar to that of RFL dips, assuming that a significant level of resorcinol-formaldehyde condensate can find its way into the interface.

In the elastomer domain, zinc oxide is important and a fatty acid such as stearic is generally included in the formulation. Solubilization of the zinc oxide and subsequent reaction of the metal with resorcinol and accelerators appear to be preliminary steps to the actual R-F condensation. Lack of kinetic response through the silica suggests that it may contribute to pre-adsorbing the rubber-insoluble reactants and by uniformly distributing them without actually migrating or carrying them.[11]

Another system that has provided useful information for mechanistic speculation is a proprietary adhesive that has been under investigation recently in the Hughson laboratories. This is a solvent-based formulation with considerable versatility for bonding various textile fibers to the important industrial elastomers. Its principal ingredients include a mixture of di- and tri-functional polyisocyanates, a crosslinker capable of rapid reaction with elastomers at their vulcanization temperature, a halogen-containing film former, and carbon black.

Figures 16 and 17 show the relationship between the amount of adhesive applied to strips of nylon and polyester fabric and the peel strength levels that were achieved after vulcanization to a standard SBR stock. These results tell us little of mechanism but they do show the critical dependency of peel values over a relatively narrow range of adhesive add-on.

For purposes of comparison, a similar study was made for the same nylon fabric and SBR elastomer using a typical RFL formulation (Figure 18). The add-on versus peel strength curves were not greatly dissimilar.

The next two experiments were designed to show the vital roles played by both the crossbridging agent and the mixed polyisocyanates in providing maximum peel strength values.

For Figure 19, increasing levels of elastomer reactant (or crossbridging agent) were accompanied by proportionate reductions in the isocyanate level, so that total adhesive solids (and add-on) were reasonably constant. The peel strength data for increasing polyisocyanate levels were also determined after dipping the fabric strips in formulations of roughly constant non-volatile contents (Figure 20).

ELASTOMER-TO-TEXTILE BONDING

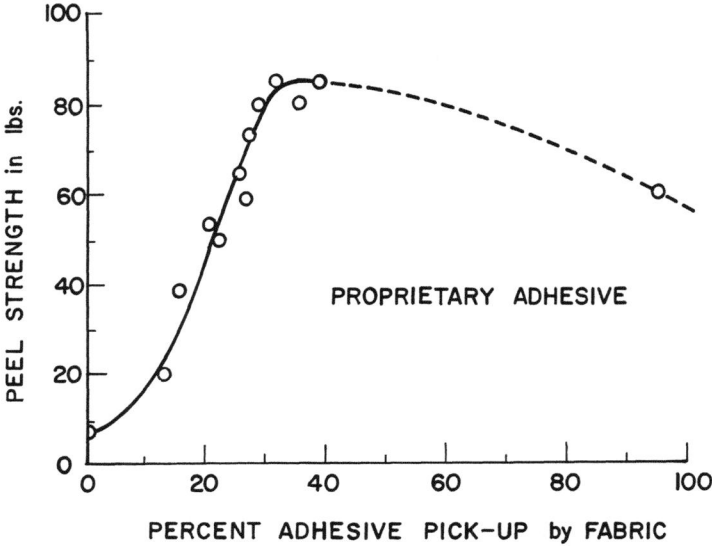

Figure 16. Nylon to SBR.

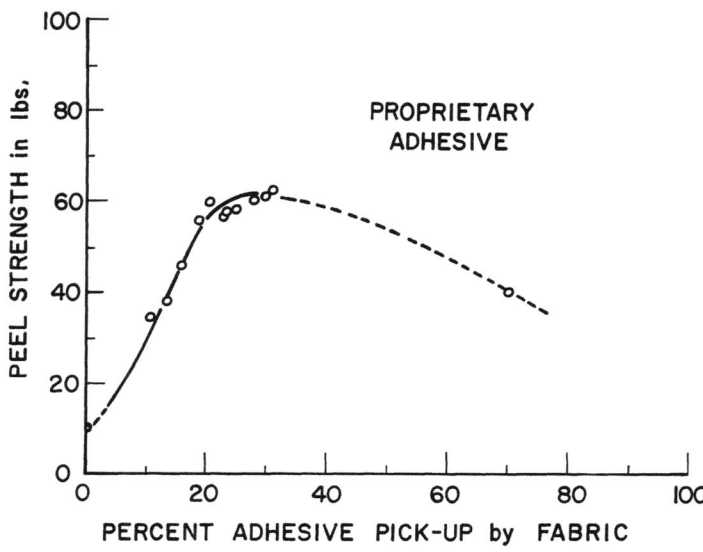

Figure 17. Polyester to SBR.

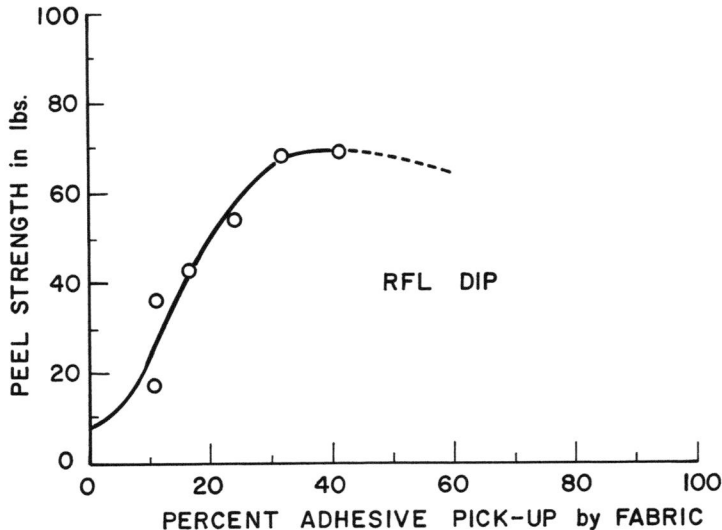

Figure 18. Nylon to SBR.

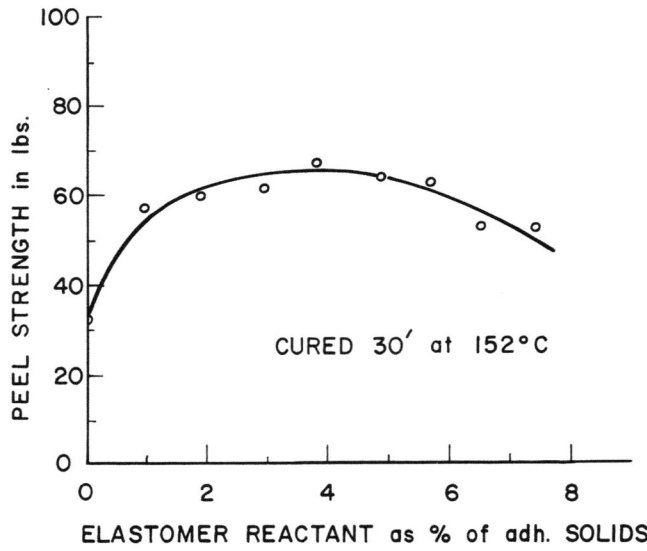

Figure 19. Nylon to SBR.

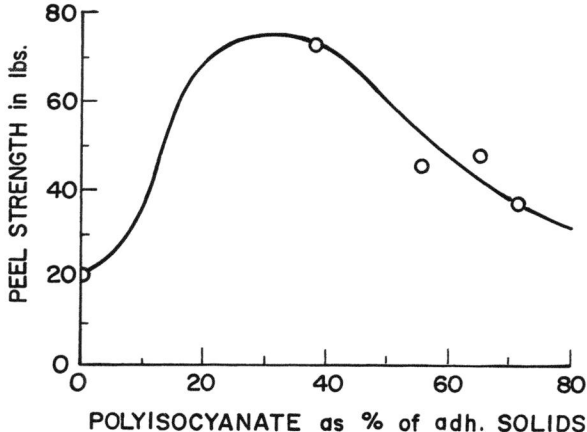

Figure 20. Nylon to SBR.

Each of the curves showed an optimum, indicative of critical contribution from both reactive species. It is reasonable to assume that the mixed polyisocyanates exerted their special influence at the nylon end, while the crossbridging agent was forming linkages through the SBR interface. These progressive changes in adhesive make-up were responsible also for shifts in the mode of failure as the nylon-SBR composites were ruptured. The boundary formulation free of polyisocyanates resulted in cement-to-fabric failure when its laminate was pulled. Conversely, the experimental adhesive devoid of crossbridging agent clearly resulted in cement-to-elastomer failure.

The contrasting modes of failure were photographed for Figure 21. Adhesive retention by the fabric caused the glossy blackness of the right hand sample. "Prop. Add." is an abbreviation for proprietary additive and refers to the crossbridging agent.

The final data of interest for the proprietary bonding agent were in regard to peel strength development as a rate process. Here we compared the solvent based adhesive again with a typical RFL dip for laminating nylon strips to SBR (Figure 22).

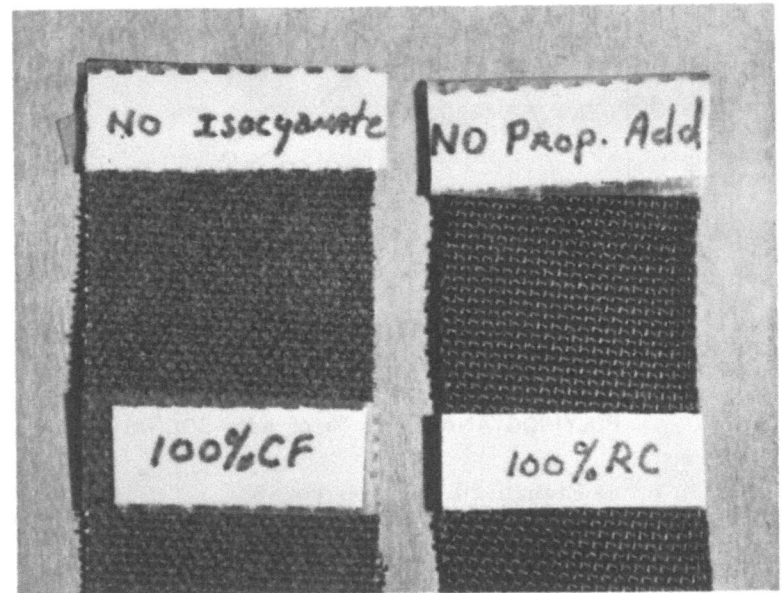

Figure 21. Nylon to SBR Modes of Failure.

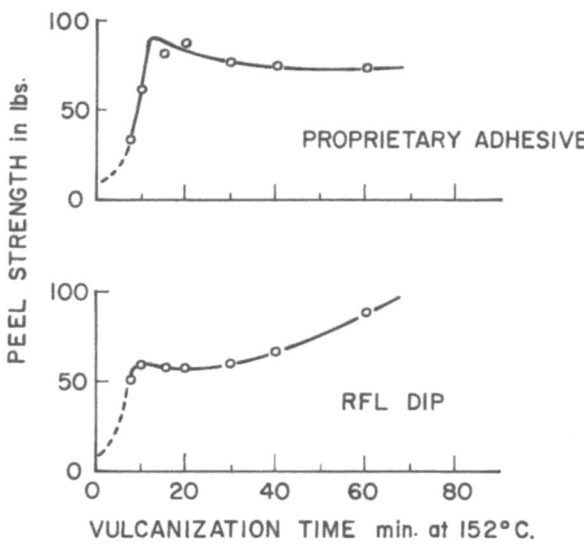

Figure 22. Nylon Fabric to SBR.

Peel strength developed very rapidly, defined a slight peak, and then leveled out. This is characteristic of vulcanization bonding formulations which contain the strong elastomer-reactive ingredient. In fact, the total curve shape was in rough parallel to changes in the bulk tear strength property of the vulcanizing SBR. This implies that the adhesion processes occurred more rapidly than the stock itself vulcanized.

With the RFL adhesive, a somewhat different kinetic profile was recorded. Certainly, a similar peak in peel strength was observed after around 10 minutes of vulcanization. But pull values were still increasing significantly after 40 or 60 minutes. This slower buildup in peel strength can be attributed to sulfur migration from the SBR into the RFL adhesive layer. The net physical effect could be a smoother gradient of crosslink densities for better stress distribution prior to rupture -- a process not unlike annealing.

It seems appropriate to close this discussion of textile-to-elastomer vulcanization bonding with the same type of schematic portrayal as was shown earlier for elastomer-to-metal adhesion (Figure 23). Not all of these phenomena or processes can be operative for any single bonding system, but each would appear to be a possibility in one or more of the material approaches that have been cited.

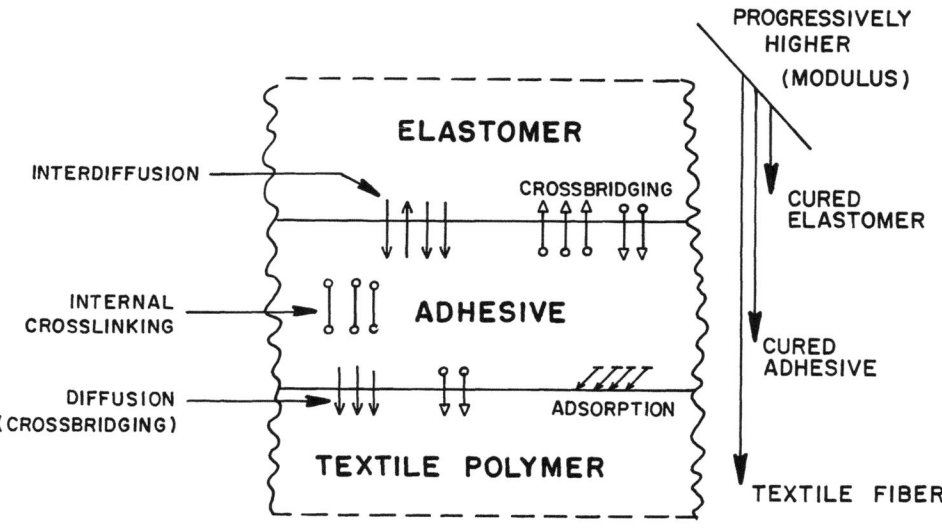

Figure 23. Elastomer-to-Textile Bonding Processes.

For vulcanization bonding generally, processes of migration and chemical reaction appear to be of greatest significance in the mechanisms of adhesion.

REFERENCES

1. F. H. Sexsmith, Adhesives Age 13, No. 5, 21, No. 6, 31 (1970).
2. F. H. Sexsmith, Mechanisms of Elastomer-to-Metal Adhesion, paper delivered at Gordon Research Conference on Elastomers, July 1973.
3. S. Buchan, *Rubber to Metal Bonding*, Palmerton Publishing Co., Inc., New York (1959).
4. A. E. Hicks and F. Lyon, Adhesives Age 12, No. 5, 21 (1969).
5. C. Frank, G. Kraus and A. Haefner, Ind. and Eng. Chem. 44, 1600 (1952).
6. F. H. Sexsmith and R. D. Sites, "The Kinetics of Rubber-to-Metal Bonding", Conference International Du Caoutchouc, (proceedings, page 53), Paris (June 1970).
7. W. C. Wake, Rubber Ind. 10, 242 (1973).
8. C. R. Parks and R. J. Brown, J. Appl. Polym. Sci. 18, 1079 (1974).
9. Klaus-Dieter Albrecht, Rubber Chem. Technol. 46, 981 (1973).
10. C. S. L. Baker, D. Barnard and M. Porter, Rubber Chem. Technol. 43, 510 (1970).
11. N. L. Hewitt, Rubber Age 104, 59 (1972).

APPENDIX

Preparation and Testing of Fabric-Elastomer Composites

1. Elastomer

A compounded styrene-butadiene recipe, based upon the 1500 series (International Institute of Synthetic Rubber Producers Inc.) was freshly milled on a 6" laboratory mill, sheeted off to 125 mils (0.32 cm), and two 5" x 7" (12.7 cm x 17.8 cm) pads were made. An area of approximately one square inch along the top of each pad was blocked off with masking tape to provide a tab for attaching the grips of the peel testing machine.

2. Resorcinol-Formaldehyde Latex Adhesive

The following formulation was employed as a typical RFL dip.

A. Deionized water   9.6 (% by weight)
    Aqueous KOH (5%)   9.6
    Pennacolite Resin R2170 (75%)[1]   4.1

B. Goodrite 2528 (40%)[2]   71.6

C. Formaldehyde (37%)   1.7
   Deionized water   3.4

Part A was added to B, and then C to (A+B). The final dip, at 32% solids, had a pH of approximately 10.

[1] Pre-condensed resorcinol-formaldehyde resin, Koppers, Inc.
[2] Styrene-butadiene-vinyl pyridine latex, B.F. Goodrich Chemical Co.

3. Fabric

All fabric specimens were cut (5" x 7"), weighed, dipped into the adhesive and hung to dry.

	Nylon*	Polyester*
Weight	15 oz./sq. yd.	15 oz./sq. yd.
Thread Count	48 x 28	48 x 28
Permeability	3-8 cfm	3-8 cfm
	non-scoured	non-scoured
	non-heat-set	non-heat-set

*Source: National Filter Media Corp.

The add-on was determined in the usual way,

$$\frac{\text{wt. of dried, dipped fabric} - \text{wt. fabric}}{\text{wt. of fabric}} \times 100.$$

4. Bonding and Testing

After drying and re-weighing, the adhesive coated fabric was placed between the two pads of rubber. This assembly was then vulcanized in a compression mold (same as used for ASTMD429-B). After vulcanization (e.g., 30' at 307°F) the sandwiched pad was cooled and 1" wide specimens were cut for peel testing.

The 1" x 5" bonded samples were peeled at 180° on a Scott Tester or similar machine at 2"/minute. The peel value was recorded and the type of failure was noted [e.g., 80% R, 15 RC, 5 CF is failure within the stock (80%), some failure between the rubber and the cement (15%), and a slight amount of failure (5%) between the cement and fabric].

On Bonding Rubber to Glass

A. Ahagon, A. N. Gent and E. C. Hsu

Institute of Polymer Science

The University of Akron

Akron, Ohio   44325

   This paper deals with the relationship between strength of adhesion and density of chemical bonding at the interface between an elastomer and a glass substrate. Reactive sites of varying surface density were obtained by treating glass plates with mixtures of vinyl and ethyl silanes in varying proportions. A layer of polybutadiene or poly(ethylene-co-propylene) containing dicumyl peroxide (DCP) was then pressed into contact with the treated glass surface and crosslinked in situ by heating to the decomposition temperature of DCP. Interfacial chemical bonding via vinyl groups (when present) in the silane-treated surface was assumed to occur simultaneously with the crosslinking reaction. Evidence for this was obtained from near-infrared spectroscopy of model systems, using finely-divided silica in place of glass.

   Measured strengths of adhesion are reported for rubber-glass combinations prepared in various ways, and compared with results obtained using untreated glass. They are also compared with the cohesive rupture strength of the elastomers and with theoretical estimates of adhesive strength for unbonded and covalently-bonded interfaces.

INTRODUCTION

In order to improve the performance of composites of silicate glass and polymers, the glass is commonly treated with silane coupling agents or adhesion promotors. Because of their dual functionality, these silanes are capable of chemical bonding both with the glass and the polymer, and improvements in performance with silane-treated glass are generally attributed to such chemical bonding. Unfortunately, however, a direct characterization of adhesion has not been attempted in the past, and interpretations of the role played by interfacial bonding are rather divergent.[1-3] An examination of the role of chemical bonding in adhesion thus seems desirable to elucidate the true nature of the system.

The existence of chemical bonds at the glass-polymer interface has been sometimes questioned. Although spectroscopic and extraction techniques have suggested that chemical bonding occurred when appropriate silanes were used[4-7], definite proof is lacking. Moreover, it is not known to the present authors whether any relation has been deduced between the extent of interfacial chemical bonding and the strength of adhesion. An attempt has therefore been made to produce adhesive joints with varying degrees of interfacial chemical bonding and to determine their strength under quasi-equilibrium conditions, when the elastomer is completely elastic and free from viscous effects.

EXPERIMENTAL

(i) Interfacial Bonding[8]

Chemical reactions at the glass-silane-polymer interfaces were studied using finely-divided fumed silica (Cab-o-sil, grades M-5 and EH-5, Cabot Corporation) as a model for the glass, and a simple rubbery polymer (Vistalon 404, a random copolymer of ethylene and propylene, ca. 50/50 molar ratio, Exxon Chemical Company). The silanes chosen for this experiment were vinyl dimethylethoxysilane and trimethylethoxysilane. These silanes, having a single hydrolysable group, were convenient for studying chemisorption because they did not form a polysiloxane layer on the silica surface, whereas other silanes having three hydrolysable groups appeared to do so.

Silica powder was pressed to form a pellet, and then treated with the silanes for periods of 10 to 20 hours at room temperature. The liquids were then decanted and the pellets were dried under vacuum. Material remaining on the silica surface can be regarded as chemisorbed, because the drying process removes substantially all material physically adsorbed on the surface.[9]

The treated silica was then examined by near infrared spectroscopy (Fig. 1). Terminal vinyl groups exhibit characteristic absorption peaks at 1.6 and 2.1 μm[10-12], and the latter, somewhat stronger, peak was used as a measure of the glass-silane reaction. The extent of surface coverage of the silane on the silica was calculated from the surface area of the silica and the intensity of the vinyl peak to be approximately one molecule per 100 $Å^2$ of surface. This figure is similar to that reported by Evans and White.[9] It was found to be independent of the particle size of silica. Moreover, for treatment times of over ten hours, the extent of surface coverage did not change. The surface coverage obtained here can thus be regarded as the equilibrium coverage under present treatment conditions.

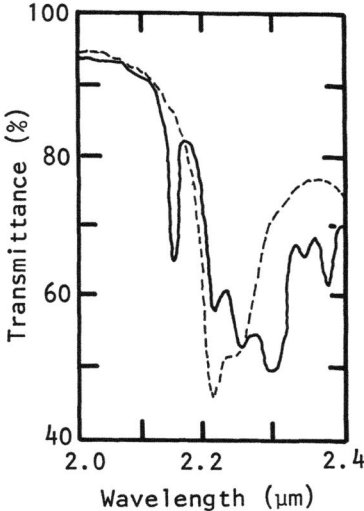

Figure 1. NIR Spectra of Silica (broken curve) and Vinylsilane-Treated Silica (full curve)[8].

Silica was also treated with mixtures of vinyl and ethyl silanes in varying proportions. In this case, the surface coverage by vinyl silane was found to be in approximately the same proportion as the amount of vinyl silane in the mixture. That is, treatment with a 50/50 (vinyl/ethyl) mixture yielded 0.5 vinyl groups per 100 $Å^2$ silica surface, and with 10/90 (vinyl/ethyl) mixture 0.09 vinyl groups per 100 $Å^2$. Consequently, it is concluded that vinyl and ethyl silanes have about equal reactivity toward silica, and presumably toward glass.

Subsequent reaction between vinyl groups on the silica surface and the polymer was also examined. The rubbery polymer was mixed with 2.7% by weight of dicumyl peroxide, and 0.32% by weight of sulfur, and 50 parts by weight of the silane-treated silica. This mixture was then heated in a press at 160°C for one hour. After this treatment, the vinyl peak in the NIR spectrum decreased in intensity by about 40 per cent. When a similar sample without crosslinking agents was examined, however, substantially the same intensity for the vinyl peak was obtained before and after the heating treatment. This indicates that vinyl groups on the silica surface react during crosslinking of the polymer, probably to yield covalent bonding between the silane and the polymer.

(ii) Adhesion to Silane-Treated Glass[13,14]

The strength of adhesion of elastomers to silane-treated glass was characterized by measuring the work $\underline{W}$ required to peel off a thin elastomer layer, per unit area of the interface (Fig. 2). Measurements of $\underline{W}$ were carried out in this way over wide ranges of tem-

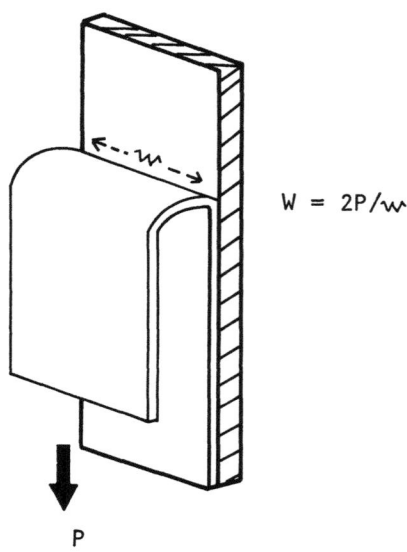

Figure 2. Test Method

perature and rate of crack propagation, and the results were found to obey the WLF rate-temperature superposition principle, as before[15,16]. The observed work of detachment was then extrapolated to very low reduced rates of crack propagation (Figs. 3 and 4). In this way, the strength of the joint was obtained at vanishingly-low rates of detachment, when the elastomer was completely elastic and free from viscous effects. This threshold value, denoted W_o, is regarded as a measure of the intrinsic strength of the joint.

Experimental results are given in Table 1 for adhesive joints having varying degrees of interfacial bonding. They were prepared by treating glass plates with mixtures of ethyl and vinyl silanes in varying proportions to give substrates having different surface densities of vinyl groups, while remaining substantially hydrocarbon in nature. A thin layer of polybutadiene was then applied and crosslinked in situ by means of a free-radical crosslinking agent, dicumyl peroxide (DCP).

Interfacial chemical bonding is inferred to take place during this crosslinking process, between surface vinyl groups and the elastomer, from the corresponding increase in the threshold value of the work $\overline{W}$ of detachment, Table 1. In contrast, surface ethyl groups were found to be inert; no significant changes in adhesion being observed in this case on crosslinking the elastomer layer.

It is noteworthy that the maximum enhancement of strength shown in Table 1, about 40-fold, is comparable to the ratio of the dissociation energies for C-C covalent bonds, about 80 kcal/mole, to that for Van der Waals bonds, 1-2 kcal/mole. Furthermore, the maximum

Table 1

Threshold Values of the Work of Detachment for Polybutadiene Crosslinked in Contact with Silane-treated Glass[13]

Ratio of vinyl to ethyl silane on the glass surface	Work of detachment W_o (j m^{-2})
0 : 1.0	1.3
0.05 : 0.95	3.4
0.1 : 0.9	6.3
0.2 : 0.8	9.5
0.3 : 0.7	13.0
0.5 : 0.5	17.5
0.7 : 0.3	28.5
0.8 : 0.2	41.0
1.0 : 0	51.0

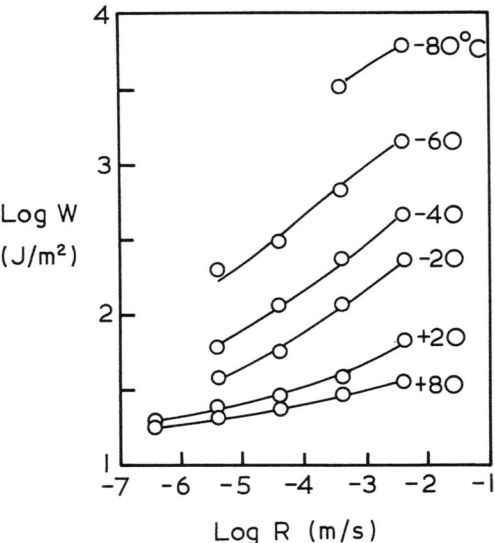

Figure 3. Work W of Detachment of a Polybutadiene Layer from a Glass Substrate Treated with 50 percent Vinyl and 50 percent Ethyl Silane[13]. R Denotes Rate of Peeling.

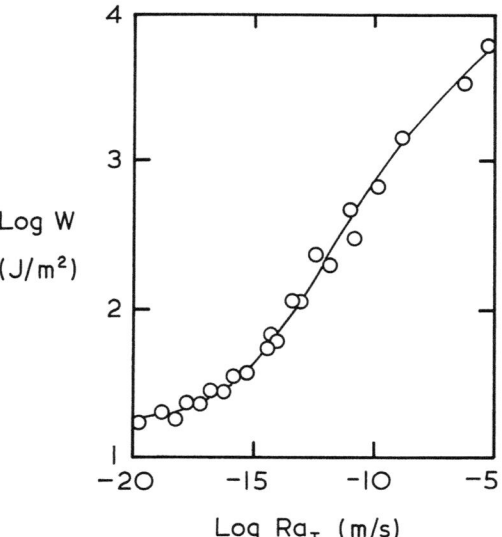

Figure 4. Results shown in Figure 3 Replotted Against the Effective Rate of Peeling Ra_T at T_g (-90°C) of Polybutadiene[13].

joint strength obtained is closely similar to the tear strength of the elastomer itself, about 50 j m^{-2} under zero-rate conditions.[17] Thus, as the substrate changes from a hydrocarbon non-reactive surface (ethyl) to a hydrocarbon surface capable of entering into coupling reactions with the elastomer layer (vinyl) the strength of adhesion increases in the ratio of covalent bond strengths relative to dispersion bond strengths, to reach the same level as the cohesive strength of the elastomer.

Now the elastomer is surprisingly strong, even under zero-rate conditions. This feature has been attributed by Lake and Thomas[18] to the polymeric character of the elastomer: many molecular bonds must be stressed in order to rupture one. In consequence, the observed strength is about 20 times as large as that calculated on the basis of C-C bond strengths alone. A result of this interpretation of cohesive strengths is that the threshold tear strength is predicted (and found) to decrease as the degree of crosslinking is increased. Similarly, it now appears that the adhesion strength of an elastomer to a substrate via covalent bonding at the interface also decreases as the degree of crosslinking (and simultaneously the degree of coupling) increases. Furthermore, the anomalously high strength of adhesion found for an inert substrate, Table 1, about 1.3 j m^{-2} in place of the value of about 0.05 j m^{-2} predicted by surface energy considerations, may now also be attributed to the polymeric nature of the adhering elastomer, in an analogous way to the cohesive strength and adhesive strength under covalent bonding conditions.

We conclude that deformable polymeric adhesives will be unusually strong, either as wetting materials or as covalently-bonded adhesives. (In the latter case, however, the strength will be as much as 40 times greater than in the former.) Under non-equilibrium conditions, for example at high rates of detachment or at low temperatures, the work of detachment is much larger due to dissipative processes in the adhesive, and the difference between different types of bonding or extents of crosslinking is smaller. Indeed, at extremely high rates of detachment or, equivalently, at low temperatures when the adhesive layer becomes glasslike, these differences tend to disappear.

ACKNOWLEDGEMENTS

This work was supported by a research grant from the Engineering Division of the National Science Foundation. The authors are also indebted to Dr. E. M. Dannenberg of Cabot Corporation for helpful suggestions, to Dr. M. W. Ranney and Mr. K. J. Sollman of Union Carbide for the supply of silane coupling agents, to Professor J. L. Kardos and Dr. R. L. Kaas for advice on the catalysis of silane reactions, and to Dr. Y. Kesten for other possible coupling reactions.

REFERENCES

1. P. E. Cassidy and B. J. Yager, J. Macromol. Sci.-Revs. Polymer Technol., $\underline{D1}$, 1 (1971).
2. P. W. Erickson, J. Adhesion $\underline{2}$, 131 (1970).
3. D. H. Kaelble, *Physical Chemistry of Adhesion*, Wiley, N.Y., Chap. 13 (1971).
4. R. L. Kaas and J. L. Kardos, Polymer Eng. Sci. $\underline{11}$, 11 (1971).
5. W. D. Bascom, Macromolecules $\underline{5}$, 792 (1972).
6. J. L. Koenig and P. T. K. Shih, J. Coll. Interface Sci. $\underline{36}$, 247 (1971).
7. O. K. Johanson, F. O. Stark, G. E. Vogel and R. M. Fleischmann, J. Composite Materials $\underline{1}$, 278 (1967).
8. A. N. Gent and E. C. Hsu, Macromolecules $\underline{7}$, 933 (1974).
9. B. Evans and T. E. White, in *Fundamental Aspects of Fiber Reinforced Composites*, ed. by R. T. Schwartz, Interscience, N. Y., p. 177 (1968).
10. W. Kaye, Spectrochim. Acta. $\underline{6}$, 257 (1954).
11. R. F. Goddu, Anal. Chem. $\underline{29}$, 1790 (1957).
12. R. T. Conley, *Infrared Spectroscopy*, 2nd ed., Allyn and Bacon, Boston, p. 246 (1972).
13. A. Ahagon and A. N. Gent, J. Polymer Sci.: Polymer Phys. Ed., in press.
14. A. N. Gent and E. C. Hsu, in preparation.
15. A. N. Gent and R. P. Petrich, Proc. Roy. Soc. (Lond.), $\underline{A310}$, 433 (1969).
16. A. N. Gent and A. J. Kinloch, J. Polymer Sci., $\underline{A-2}$, $\underline{9}$, 659 (1971).
17. A. Ahagon and A. N. Gent, J. Polymer Sci.: Polymer Physics Ed., in press.
18. G. J. Lake and A. G. Thomas, Proc. Roy. Soc. (Lond.), $\underline{A300}$, 108 (1967).

The Microscopy of Polymeric Adhesive Systems

Ronald W. Smith

The B.F. Goodrich Company Research & Development Center

Brecksville, Ohio   44141

When polymeric materials are joined together, either by using an adhesive bond or by direct contact, critical information can be gained by examining the resulting bond with optical and/or electron microscopes.

Two preparative techniques have been proven to be invaluable for the microscopic examination of bonded polymeric materials: the "Ebonite" and "Cyro" methods. By using these methods to harden soft and rubbery materials, it is possible to cut thin sections for either optical or transmission electron microscopy.

Information obtained from the microscopical examination of adhesive bonds can be reduced to two broad categories. First, geometrical considerations involving gross morphological characteristics such as bond thickness and uniformity, surface roughness of substrates and matters relating to flow of materials during forming and curing. Second, compositional consideration such as pigment and polymeric dispersions and interactions between the adhesive solvent and the substrates prior to curing.

INTRODUCTION

To say that adhesion plays an extremely important role in the performance and durability of multi-component polymeric systems is a gross understatement. And even though everyone recognizes the importance of adhesion, it is paradoxical how little attention has been paid in the literature to the subject of the microscopical examination of adhesive systems. I suspect that there are two reasons for the scanty literature on the subject: 1) Adhesion values cannot be "seen" in a microscope. There is no way that the microscope can evaluate the strength of an adhesive bond and generate the numbers that are so important to adhesive technology. Only in a qualitative way can one say that an adhesive bond is good or bad by merely looking at it through a microscope. And the criterion used for this evaluation is whether or not the adhesive bond survived the method of sample preparation; 2) The second reason may be that not very much work is done in this area in the first place. Adhesive bonds are dimensionally very small, and the techniques useful for the general microscopy of bulk specimens become very tedious when applied to micron-sized thin layers of adhesive bonds. Sample preparation is further complicated by the diverse types of substrates one can encounter. Rubbers, plastics, fabrics, metals--all glued together in every possible combination. All microscopists know that complex structures comprised of greatly varying moduli materials are much harder to deal with than similar-moduli constructions.

In his excellent review on the subject of rubber microscopy, J. Kruse[1] did not elaborate on adhesive systems or the preparation of specimens for the study of finished polymeric products. The reason for this, he stated, was that there were too many objects for this type of investigation. Indeed there are, but perhaps some principles and guidelines for the application of microscopy to adhesive systems would be of value to the adhesive chemist who would like to know more about his product.

The objective of this presentation is to show how all types of microscopy--optical and electronic--can be of great value in studying polymeric products with respect to their cement lines, splices, and junctions; and to adhesive systems alone prior to fabrication. Numbers will not be generated, but the importance and occurrence of certain morphological characteristics will be stressed. In addition, various specimen preparations I have found useful will be presented.

PROCEDURES

Basic to the microscopical examination of polymeric systems is the need to cut thin sections, and this is called the science of microtomy. But before soft rubber specimens can be microtomed they must be hardened in some manner. The two methods used in this study are the freezing, or cyro, method, and chemical hardening of unsaturated rubbers with molten sulfur, called the ebonite method.

Cryomicrotomy

In 1920, Depew and Ruby[2] reported a method for cutting rubbers in the frozen state. Since then, the procedure has been refined by Latta[3] who used glass knives, and Leigh-Dugmore[4] who quantified the procedure for measuring carbon black dispersions. Since then, the procedure has been incorporated in ASTM D-2663 "Methods for Determining the Dispersion of Carbon Black in Rubber Compounds". The method involves freezing a small block of the rubber in liquid nitrogen and cutting approximately 2 micron thick sections from the block using a glass knife and a sledge type microtome. Detailed procedures are described in ASTM D-2663.

Ebonite Method

Roninger[5] used molten sulfur to transform rubber to ebonite that could then be polished for microscopical examination of surfaces. Folt and Smith[6] and Smith and Andries[7] reported that this ebonite could be easily cut into ultrathin sections suitable for transmission electron microscopy. It has the advantage of maintaining accurate geometric relationships within the specimen.

A small rectangular block (10x5x3 mm) is excised from the specimen containing the feature of interest and immersed in molten sulfur at 125°C for 24 hours. At the end of this period, the specimen is obtained in the form of a hard ebonite-like solid suitable for ultrathin sectioning in the usual manner. In this study, the rubbers were sectioned in the usual manner with a Reichert OMU-2 ultra-microtome and subsequently examined using transmission electron microscopy.

RESULTS AND DISCUSSION

There are probably several ways to present the data in a paper such as this, and the following outline will be followed:

A. Optical Microscopy
   1. Uncemented Junctions: Non-Black Systems
   2. Uncemented Junctions: Black Loaded Systems
   3. Cemented Junctions
   4. Cements Prior to Application
   5. Analysis of Adhesion Testing
B. Transmission Electron Microscopy
C. Scanning Electron Microscopy

A. Optical Microscopy

In this section, samples have been prepared by freezing the rubber and microtoming it to produce thin sections as described above. There are four exceptions that are noted.

1. Uncemented Junctions: Non-Black Systems

Figure 1 is an optical micrograph of the junction between two identical natural rubber compounds. The substrates were cured together while they were very fresh and complete fusion has occurred, leaving no trace of the boundary between the two. This is as we hope the condition to be for a "perfect" bond. In fact, it is almost axiomatic that the harder a bond, or cement line is to see in a microscope, the better it is.

In contrast to this, Figure 2 shows a distinct bond line, and some flaws, between two identical natural rubber stocks. In this case, the rubber surfaces were exposed to ultraviolet light in laboratory air for 6 hours before they were assembled and cured together. The surfaces, changed by crosslinking and oxidation, have not fused together completely, and some change in refractive indices of the surface rubber makes the junction very visible.

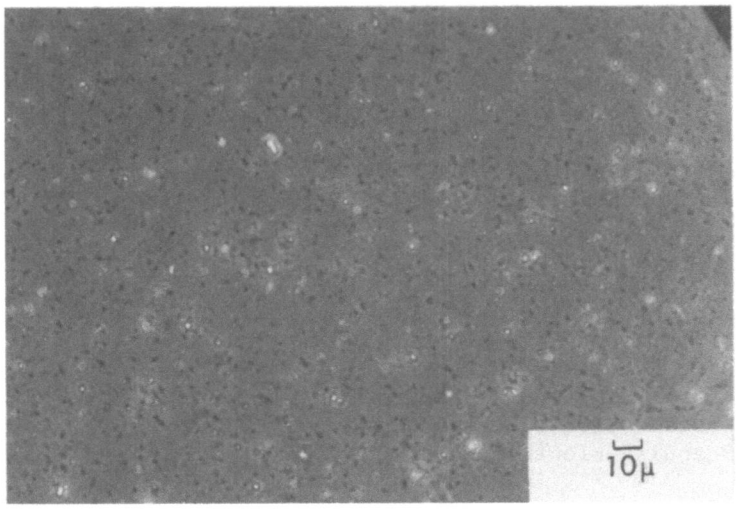

Figure 1. Bond Between Identical Natural Rubber Compounds.

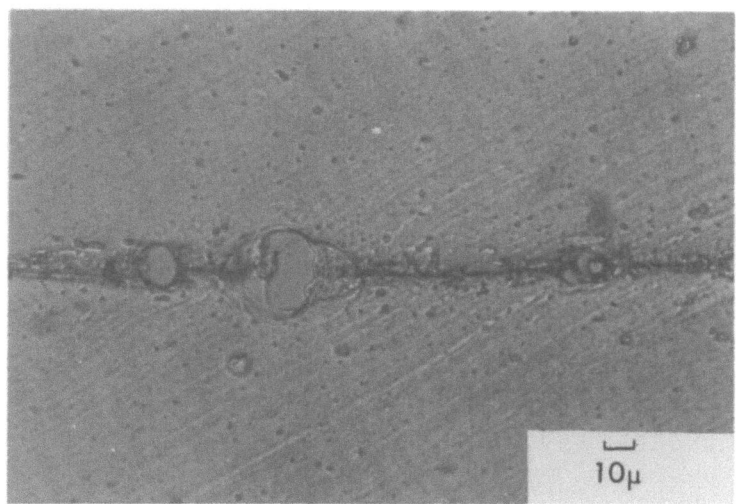

Figure 2. Surfaces Exposed to UV Light Before Assembly.

The phase contrast microscope is invaluable for examining other junctions formed between rubbers differing only in refractive indices. Figure 3 shows the junction between natural rubber and SBR. There is enough refractive index difference to show the bond line in the phase contrast microscope. Bond lines can be observed even when identical rubbers are used if they have been cured differently. For instance, one can see the bond between peroxide and sulfur cured natural rubbers.

Indeed, when different cure systems are used, there may be changes in the bond line rubber other than a mere step-off in refractive index. Figure 4 is a scanning electron micrograph of two gum natural rubber compounds. One was cured with a standard sulfur/accelerator system (positioned on the left side of the micrograph) and a peroxide cured rubber on the right. This bond was exposed by merely cutting the construction with a sharp razor. The cutting characteristics of the surfaces that have been joined are different, and a nubbly-appearing zone about 50 microns wide appears along the path of the junction. The rubber in this area has probably seen two types of curatives, leaving a network that has different physical properties than the bulk rubbers on either side. Figure 5 is a higher magnification view. The inhomogeneous nature of that zone is easily seen to be due to something that is about 5 microns in diameter. These could be zones of extraordinarily high crosslink density. Interactions between phases caused by diffusion of materials is a very important subject area, and other examples will be shown later on this same effect.

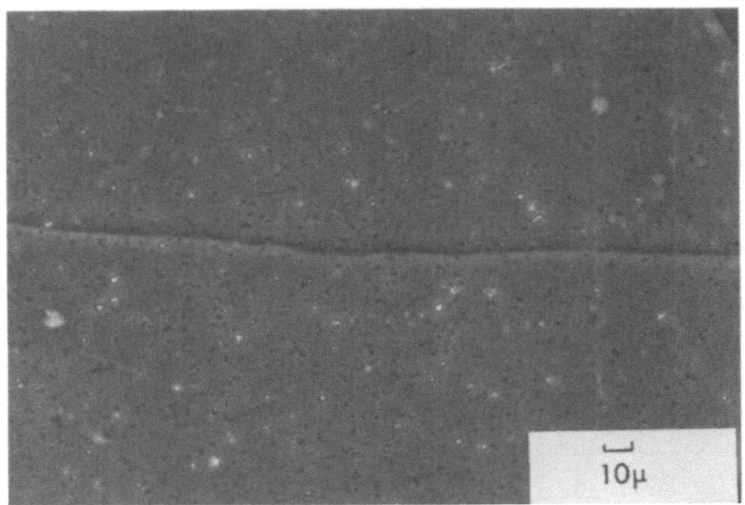

Figure 3. Bond Between Natural and SBR Rubbers.

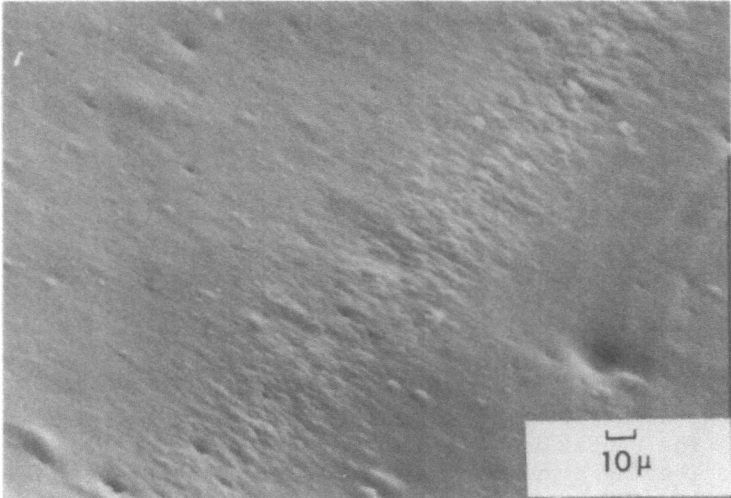

Figure 4. Bond Between Natural Rubbers of Different Cure.

MICROSCOPY OF ADHESIVE SYSTEMS

2. Uncemented Junctions: Black-Loaded Systems

The presence of reinforcing carbon blacks makes the use of the phase contrast microscope impossible, and one must rely upon the pigmentation of the system to produce the optical effects needed to see junctions.

Figure 6 is a high magnification optical micrograph of the junction between two similar black reinforced rubbers. The fact that one can see anything at all of the bond is due to the pigmentation. Carbon black is "wet" with the polymer and fully dispersed before assembly. When a free surface of the compound is formed, either by cutting, tearing, or molding there will always be a very thin layer of rubber at the surface, and when two such surfaces are joined, there is a discontinuity of the random carbon black dispersion, and this is what we see as a bond line. This is a very difficult bond to see, but generally there are areas that can be discerned. Of course, the situation is much different when dissimilar types of carbon black are present. Figure 7 is similar to Figure 6 except that there are two types of carbon black, and the bond line is very visible.

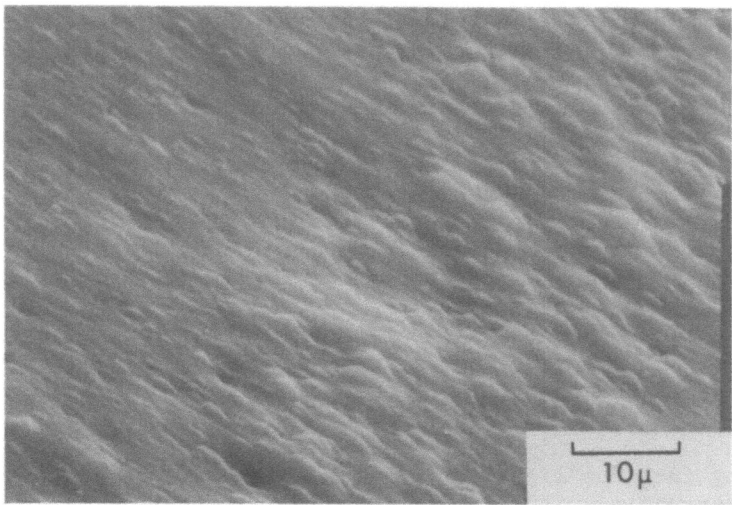

Figure 5. Bond Between Natural Rubbers of Different Cure.

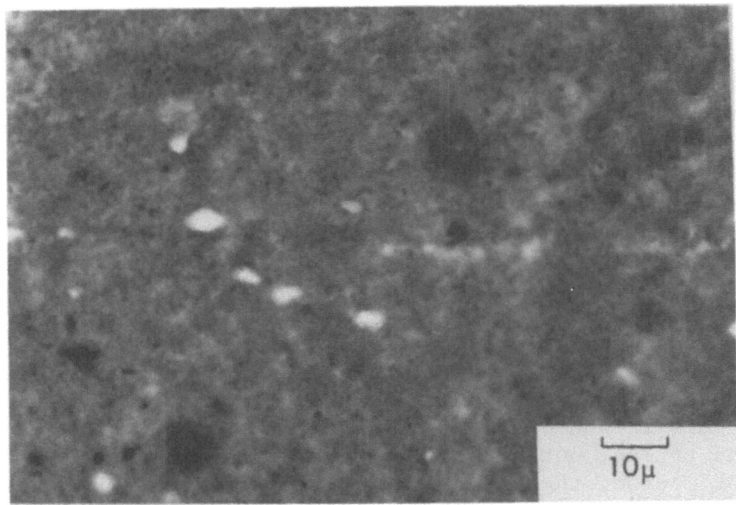

Figure 6. Bond Between Identical Black Loaded Compounds.

3. <u>Cemented Junctions</u>

In this section, we will be dealing with carbon black reinforced systems only, cemented together with carbon black loaded cements. These systems are much easier to see microscopically than the uncemented systems.

Figure 8 is a 40X optical micrograph of a cement line joining two similar compounds. In this case, we are not looking through a thin section, but rather at a surface that was exposed with a sharp razor, and photographed with incident light. The information available here is limited due to the distortions caused by the cutting blade, but one can, by using a reticule in the microscope eyepiece, determine bond thickness, trace the bond line throughout the construction, and check for incipient failure due to extremely poor stock knitting or the presence of foreign materials. In this case, the thickness of the cement line is nominally 40 microns.

In most cases, cement lines exposed in this manner are easily observable, but when the cement used has a loading similar to, or identical to, the substrates, it may be necessary to examine thin sections at higher magnifications.

Using incident light for viewing surfaces such as these is not satisfactory at magnifications above 100X. This is the area where the scanning electron microscope (SEM) is indispensible. Figure 9 is an SEM micrograph of the same surface as shown in Figure 8, but

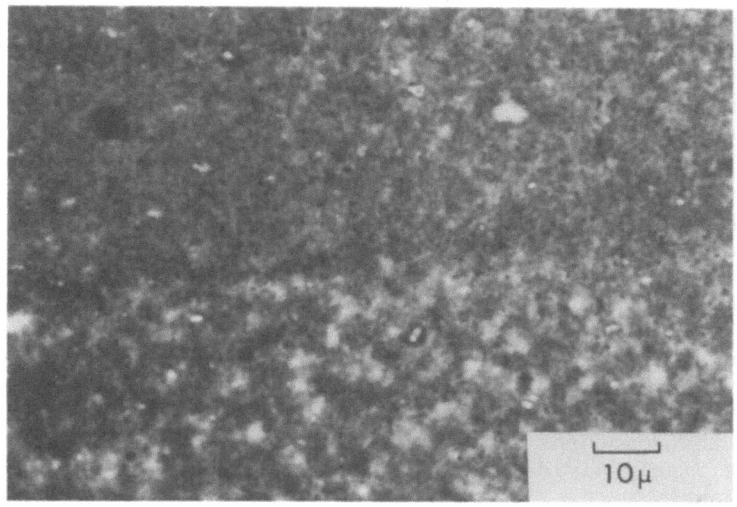

Figure 7. Bond Between Dissimilar Black Loaded Compounds.

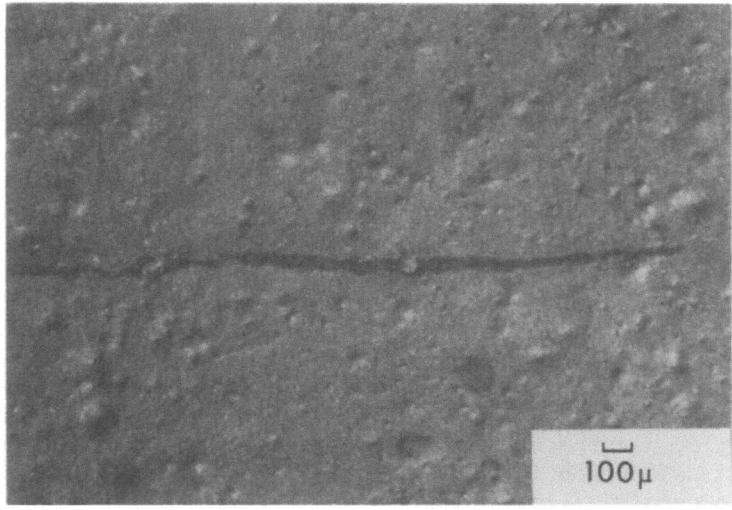

Figure 8. Cemented Bond.

at a magnification of 500X. The cutting marks left by the blade are more pronounced and the higher magnification does not produce any more accurate measurements, but one additional bit of information is available, and that comes from the nature of the cutting marks themselves. The hardness of a black reinforced rubber has a definite effect on the magnitude of cutting marks or abraded surface characteristics: the softer the compound, the more pronounced are these markings. Thus, in Figure 9, we can say that the hardness (hence, modulus) of the compound used as an adhesive is higher than the adherend compound because the cut surface characteristics of the adhesive are finer in detail.

A great wealth of information is available about these cemented systems by examining microtomed thin sections in transmission, thus getting a view of the interior of the system. Figure 10 is a microtomed thin section of two dissimilar compounds joined by a cement having a third composition. The reason for the different colors exhibited by the compounds is due to the light scattering characteristics of the finely divided carbon blacks, and one can use the color effects to analyze the system. Use of these color effects was shown to be valuable in identifying grades of carbon black by Allen[8]. Briefly, the finer the particle size of the black, the more brown will be the color exhibited in transmission optical microscopy. The three different shades of brown shown in Figure 10 tell us that there are three different compounds involved. In addition to this information, one can measure the carbon black dispersion of the adhesive compound, and gain information on the presence of, or the states of dispersion of other compounding ingredients.

Too often perhaps, one thinks of a cemented splice or junction in very simple terms, when in fact, some very complicated and far reaching phenomena can exist. One has to do with the effect of the solvent that is used for the cement. The organic solvents used can, and will, have an effect on the substrate itself, and cannot be ignored. The effect on an uncured substrate by a solvent can be easily demonstrated. Figure 11 is a bond line formed by the junction of two identical black loaded, SBR substrates. Before the bond was assembled, each substrate was "freshened" with benzene. The freshened surface was allowed to dry, the junction assembled, and cured. Upon examination of a section microtomed through the junction, one can see a thin band of unpigmented polymer that had been extracted from the substrates and deposited on the surfaces. This is an extreme example, but the effect can be seen in real systems. Figure 12 is a real situation in which a thin band of unpigmented polymer can be seen running along the top boundary. This system was not solvent freshened, and the effect is due to extraction by cement solvent. These are regions on non-reinforcement, and represent weaknesses in the adhesive system.

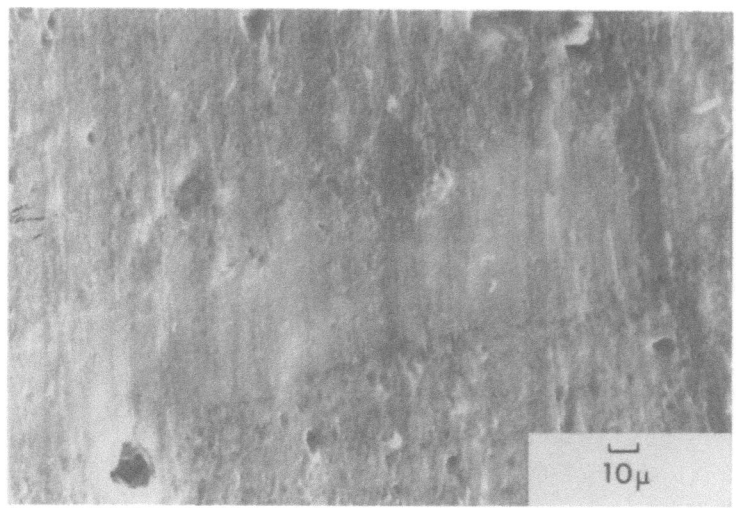

Figure 9. Cemented Bond.

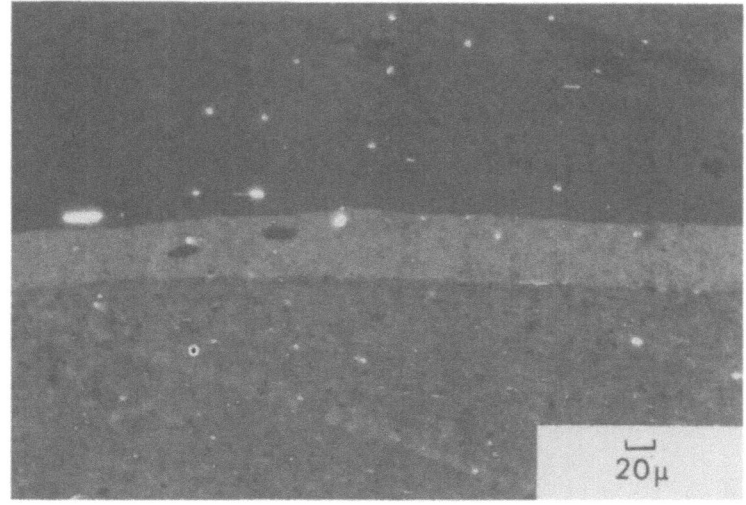

Figure 10. Cemented Bond, Microtomed Section.

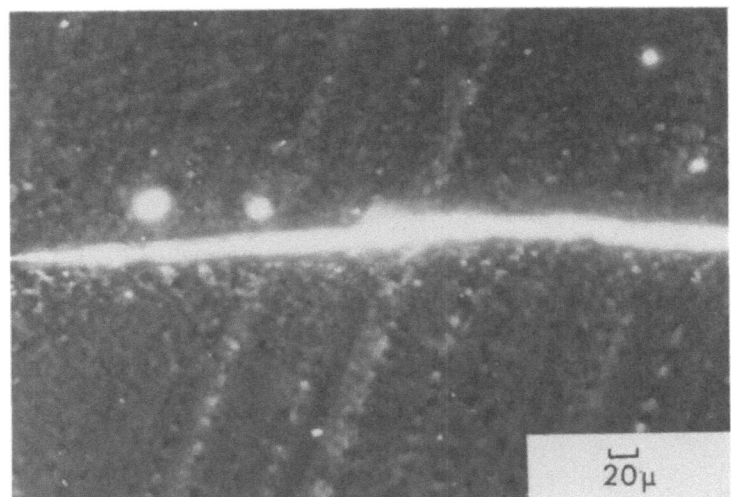

Figure 11. Extracted Polymer at Interface.

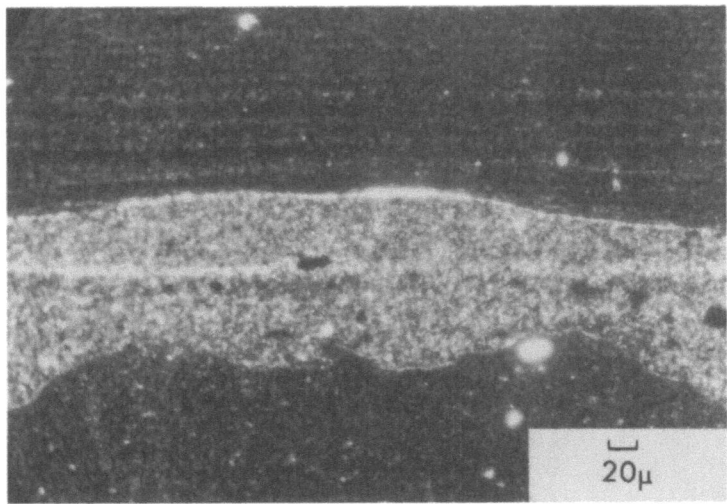

Figure 12. Extracted Polymer at Adhesive/Substrate Interface.

MICROSCOPY OF ADHESIVE SYSTEMS

Microscopy can also provide information with regard to the nature of the substrate itself. Figure 13 shows a very ragged substrate that was formed by a very dull cut-off blade. Such a surface presents problems associated with incomplete wetting by the adhesive or the entrapment of air, either of which would be potential flaws.

As mentioned earlier, every rubber surface carries with it some history of the forming process. One forming process that produces a lot of orientation to surfaces is that of calendering. Figure 14 shows an adhesive system that contains not only poor cement dispersion, but a considerable amount of graininess on the lower substrate attributed to calendering. Even though milling rubber might be less severe in producing sub-substrate orientation, it too, can have a definite orienting effect.

In Figure 15 there is a different type of substrate effect, and again, this can be due to cement solvent, and is probably the precursor to complete extraction of polymer. The normal random carbon black distribution has been effected by the solvent on either side of the cement. Obviously, for this to happen, on both sides of the cement line, the substrates would have had to have been assembled before the cement had completely dried.

Occasionally, one can see effects caused by migration of curing ingredients from one phase to the other. One of these effects is the formation of "nodules" in rubber compounds that has been described earlier.[9] Briefly, these are 10 to 30 micron sized spherical networks of gelled rubber that are caused by the diffusion, crystallization, and crosslinking of mobile curing and accelerating species such as sulfur, accelerator fragments, and zinc salts of fatty acids. Figure 16 shows a line of nodules has formed on both sides of the cement phase in the substrates. The opposite can also occur, i.e. nodules can form within the adhesive phase itself. Figure 17 shows this condition. In the first case, materials have diffused from the adhesive into the adherends, and in the second case, diffusion was from the adherends into the adhesive. The presence of these nodules in rubber compounds has never been related to poor performance of the compound. They seem to be a normal consequence of age and heat history, however, one would speculate that in an extreme situation they might serve as a locus for failure initiation.

Stored adhesive cements are prone to settling upon standing, and when this does occur, pigments drop to the bottom leaving unpigmented polymer at the top. Microscopy can reveal when this has happened to an adhesive before application. Figure 18 shows an adhesive layer that is marbled with unpigmented polymer.

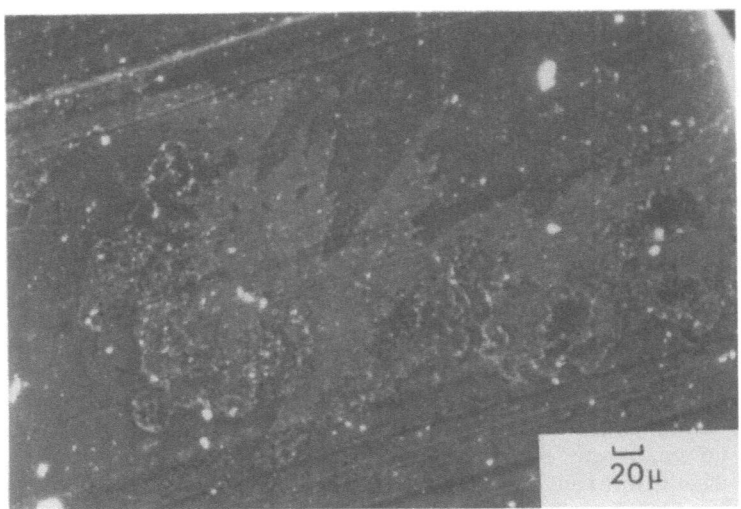

Figure 13. Ragged Substrate.

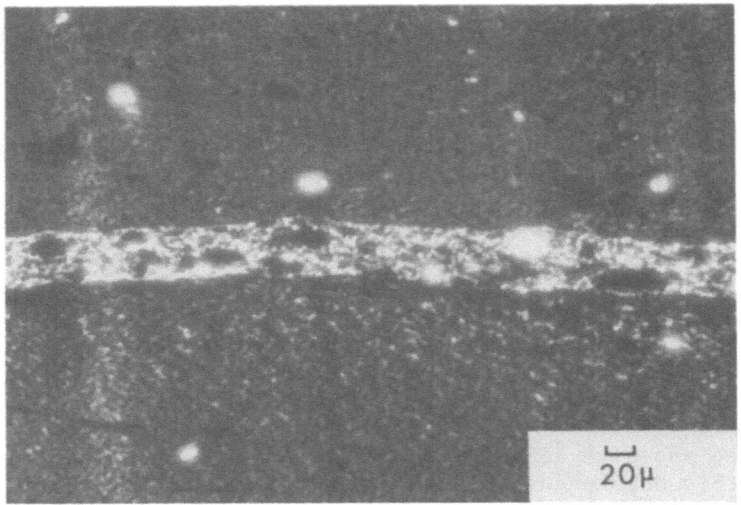

Figure 14. Orientation of Substrate by Calendering.

MICROSCOPY OF ADHESIVE SYSTEMS 303

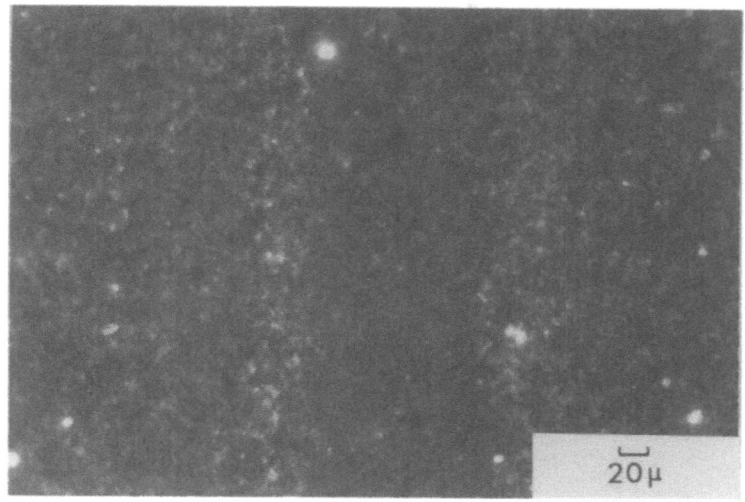

Figure 15. Alteration of Carbon Black Distribution of Sub-Surfaces.

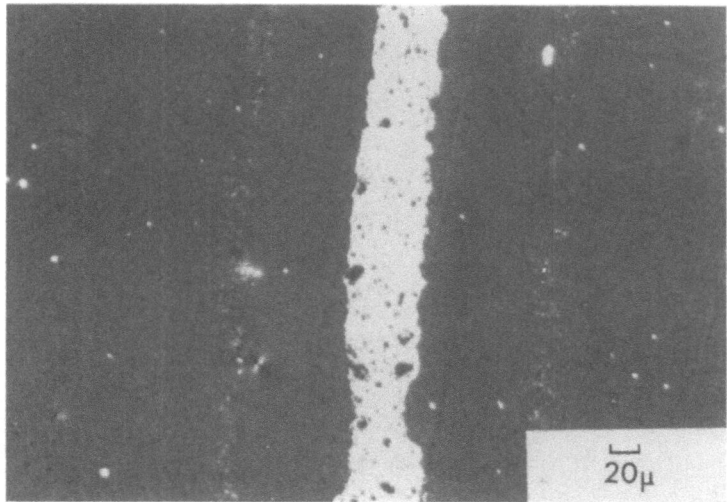

Figure 16. Diffusion of Materials from Adhesive to Substrate.

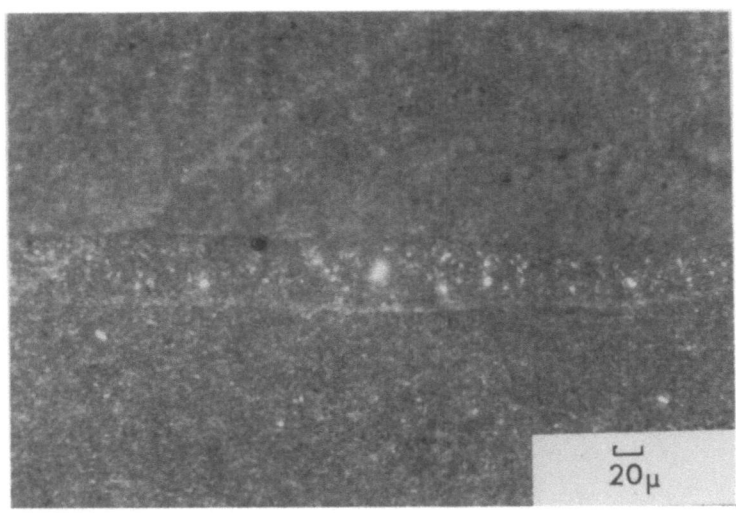

Figure 17. Diffusion of Materials from Substrates into Adhesive.

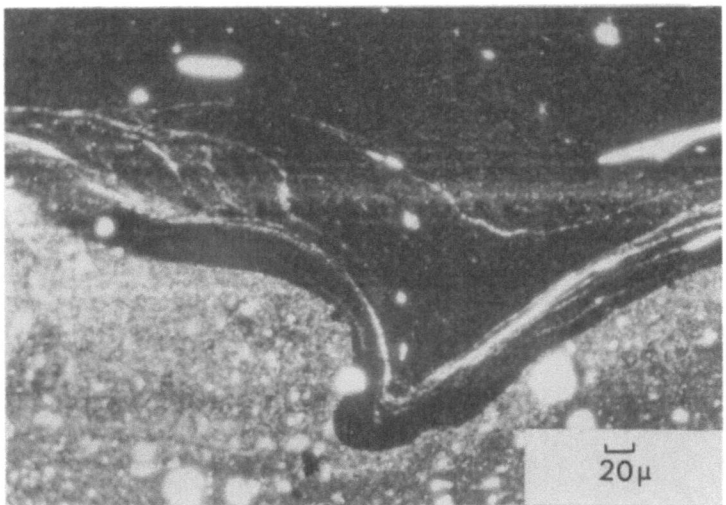

Figure 18. Phase Separation of Cement in Bonded System.

MICROSCOPY OF ADHESIVE SYSTEMS

4. Cements Prior to Application

Just as Figure 18 showed effects of phase separation, Figure 19 also shows a clear polymer phase. In this case, the adhesive cement was not applied to a substrate and cured, rather, the adhesive itself was dried on a glass microscope slide, cured, and then peeled off and microtomed in the usual manner. As long as the adhesive is compounded with curing ingredients, it too can be examined, just as any rubber sample. Using this technique, cements can be assessed for their settling characteristics.

5. Analysis of Adhesion Testing

When an adhesion test is concluded, whether it be a ply-pull, peel, or shear test, one has to know how the construction came apart, and of course, there are several possibilities. Many times it is impossible to tell whether a failure was cohesive or adhesive unless the test specimen is examined with a microscope.

Figure 20 shows a separation propagating along the interface between the adhesive and the adherend. Figure 21 shows another failure that is propagating at a location in one of the adherends so close to the interface that it would be difficult to tell whether or not the failure was interfacial.

Likewise, in adhesion testing of rubber to fabrics, it is not always easy to tell whether a failure occurred between the fabric and the adhesive dip or between the adhesive dip and the rubber. Figure 22 is a microtomed section from the failure region of a rayon fabric/rubber composite. Such a section can clearly show where the failure occurred. In this case, remnants of RFL adhesive dip remain on the rubber, leaving the imprint of the fiber filaments. The obvious conclusion is that the failure occurred between the dip and the filaments. One should not stop asking questions at this point though, and the next one would be was whether or not at a higher level of magnification this failure would still appear to be interfacial. For this type of examination, the transmission electron microscope (TEM) is utilized.

B. Electron Microscopy

In this section, all of the micrographs except Figure 23 have been taken on samples that have been hardened using the ebonite method and microtomed with an ultramicrotome.

The magnification capabilities of the transmission electron microscope, several thousand to several hundred thousand times, provide a step closer in the study of the same subjects discussed in the optical microscope section.

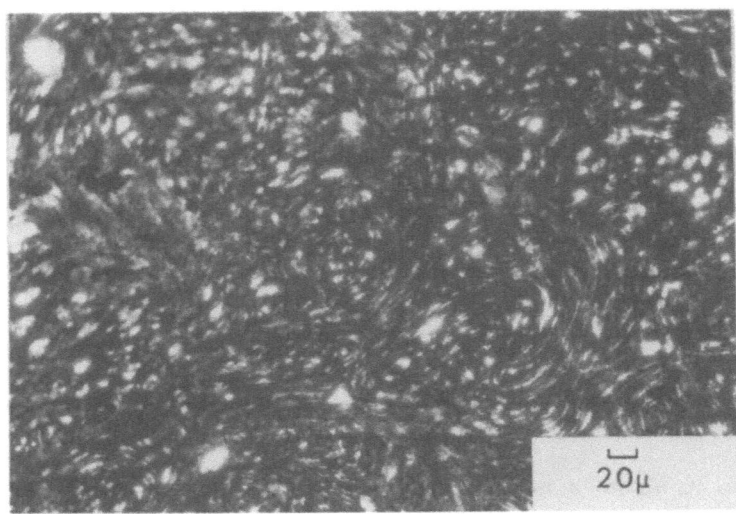

Figure 19. Phase Separation of Cement Prior to Application.

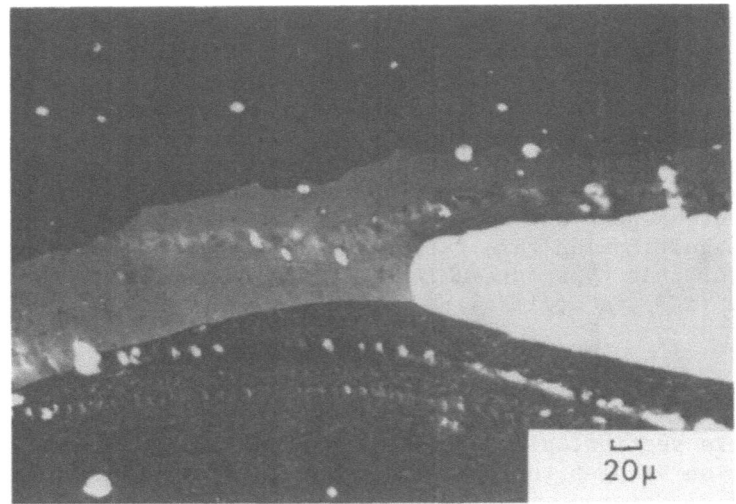

Figure 20. Separation Between Substrate and Adhesive.

MICROSCOPY OF ADHESIVE SYSTEMS 307

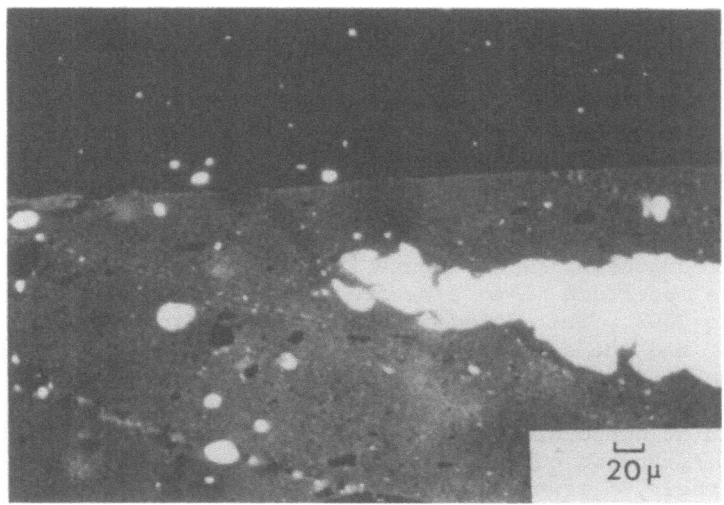

Figure 21. Separation in Substrate.

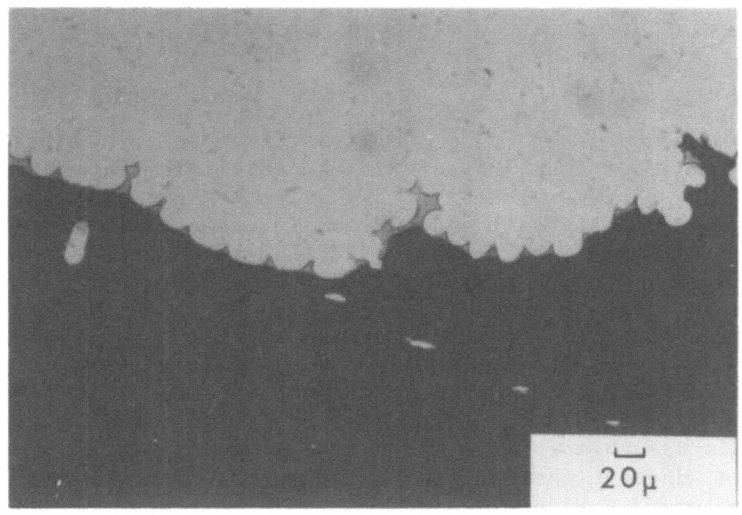

Figure 22. Fragments of RFL on Substrate.

Figure 23 shows an interface between two dissimilar compounds at 50,000X magnification. In this micrograph, the sample was prepared by microtoming a frozen specimen. This technique can also be used for electron microscopy although the sections have to be ultrathin sections in order of 500 to 1000Å. Sections such as these can be mounted in a special tensile holder and strained before examining them in the microscope. This technique does give information with regard to adhesion of the polymer to pigment or carbon black particles. The evaluation is based upon comparison only, because no tensile strength measurements can be taken. This method was used to relate vacuole formation (or, dewetting) to hysteresis in various compounds.[10] In Figure 23, the section was microtomed through the interface of an adhesive cement and substrate, and strained to 200%. The adhesive compound is toward the top of the micrograph, and the substrate at the bottom. The adhesive has taken up most of the elongation. Using this technique, it is possible to make relative comparisons of elongation properties between micro areas of compounds. The carbon black particles are fully resolved. Again, we can see the bond between the two compounds by virtue of their pigmentation. We would have a much more difficult time seeing the bond line unless there were some "tag" to one of the polymers such as pigmentation or a great difference in electron scattering effects.

In electron microscopy it is again axiomatic that the more difficult it is to see a bond, the better it is. Figure 24 is the electron microscope counterpart to Figure 2. The poor knitting of rubbers is also apparent in the electron micrograph. If the surfaces had been fresh before assembly, the interface would not have been visible.

Certain features of composite structures can be seen with the electron microscope that are not visible at all in the optical microscope. Figure 25 is a portion of two polyester tire cord filaments (light hemispherical areas) and surrounding resorcinol/formaldehyde/latex adhesive dip. With the electron microscope, one can see the rubber latex phase and how it is distributed. Figure 26 is a section through a similar composite structure, but here the parent rubber, adhesive dip, and filament sections are all seen in perspective. In this formulation, the adhesive dip was not RFL. Instead, a carbon black reinforced polymeric cement was used. The carbon black dispersion within the cement can be easily assessed. This, in turn, yields valuable information on the dispersion of all components of the adhesive whether they be pigments other than carbon black or polymers. Figure 27 is from the same test specimen, but at an area that had been pulled apart. At a higher magnification, one can get a good view of the areas on the adhesive where a polyester filament had been. It is a very smooth interface, and one must conclude that the failure was probably interfacial between the adhesive dip and the polyester. However, Figure 28 is from another area

MICROSCOPY OF ADHESIVE SYSTEMS

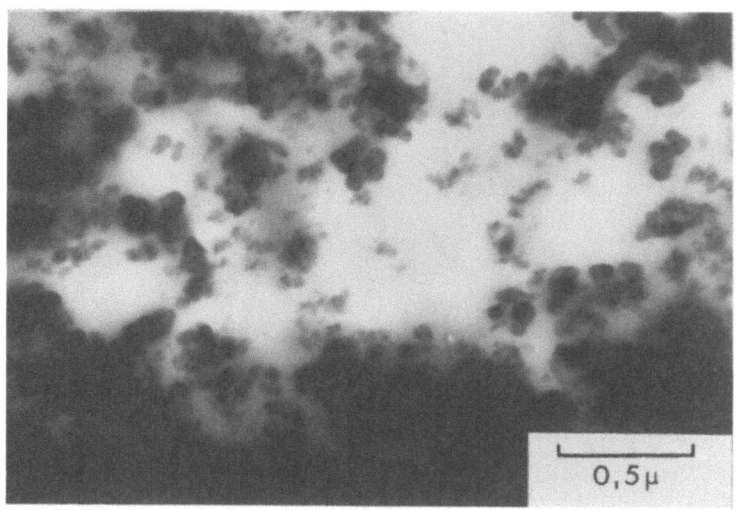

Figure 23. Strained Interface.

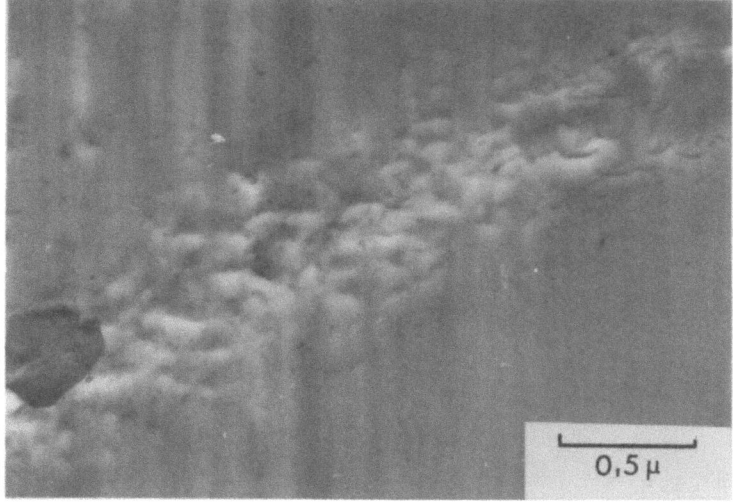

Figure 24. Poor Interface (Non-Black Compound).

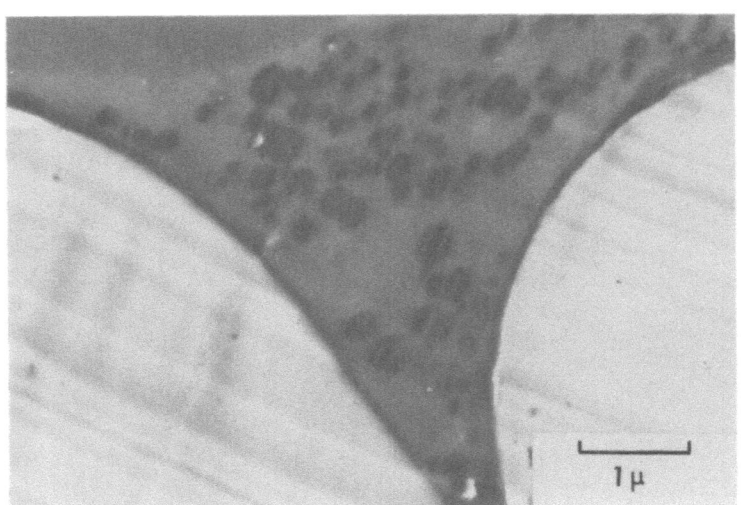

Figure 25. Polyester/RFL System.

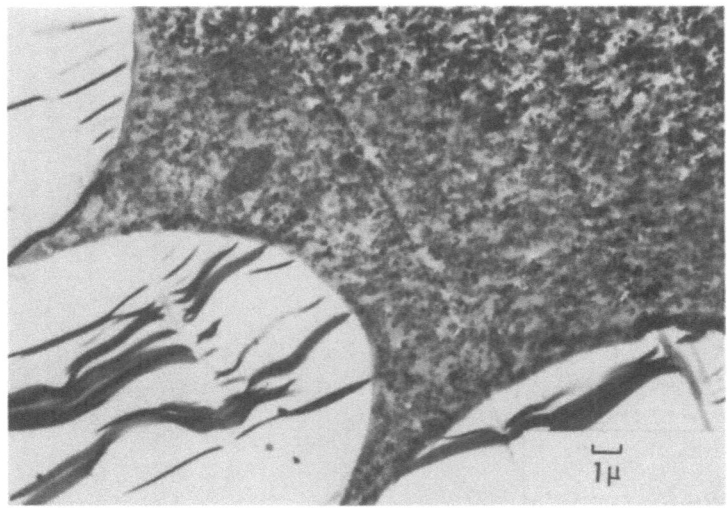

Figure 26. Polyester/Black Loaded Adhesive System.

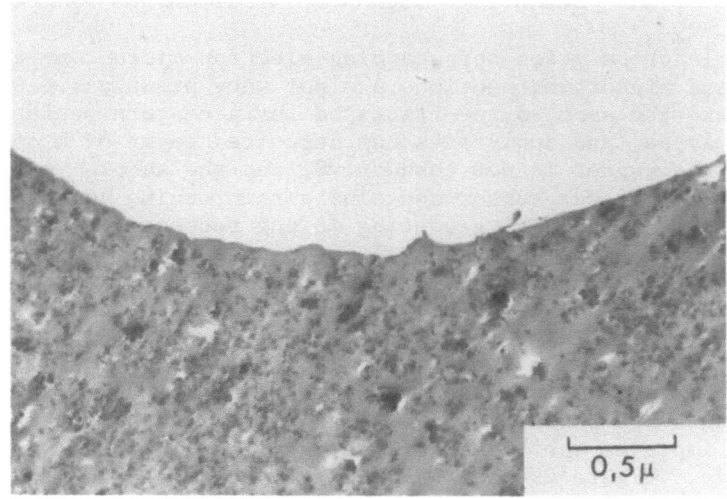

Figure 27. Separation at Dip/Filament Interface.

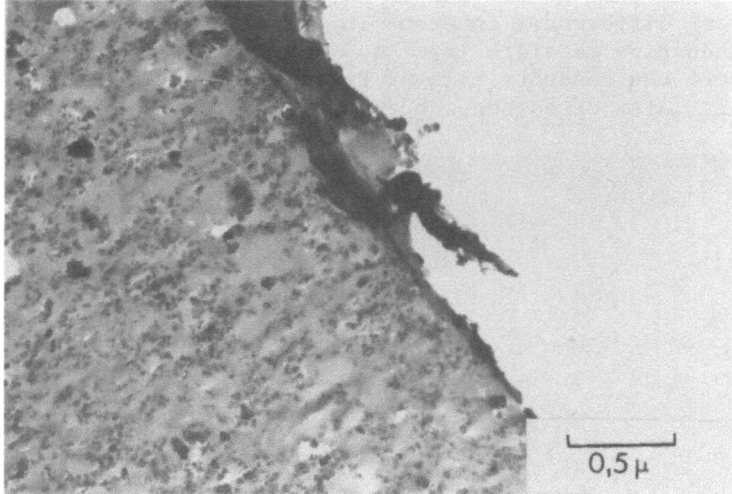

Figure 28. Remnants Adhering to Interface.

of the same specimen, and here, one can see that there are remnants of some kind on the adhesive dip, so apparently the total separation was a mixed failure--adhesive and cohesive between the two materials.

C. Scanning Electron Microscopy

Sample preparation for scanning electron microscopy is very simple, and microtomed sections are not very practical. One merely has to take the same as is--fractured surface--torn surface--whatever it may be, and apply a vacuum deposited layer of a conductive metal if the rubber is not conductive, and the sample is ready for examination. If the rubber contains a reinforcing amount of carbon black, even the conductive coating is not necessary. Adhesion can be assessed using the scanning electron microscope. This is based on whether or not one can "see air" between phases after the desired test has been conducted. This has been demonstrated many times in the fiber-reinforced resin systems.

In some areas, it is desirable to have poor adhesion. After saturating a non-woven fabric with a polymeric latex system, poor adhesion improves the "feel" of the material, making it less stiff and boardy. Figure 29 is a SEM micrograph through such a construction, and because we can "see air" between filaments and polymer, we can say that there is no adhesion there whatsoever.

Figures 4 and 5 were scanning micrographs that showed very little contrast between the two rubbers present, the only difference being the fact that sulfur was in one rubber and not in the other. Any contrast differences inherent in this system were covered over by the conductive metallic layer applied in that case. Figure 30 is a case where the conductive layer was not a heavy metal, but rather a very thin layer of carbon, and substrate effects are helpful.

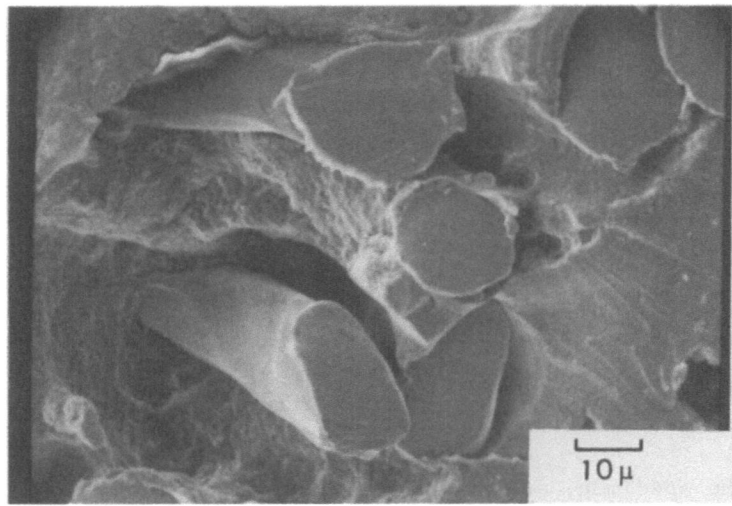

Figure 29. Saturated Non-Woven Fabric.

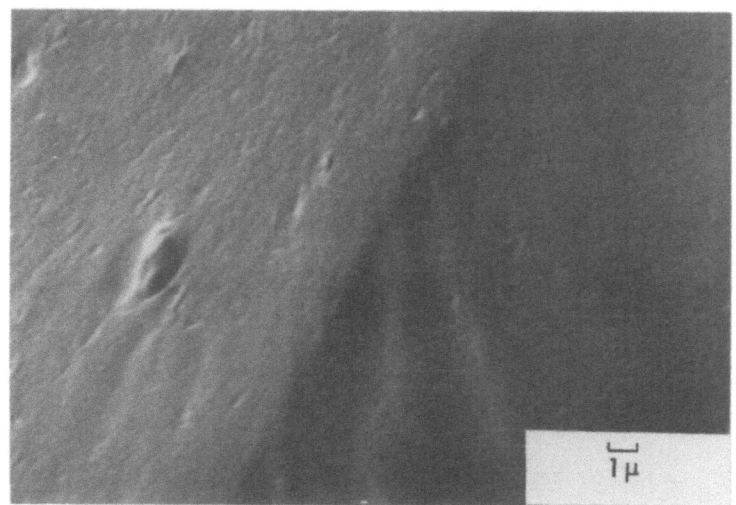

Figure 30. Natural Rubber/Neoprene Interface.

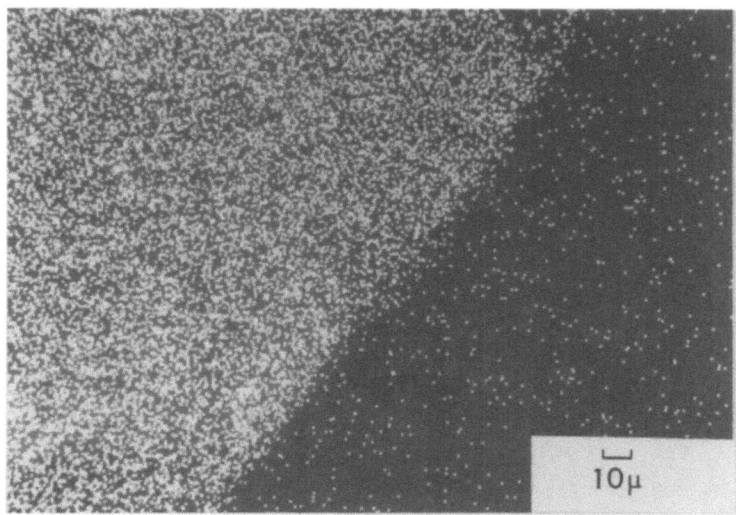

Figure 31. Chlorine Map of Figure 30.

This micrograph shows the bond line between natural and neoprene rubbers. The neoprene rubber is on the left side of the picture, and shows up brighter because it has a better secondary electron emissivity than does natural rubber that has no chlorine bonded to it. In addition, if the microscope has x-ray analyzing capabilities for mapping elements, one can use this to locate boundaries. Figure 31 is a map of chlorine from the same surface shown in Figure 30.

SUMMARY

Microscopical examination of polymeric adhesive systems shows that they can be very complex. Interactions between solvents and substrates, diffusion of materials, sub-surface characteristics of the substrate, the quality of the adhesive itself, and the true nature of separations can all be assessed through proper use of microscopes.

REFERENCES

1. J. Kruse, Rubber Chem. & Technol. 46, 653 (1973).
2. H. A. Depew and I. R. Ruby, Ind. Eng. Chem., Anal. Ed. 12, 1156 (1920).
3. H. Latta and J. F. Hartmann, Proc. Exp. Biol. Med. 74, 436 (1950).
4. C. H. Leigh-Dugmore, Rubber Chem. & Technol. 29, 1303 (1956).
5. F. H. Roninger, Ind. Eng. Chem., Anal. Ed. 5, 251 (1933).
6. V. L. Folt and R. W. Smith, Rubber Chem. & Technol. 46, 1193 (1973).
7. R. W. Smith and J. C. Andries, ibid. 47, 64 (1974).
8. R. P. Allen, ibid. 16, 219 (1943).
9. R. W. Smith and A. L. Black, ibid. 37, 338 (1964).
10. R. W. Smith, ibid. 40, 350 (1967).

High Speed Testing of Rubber to Metal Bonds

David A. Given and Raymond E. Downey

The Goodyear Tire & Rubber Company

Research Division

142 Goodyear Boulevard

Akron, Ohio 44316

With the advent of the energy absorbing bumper systems on automobiles, a need has developed for bonding and testing post cure bonded and vulcanized in place rubber-to-metal bonds.

This paper will deal with the adhesives used in both types of bonding and the technique of fabricating the rubber-to-metal bonds. A special test will be described using high speed stress strain equipment to simulate impacts in the five- to 10-miles per hour range.

INTRODUCTION

One of the recent contributions to rubber technology has been the development of methods whereby both natural and synthetic, both cured and uncured, may be bonded directly to metal surfaces. Earliest methods used to fasten rubber to metal were mechanical. Ebonite[1] was also used in early work, the ebonite surface giving a strong adhesion to the metal while uniting with the rubber on the other side. Ebonite as a bonding agent reduces the flexibility of the unit as a whole, especially when the rubber section itself is not very thick. Secondly, ebonite is thermoplastic and the tenacity of the bond falls off rapidly even at moderate temperature. Recently, other modified derivatives of rubber have become available which have excellent bonding properties without possessing the drawback of being thermoplastic. Typical of these are the chlorinated and hydrochlorinated rubbers and the functional terminated polymers.[2]

The need for bonding rubber-metal units has been most evident in the engineering industries, where they are used to reduce vibrations, absorb shock and aid in the reduction of noise. The use of rubber-metal bonded units has been of great assistance to the engineer in designing his equipment. Rubber possesses inherent properties obtainable in no other material and the ability to bond it to metal allows the designer to introduce a bonded unit at almost any point in his layout without having to rely on mechanical means for locating the rubber phase.

Although this paper does not deal with the tire cord adhesives, the use of the steel belt in the newer radial tires and in truck tires has indicated a need for better adhesive systems and methods of testing the steel cord to rubber adhesion.[3]

DISCUSSION

The development of the elastomeric energy absorbing bumper has created the need for a method of testing post cure bonded and vulcanized in-place rubber-to-metal bonds. Furthermore, this method should be capable of testing the bonds in the five- to 10-miles-per-hour range and should be capable of reproducible results. The so-called button test[4] has been modified for use as a test to determine the bond strength of the adhesive at the rubber-metal bond.[5] However, this method has been shown to give results that vary widely from low to high for the same elastomer-adhesive combination. There are three elements to consider in the development of a new test method that will satisfy the requirements of dynamic shear testing. They are:

1. High-speed test machine
2. Preparation of the samples
3. Testing of the bonded sample

High Speed Test Machine

The dynamic shear testing of rubber-to-metal bonds was done with a Plastechon Universal tester.[6] This machine is an electrohydraulic floor model with a 20,000-lb capacity. (Fig. 1.) It has additional interchangeable actuator/piston sleeve assemblies available in 2,500, 5,000 and 10,000-lb capacities. The speed range of the tester varies with the capacity. The range under 20,000-lb actuator service is from 2.0-3000 in/min, with speed range being

Fig. 1. High Speed Tester

extended to nominal 6000-8000 in/min at 10,000-lb service, to 15,000 in/min range (nominal) at 5000-lb service, and the 30,000 in/min range (nominal) at 2500-lb service.

The tester is available in either open- or closed-loop models. An open-loop model was used in our tests. Load was measured by means of force transducers of either piezoelectric or semiconductor gage type. Elongation may be followed via potentiometer, magnetic tape, microformer, strain gage, or non-contacting (capacitance, photo-optical) techniques. Load vs elongation is displayed on an oscilloscope and can be recorded with a Polaroid camera. An electronic console is separated from the tester for remote control operation. The tester is actuated by push button which initiates a "fire" sequence. The load curve is recorded by opening the camera shutter immediately prior to pushing the "fire" button.

Preparation of the Samples

A natural rubber compound that has been approved for use in an elastomeric energy absorbing bumper was selected for study. The test sample was designed to simulate the bumper as it is mounted in an automobile. Adhesives used in this study were selected from several systems that included both commercial and experimental types.

One-inch steel strips were sand-blasted to a white surface and degreased with tetrachloroethylene to remove any loose particles. A suitable metal primer was applied to the steel. After drying the primed steel received the adhesive coat. When the rubber parts were post cure bonded, they also received the adhesive coat after degreasing with tetrachloroethylene. Three metal strips were aligned with two rubber parts, compressed by 10 to 15 percent, and cured in an air circulating oven for 90 minutes at $280°F$. (Fig. 2.) Vulcanized-in-place samples were prepared in a similar manner. However, the rubber did not receive an adhesive coat. The rubber and the coated metal strips were aligned in a mold and cured in a press for 30 minutes at $300°F$.

All test samples were bonded with a two-coat adhesive system consisting of a metal primer and a cover adhesive. A metal primer is necessary because the cover adhesive exhibited very little inherent metal adhesion. Metal primers used in this study included one commercial and three experimental primers. One commercial and six experimental cover adhesives were included in the testing.

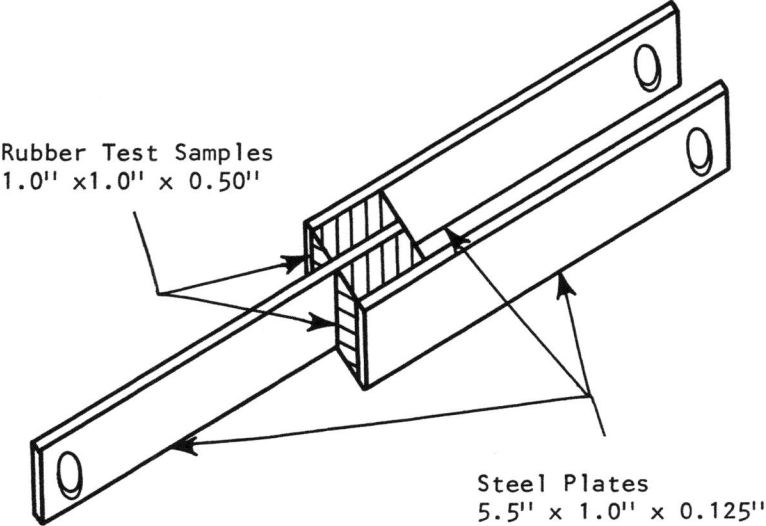

Fig. 2. Assembly for High Speed Test

Testing of the Bonded Samples

The Plastechon Universal Tester was used for all of the high speed testing and was mounted with the 2500-lb actuator/piston sleeve assembly which has a nominal range of 30,000 in/min. The speed was set at 5100 to 5300 in/min to simulate the five miles per hour impact requirement of the elastomeric energy absorbing bumper. Test samples were mounted on the piston sleeve assembly and attached to a ram. The ram can be brought to the operating speed of 5280 in/min over a time span of a few milliseconds. A plot of the force versus the deflection is displayed on the oscilloscope and is recorded by a Polaroid camera that is fired in conjunction with initiating the "fire" sequence of the tester. Traces on the oscilloscope grid show the force in pounds and the deflection in inches. (Fig. 3.) The amount of rubber coverage remaining on the metal is also recorded for each sample.

Our first use of this test procedure was the evaluation of rubber compounds to determine the effect of compounding ingredients on post cure bonding (Table 1). Materials for this study were

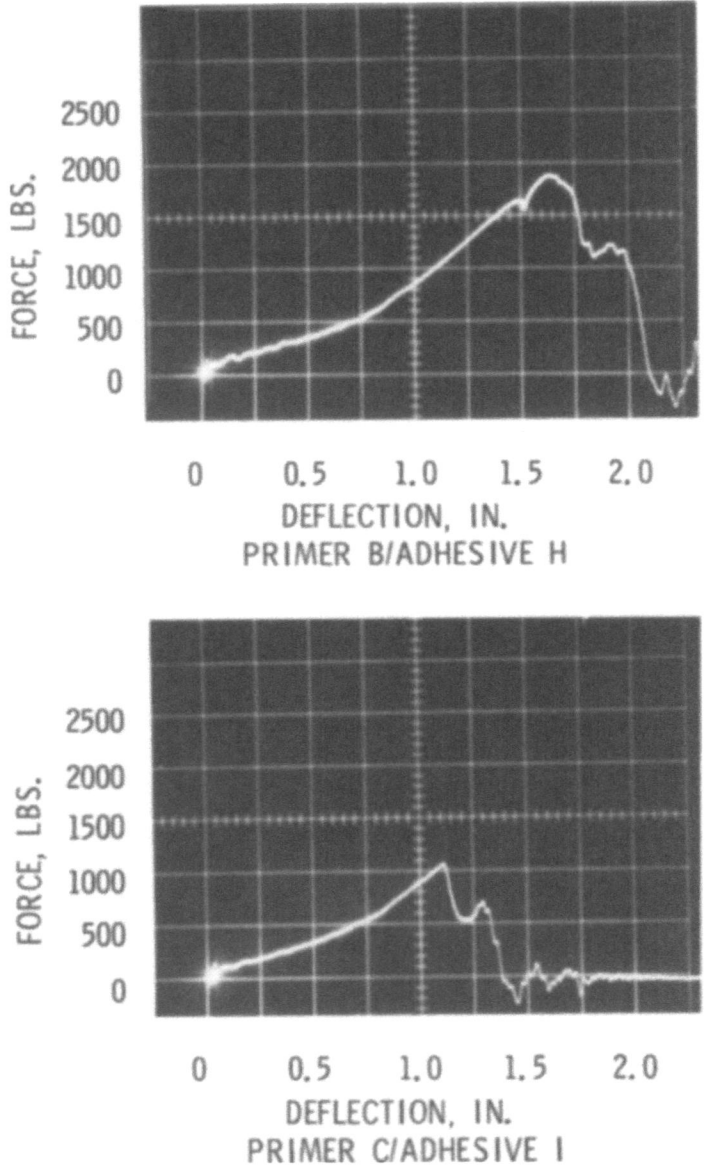

Fig. 3. Oscilloscope Trace of Force vs. Deflection

Table 1

Effect of Compounding Ingredients on Post Cure Bonding

Compound	Average Force (lbs)	Average % Rubber Failure	% Deviation
R168x601	975	87.5	12.8
R168x602	860	100.0	7.0
R168x603	580	100.0	3.4
R168x604	690	100.0	10.1
R168x605	840	82.5	2.4
R168x606	1320	100.0	6.1
R168x607	325	22.5	7.7
R168x608	530	100.0	5.7
R168x609	1070	100.0	0.93
R168x610	993	100.0	0.75
R168x611	920	100.0	13.0
R168x612	785	100.0	0.64
R168x613	1220	100.0	1.6
R168x614	740	100.0	2.7
R168x615	940	75.0	6.7
		av	5.4

selected from different classes of compounding ingredients commonly used in rubber-metal applications. A linear screening design was chosen since it allowed 11 different variables to be screened with only 16 experiments. Rubber-metal bonds were tested by three methods: the high-speed test and two variations of the button test (ASTM D429). The first button test (A) called for a sample that was 2.0 in. x 2.0 in. x 1.5 in. bonded to two metal disks with a bond area of 2.0 sq. in. (Fig. 4). The second test (B) called for a sample that was 1.0 in. x 1.0 in. x 0.5 in. bonded to two metal disks with a bond area of 2.0 sq. in. (Fig. 5). Both tests call for a tensile pull of the sample at 2.0 in/min (Table 2). Results of the three tests indicated that the level of carbon black and the cure system are the most important compounding materials that affect the adhesion. High carbon black level and the low sulfur cure system produced the best results. Linear regression analysis to determine the percent effects of compounding ingredients on the test results showed that the high speed test and test B results were statistically significant. Although test A results agreed with the other tests on the most important materials, the results were not statistically significant and only indicated trends in improved adhesion (Table 3).

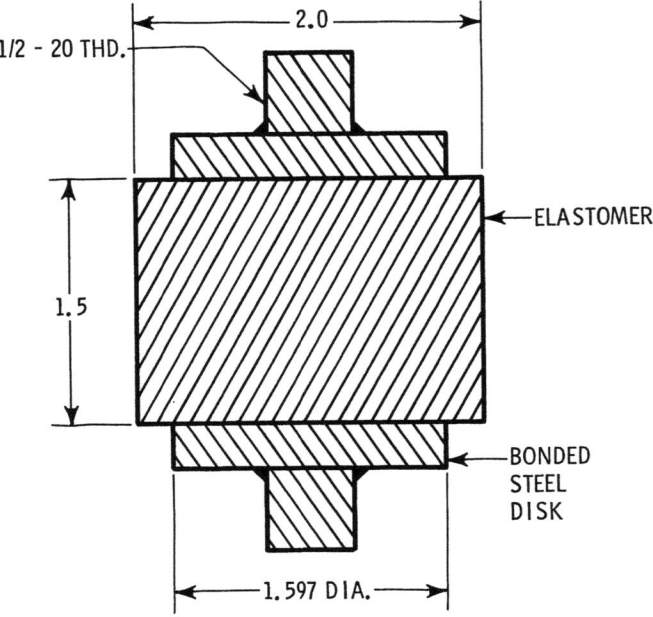

Fig. 4. Assembly for Test A

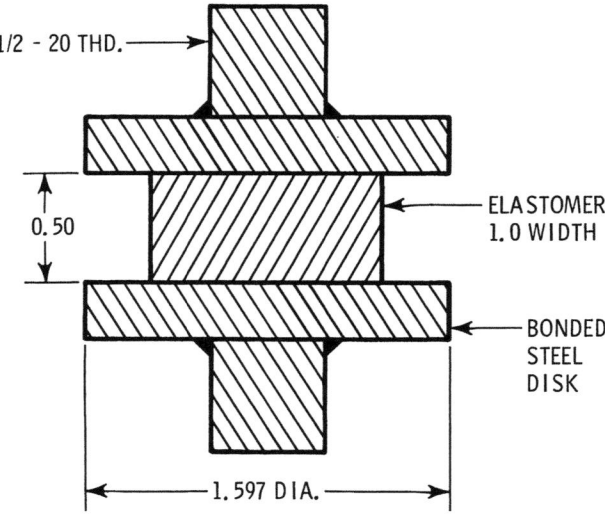

Fig. 5. Assembly for Test B

Table 2

Button Test Results

Compound	Test A (psi)	Test B (psi)
R168x600	91	385
R168x601	105	665
R168x602	275	560
R168x603	269	450
R168x604	328	620
R168x605	140	940
R168x606	229	870
R168x607	154	475
R168x608	180	460
R168x609	149	725
R168x610	369	650
R168x611	140	750
R168x612	355	735
R168x613	196	645
R168x614	303	650
R168x615	157	710

These high speed test results were found to be reproducible when duplicate samples were tested. Percent deviation of the measured force from the average force varied from 0.64 to 13.0 percent indicating fairly close agreement in the repeat tests. Average deviation of all of the test samples was 5.4 percent (Table 1). The results of tests A and B were taken from tests run by outside concerns and reported as the average of several runs. As a result the percent deviation could not be calculated. Differences in the results of these two tests can be attributed to the size of the rubber sample in each test. Samples in Test A were larger in width and thickness than the Test B samples. For this reason, the initial failure occurred at the periphery of the metal disk in Test A. Since the samples in Test B were smaller, the failure occurred in the body of the rubber and produced higher test results.

A subsequent study involved the screening of various adhesive systems that could be used to bond a natural rubber compound that was approved for an elastomeric energy absorbing bumper. The scope of this study was greater than the compounding study in that the number of duplicate samples was higher for many of the systems that were evaluated. Results of this study are found in Table 4. Percent deviation of the measured force from the average force varieed from 0 to 26.2 percent. The magnitude of the deviation is very likely due to the greater number of repeat runs. The deviation

Table 3

Linear Screening Design for Post Cure Adhesion Study
(% effects of compounding ingredients on test results)

Compounding Ingredient	phr Range	Test A	Test B	High Speed Test
ZnO	5-15	- 7.14	+ 0.01	- 3.57
SRF Black	40-80	+18.50	+27.99*	+28.81*
Stearic Acid	1-3	- 0.37	- 0.90	- 8.91*
Polyethylene	1-3	+ 0.25	- 5.50	- 0.39
Wax	1-3	+ 0.33	0	- 1.27
Benzoic Acid	0.5-1.5	+ 3.25	+10.20*	+13.86*
Octamine	0.5-1.5	+ 2.33	- 0.25	- 0.54
UOP 88	1-3	-11.56	- 3.99	-17.94*
Paraffinic Oil	4-8	- 8.70	- 1.21	+ 0.94
Aromatic Oil	4-8	+ 4.55	- 6.23	+ 1.43
Cure System	A or B**	+30.81	-42.10*	-18.35*
Total % effects		87.78	98.39	96.01

** Cure Systems

A (+1)		B (-1)	
Sulfur	2.75 phr	Sulfur	0.3 phr
MBTS	0.75	MBTS	2.0
TMTD	0.15	TMTD	1.0

* Statistically significant @ 95%

was usually less for samples with fewer repeat runs (Table 4). Results of two adhesive systems tested with the modified button test (A) had a percent deviation that varied from 24.4 to 57.4 percent (Table 5). A second run of the system with the higher deviation was found to have a percent deviation of 45.8 percent. Average deviation of the tensile test was 42.5 percent compared to 11.9 percent for the high speed test. From this comparison, we can say that the high speed test has given better agreement between duplicate runs and is more reproducible than the modified button test. Principal values to receive consideration in the study were the force required to cause rubber or adhesive failure and the amount of rubber retained on the metal.

Various adhesive systems were screened using these criteria and the best systems were selected for more extensive evaluation. Based on the data in Table 4, the best primer would be B and the best cover adhesive would be adhesive H. Primers A and C were also

Table 4

Adhesion Study - Post Cure Bonding

Adhesive System	Average Force (lbs)	Average % Rubber Failure	% Deviation	Number of Runs
A/E	1736	97.3	12.3	7
B/E	1580	90.0	20.9	4
B/G	1897	100.0	26.2	7
A/G	1850	100.0	13.5	4
A/F	2133	91.4	18.0	6
B/F	1950	98.4	16.9	4
C/H	2114	96.8	13.5	5
B/H	2073	100.0	13.4	3
D/H	1875	95.0	14.7	2
C/I	1175	50.0	8.9	2
B/J	1150	80.0	2.6	2
D/J	1850	87.5	2.7	2
B/K	1120	54.0	3.6	2
C/G	1740	100.0	0	2

av 11.9

Table 5

Adhesion Study - Button Test B

Adhesive System	Average Tensile (psi)	% Deviation	Number of Runs
A/F	291	24.4	5
B/G	502	57.4	5
B/G	554	45.8	3

av 42.5

found to be effective with some of the cover adhesives. Cover adhesives F and G were also reasonably effective because they caused 90 to 100 percent rubber failure even at lower force readings.

A study of vulcanized in-place bonds was more limited in scope and may need further work to confirm the results obtained thus far (Table 6). There was better agreement of the results of duplicate samples than with the post cure bonded samples.

Table 6

Adhesion Study - Vulcanized In-Place

Adhesion System	Average Force (lbs)	Average % Rubber Failure	% Deviation
A/F	2795	82.5	2.0
B/F	2875	92.5	6.1
A/G	2890	60.0	3.1
B/G	2800	67.5	3.6
			av 3.7

Table 7

Identification of Adhesives

Adhesive Number	Type of Adhesive
A	Experimental metal primer #1
B	Commercial metal primer
C	Experimental metal primer #2
D	Experimental metal primer #3
E	Experimental cover adhesive #1
F	Experimental cover adhesive #2
G	Commercial cover adhesive
H	Experimental cover adhesive #3
I	Experimental cover adhesive #4
J	Experimental cover adhesive #5
K	Experimental cover adhesive #6

Percent deviation was 2.0 to 6.1 percent and the average deviation of all the samples was 3.7 percent. One result of this evaluation was the discovery that the force required to cause rubber or adhesive failure in vulcanized in-place bonds was up to 56.2 percent higher than the force required for post cure bonds using the same adhesive system. Testing also revealed that the partial failure of the adhesive system was more likely to occur with the vulcanized in-place bonds. The best primer was found to be primer B and the best cover adhesive was adhesive F (Table 7). Primer A also gave good results when used with cover adhesive F. Cover adhesive G exhibited the highest force reading with primer A but exhibited much lower retention of rubber than adhesive F with both primers. These results are in agreement with the post cure bonded results. In both cases primer B was the best primer. Cover adhesive H was not tested with the vulcanized in-place samples. However, adhesive F was found to be reasonably effective in the post cure bonds as well as the vulcanized in-place bonds.

SUMMARY

In summary, we have presented a method for the dynamic shear testing of post cure bonded and vulcanized in-place rubber-to-metal bonds at high speeds. This method is applicable for the testing of elastomeric energy absorbing bumper systems on automobiles in the five miles per hour range. This method has the capability of reproducible results with greater consistency than previous methods.

REFERENCES

1. S. Buchan, *Rubber to Metal Bonding*, Second Edition, Lockwood, London, 1959.
2. I. Skeist, *Handbook of Adhesives*, Reinhold Publishing Corporation, 1962.
3. K. D. Albrecht, Rubber Chem. Technol. 46, 981 (1973).
4. ASTM Test D-429, "1973 Annual Book of ASTM Standards, Part 28, Rubber, Carbon Black, Gaskets", American Society of Testing and Materials, Philadelphia, p. 219.
5. K. C. Rusch and J. M. Slessor, Rubber Chem. Technol. 46, 862 (1973).
6. Plas-Tech Equipment Corporation, Natick, Mass.

The Nature of Rubber Reinforcement by Silane-Treated Mineral Fillers

E. P. Plueddemann and W. T. Collins

Dow Corning Corporation

Midland, Michigan 48640

Silane treated clays were compared as reinforcing fillers in typical rubber formulations. Properties of the rubbers showed that some silane treatments were generally much more effective than others, although better silane treatments were somewhat specific for each rubber. Silanes were relatively ineffective as adhesive primers on plane glass surfaces, but relative peel strengths could be related to silane effectiveness on the filler. Recognizing that vulcanized rubber in the interphase region adjacent to silane treated fillers may be resinous in nature, it appears that rubber adhesion and rubber reinforcement fit into the general mechanism of bonding that has been proposed for reinforced plastics. This makes it possible to predict from simple tests that certain new silanes should be very effective as treatments on mineral fillers to convert them into reinforcing fillers for rubbers.

INTRODUCTION

Higher costs and potential shortages of reinforcing carbon black have stimulated interest in the use of particulate fillers in rubber. Silica, clay, and other silicates are available at low cost in virtually unlimited supply, and in the presence of appropriate silane coupling agents, show rather creditable properties as reinforcing fillers.[1] The nature and extent of adhesion between rubber and filler, however, has been somewhat of a mystery and clouded by contradictions:

(a) Silanes that convert silicates like clay into reinforcing fillers in rubber do not promote good adhesion between similar rubbers and plane mineral surfaces.

(b) Silane treatments on clay that are effective in rubber do not improve mechanical properties of clay-filled epoxies or polyesters.[2] The poor showing of treated clay in rigid resins has been attributed to cleavage within the clay particle when stressed in a rigid matrix.

(c) Practical experience in rubber bonding indicates that vulcanized rubbers have poor direct adhesion to hydrophilic mineral surfaces. Good adhesion is obtained only by providing a resinous primer layer on the substrate. Thermoplastic rubbers may be bonded through a tacky primer, while vulcanized rubbers are invariably bonded through a crosslinked resinous primer.[3]

(d) Theoretical concepts of adhesion suggest that total performance of organic-inorganic composites may be related to polymer morphology (in the interface region) according to a master curve (Figure 1) that has a minimum in the rubbery range even with optimum chemical modification of the interface.

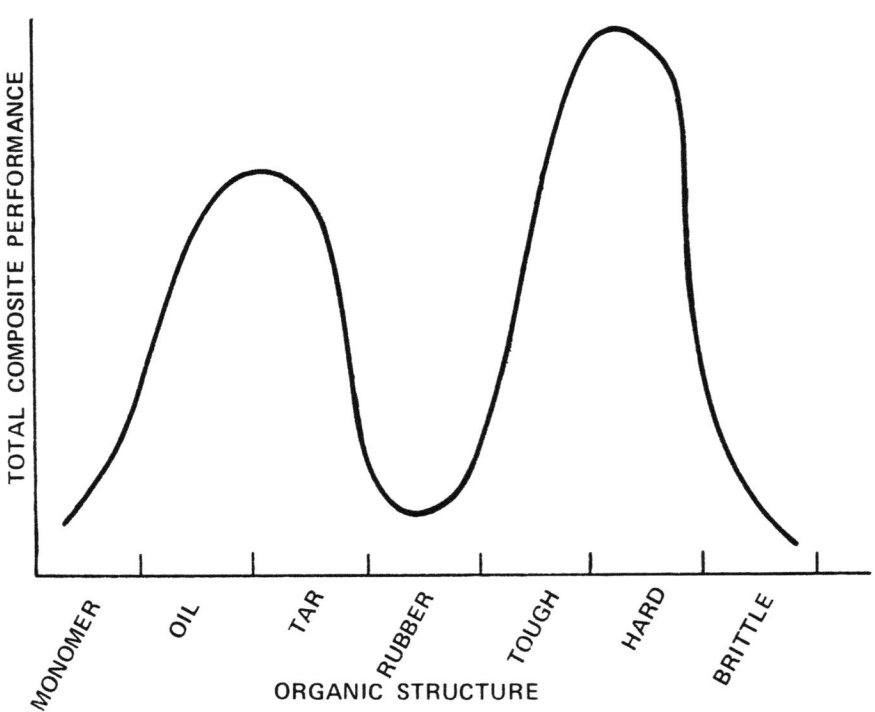

Figure 1. Variation in total performance of organic-inorganic composites with change in polymer morphology at the interface.

The apparent contradictions of rubber adhesion as related to rubber reinforcement are resolved, if it is recognized that a large concentration of high energy surfaces may modify adjacent rubber molecules to reduce polymer mobility and provide a resinous layer for bonding at the interface.[4] Under such conditions, concepts that have been used in modifying the interface in mineral reinforced thermosetting resin composites may be applied to mineral reinforced vulcanized rubbers. Limited studies of silanes with vulcanized rubber show enough similarity with thermosetting resins that certain recommendations may be made:

(a) A silane should be selected that will combine chemically with the rubber during vulcanization.
(b) Performance will improve with increased crosslinking at the interface.
(c) The silane may be applied separately to the mineral or added as an integral blend during compounding of the rubber.
(d) Hydrolyzable groups on silicon will provide silanol groups for bonding to the mineral, and will be effective on the same minerals that respond to silanes in reinforced plastics.
(e) Although only modest improvement in adhesion to vulcanized rubber is provided by thin layers of silanes on plane surfaces, there should be some correlation between peel strengths or vulcanized rubbers on silane-treated glass surfaces and the performance of the same silanes on reinforcing mineral fillers.

This work compares the performance of silanes in rubber adhesion and rubber reinforcement so that simple tests may be used to predict the performance of a silane on mineral fillers in any rubber formulation.

EXPERIMENTAL

Adhesion to Microscope Slides

Several typical rubber formulations were vulcanized against silane-treated glass microscope slides and tested for peel strength. The rubbers were commercial compounds of undisclosed formulation: natural (tire tread), SBR (tire tread), nitrile (hose and belts), neoprene (hose, belts and gaskets), EPDM (peroxide) and (sulfur) (ignition wire insulation) and Hypalon (DuPont) (ignition wire jacket).

Microscope slides were dipped in 1% aqueous silanes and dried at room temperature. Non-reactive silanes were selected to contribute varying degrees of surface energy for bonding through dispersion forces (except for polar carboxyphenylsilane). Sulfur-cured SBR and peroxide-cured EPDM showed slight increase in peel strength with increasing surface energy of the silane-treated surfaces, but none developed significant adhesion (Table 1). Sulfur-cured EPDM showed fairly good adhesion to almost all surfaces.

Table 1

Adhesion of Vulcanized Elastomers to Glass
(Glass treated with 1% aqueous silane)

Silane Functionality (cure)	Peel Strength (ppi)		
	SBR (Sulfur)	EPDM (Peroxide)	(Sulfur)
No silane (control)	nil	nil	1.1
Non-reactive silanes			
Propyl	nil	0.1	0.7
3-Chloropropyl	0.1	0.1	1.1
Phenyl	0.7	0.2	0.9
Chlorophenyl	0.4	0.2	1.3
Dibromophenyl	0.1	nil	1.5
Carboxyphenyl	0.4	0.2	2.4
Reactive silanes			
Vinyl	0.9	1.1	2.2
Methacryloxypropyl	0.1	2.9	4.6
Aminopropyl	1.1	2.1	5.7
Diamine	1.2	1.3	3.5
Mercaptopropyl	3.0	1.8	8.5
iso-Thiuronium Cl.	6.3	1.0	2.9
Cationic styryl	2.0	>20(c)	>20(c)
Anionic styryl	2.1	>20(c)	>20(c)

(c) = cohesive failure in the rubber

Reactive silanes were tested on glass with the same rubbers (Table 1), as well as with natural, nitrile, Neoprene, and Hypalon compounds (Table 2). In addition to simple mono-organofunctional silanes these tests included polyfunctional silanes:

diamine (D.C. Z-6020) $\quad -CH_2CH_2CH_2NHCH_2CH_2NH_2$

iso-thiuronium Cl. (D.C. QZ-8-5456) $\quad -CH_2CH_2CH_2-\overset{+}{S}-C\underset{NH_2}{\overset{\displaystyle NH}{\diagup\!\!\!\!\diagdown}} \quad Cl-$

cationic styryl
(D.C. QZ-9-5069)

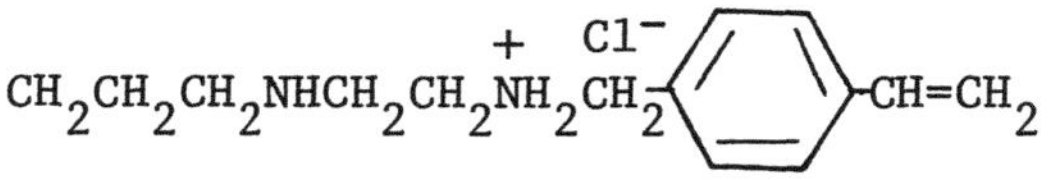

$CH_2CH_2CH_2NHCH_2CH_2\overset{+}{N}H_2CH_2$-C$_6H_4$-CH=CH$_2$ · Cl$^-$

anionic styryl
(experimental)

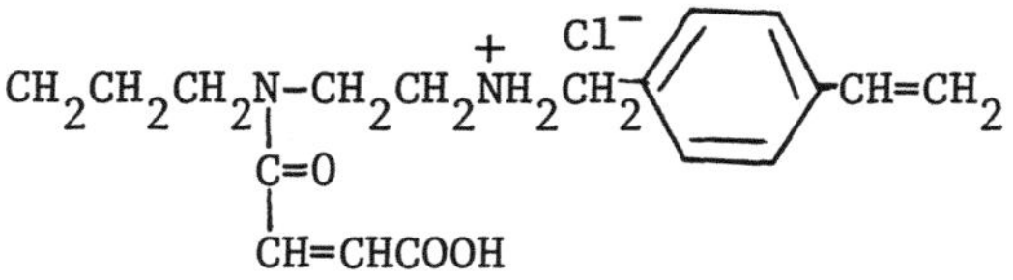

$CH_2CH_2CH_2\underset{\underset{CH=CHCOOH}{\overset{|}{C=O}}}{\overset{|}{N}}-CH_2CH_2\overset{+}{N}H_2CH_2$-C$_6H_4$-CH=CH$_2$ · Cl$^-$

Chloropropyl, vinyl, and methacrylate-functional silanes on glass were only slightly better than untreated surface for adhesion to most of the rubbers. The mercaptan and amine-functional silanes have been the preferred silanes on fillers in sulfur-vulcanized rubbers. The amine gave better adhesion to natural, nitrile, and Hypalon rubbers, while the mercaptan was better with SBR, Neoprene and EPDM. The iso-thiuronium-functional silane appears to be fairly effective in bonding all rubbers--but especially SBR. Anionic or cationic styryl coupling agents were the best unsaturated silanes and contributed true adhesion to EPDM and Hypalon.

Table 2

Adhesion of Vulcanized Elastomers to Glass
(Glass treated with 1% aqueous silane)

Silane functionality	Peel Strength (ppi)			
	Natural	Nitrile	Neoprene	Hypalon
No silane	0.1	nil	nil	1.2
Chloropropyl	0.6	0.5	0.1	1.2
Vinyl	0.7	0.1	0.1	1.8
Methacryloxypropyl	0.5	0.1	0.1	2.0
Aminopropyl	1.4	0.1	0.1	6.0
Diamine	4.4	8.0	1.1	>20(c)
Mercaptopropyl	2.0	1.2	4.0	2.6
iso-Thiuronium Cl.	3.9	2.0	2.0	2.7
Cationic styryl	4.8	5.5	2.2	>20(c)
Anionic styryl	5.2	5.2	0.6	>20(c)

(c) = cohesive failure in the rubber

Table 3

Clay-filled Rubber Formulations

Formulation	Natural (SMR-5)	SBR (1502)	Nitrile (FRN-502)
Rubber	100	100	100
Clay	50	60	70
Zinc Oxide	5	4	5
Stearic Acid	1.5	2	2
Sulfur	2.5	1.4	0.2

Other:
  <u>Natural</u>: Plastogen 5, Sunolite-240 1.5, Thermoflex-A 1, TMTD 0.5, Santocure-NS 1.0
  <u>SBR</u>: Circolite Oil 10, PBNA 1, Flexamine-G 1, MBT 0.8, Di-o-Tolylguanidine 0.3
  <u>Nitrile</u>: DOP 10, Sunolite-240 1, PBNA 1, TMTD 3

Rubbers with Treated Clay Fillers

Kaolin clay (Superex, J.M. Huber Co.) was treated in aqueous slurries with silanes and dried. All of the rubber clay compositions were prepared by mill-mixing the clays into the rubber before adding the remaining ingredients. Recipes shown in Table 3 were selected from data from suppliers to the rubber industry; however, since a control was included with each rubber, the treated samples can best be evaluated in comparison with each other.

Selected mechanical properties were determined by standard ASTM methods on samples prepared at optimum vulcanization time as determined using a Monsanto Rheometer at 60 cpm and three degrees of arc. Scorch time was taken at a two point rise in the curve, and cure time at 90% of the time for maximum point in the cure curve.

Mooney viscosities were determined with the large rotor after a one minute warm-up, followed by four minutes at 212°F. Mooney scorch tests were made with the small rotor at 250°F. Time to scorch was three units above the minimum and cure time was taken at eighteen units above the minimum.

Little variation in scorch or cure-times were observed with the various clays in Natural or Nitrile rubbers. SBR showed a much wider variation in processing characteristics as summarized in Table 4.

Two levels of silane treatment on clay are compared in natural rubber in Table 5. Differences in silane reactivity as indicated by peel tests are accentuated by increasing the level of silane

Table 4

Vulcanization Characteristics of SBR Recipes

Silane on Clay	Mooney Visc. Data			Rheometer Data	
	Visc.	M_3	M_{18}	T_2	T_{90}
Control (no silane)	38	120	149	14	46
0.5% Chloropropyl	29	125	161	16	52
1.0% Chloropropyl	32	132	168	18	50
0.5% Amine	36	94	110	13	34
1.0% Amine	32	66	75	11	32
0.5% Mercaptan	31	103	130	14	48
1.0% Mercaptan	33	108	135	15	47
0.5% i-Thiur. Cl	33	101	121	13	33
1.0% i-Thiur. Cl	32	90	112	14	40

M_3 = Mooney Scorch at 250°F (min)

M_{18} = Mooney Cure at 250°F (min)

T_2 = Rheometer Scorch at 302°F (min)

T_{90} = Rheometer Cure at 302°F (min)

Table 5

Mechanical Properties of Natural Rubber Compounds
(Clay fillers treated with organofunctional silanes)

Silane on Clay	Adhes. (ppi)	300% Mod. (psi)	Tensile St.(psi)	Elong. (%)	Tension Set(%)
None (Control)	0.1	1040	3435	585	43
0.5% Cl-Pr	0.4	1080	3725	580	43
1.0% Cl-Pr	0.6	985	3690	585	43
0.5% Mercaptan	1.1	1265	3620	555	37
1.0% Mercaptan	2.0	1480	3885	540	41
0.5% i-Thiur. Cl	2.8	1430	4110	560	37
1.0% i-Thiur. Cl	3.9	1575	3750	515	37
0.5% Amine	2.0	1505	3760	540	32
1.0% Amine	4.4	1655	3925	520	38

(Continued on next page)

Silane on Clay	Adhes. (ppi)	300% Mod. (psi)	Tensile St. (psi)	Elong. (%)	Tension Set (%)
	(Aged 70 Hours at 70°C)				
Control		1465	3835	545	40
0.5% Cl-Pr		1470	3810	515	40
1.0% Cl-Pr		1395	3600	505	29
0.5% Mercaptan		1650	3840	505	29
1.0% Mercaptan		1860	3985	485	35
0.5% i-Thiur. Cl		1875	4035	495	35
1.0% i-Thiur. Cl		2000	3825	465	30
0.5% Amine		1940	2885	415	22
1.0% Amine		2165	3840	450	27

treatment from 0.5% to 1.0%. Modulus and tensile strength of the rubbers generally increase with increased adhesion (Figure 2) although tensile strength values show considerable scatter. Heat aging causes more severe change in properties with increased adhesion.

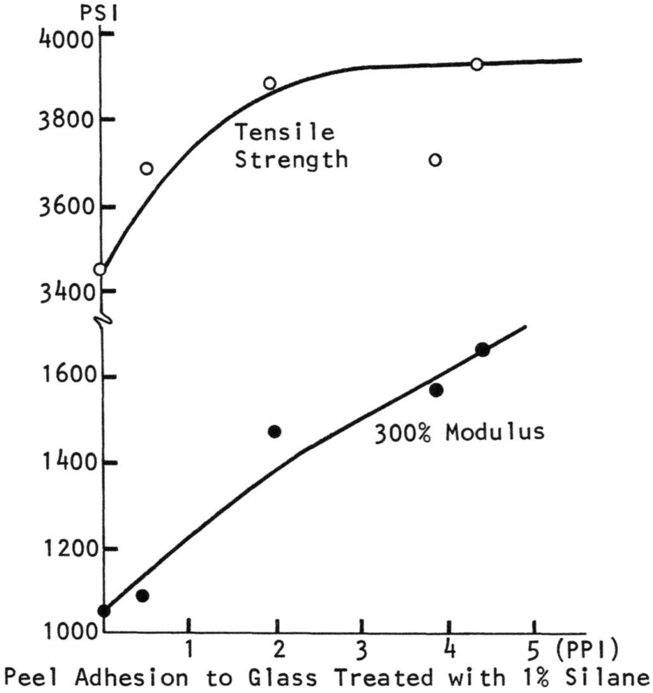

Figure 2. Relationship of rubber reinforcement with rubber adhesion to silane-treated surfaces.

Table 6

Mechanical Properties of Clay-Filled Natural Rubber
(Clay treated with 1% aqueous silane)

Silane on Clay	None	Cl-Prop.	Mercap.	i-Th-Cl	Amine
Peel Adhesion (ppi)	nil	0.6	2.0	3.9	4.4
300% Modulus (psi)	1040	985	1480	1575	1655
Tensile Str. (psi)	3435	3690	3885	3750	3925
Elongation (%)	585	585	540	515	520
Tension set (%)	43	43	41	37	38
Compression set (%)	8	7	6	5	6
Bashore rebound (%)	59	53	64	65	67
Tear strength (ppi)	127	138	140	106	118

Mechanical Properties of Clay-Filled SBR

Silane on Clay	None	Cl-Prop.	Amine	Mercap.	i-Th-Cl
Peel Adhesion (ppi)	nil	0.1	1.2	1.4	6.3
300% Modulus (psi)	285	280	400	405	370
Tensile Str. (psi)	1120	1380	1505	1590	1720
Elongation (%)	925	1015	885	930	975
Tension set (%)	37	41	35	36	38
Compression set (%)	13	11	10	11	9
Bashore rebound (%)	46	47	48	48	48
Tear strength (ppi)	141	150	154	162	157
Flex (JIS 10^{-2})	337	567	259	370	520
Abr. resistance	100	130	155	169	140
Heat build-up (°C)	91'	103'	90'	78'	97

Mechanical Properties of Clay-filled Nitrile Rubber

Silane on Clay	None	Cl-Prop.	Mercap.	i-Th-Cl	Amine
Peel Adhesion (ppi)	nil	0.5	1.2	2.0	8.0
300% Modulus (psi)	1230	1355	1755	1770	2125
Tensile Str. (psi)	3150	3255	3525	4060	3495
Elongation (%)	650	665	665	645	545
Tension set (%)	31	27	25	28	21
Compression set (%)	7	7	6	4	4
Bashore rebound (%)	24	23	24	24	24
Tear strength (ppi)	201	176	167	163	153
Flex (JIS 10)	49	102	104	75	36
Abr. resistance	100	122	141	137	166
Heat build-up (°C)	61'	98'	69	57'	55

(') sample blew out before completion of heat build-up test

Properties of three rubbers are described in more detail with 1% silane on the clay filler (Table 6). Again, it appears that reaction of organofunctional silanes with rubber types is specific enough that peel adhesion from glass correlates fairly well with mechanical properties obtained with the same silanes on clay filler.

Increased adhesion, as indicated by peel tests, consistently improves certain mechanical properties in all rubbers, has little effect on some properties, and varies with the type of rubber in other properties. Each trend in properties is accentuated by increasing the level of silane treatment from 0.5% to 1.0%.

Increased adhesion is accompanied by:

    300% Modulus increase
    Tensile strength increase
    Abrasion resistance increase
    Tension set decrease
    Compression set decrease
    Heat build-up decrease
    Shore A hardness-little effect
    Bashore rebound-little effect
    Mooney viscosity + Natural and Nitrile, -SBR
    Scorch-and Cure time + Natural and Nitrile, -SBR
    Elongation + SBR, - Natural and Nitrile,
    Tear strength + Natural and SBR, -Nitrile
    Flex resistance + SBR and Nitrile, -Natural
    Heat Aging, better Natural, poorer SBR and Nitrile

Additional data should be obtained with a cationic, and anionic styryl-functional silanes on a filler in a rubber that shows true adhesion (cohesive failure) in peel tests on glass to determine the ultimate effect of adhesion on reinforcement. Data of Tables 1 and 2 suggest that the iso-thiuronium chloride, and the vinylbenzyl functional silanes should be considered generally along with amine and mercaptan-functional silanes in modifying mineral fillers for rubbers.

REFERENCES

1. N. W. Ranney, S. E. Berger, and J. G. Marsden, *Interfaces in Polymer Matrix Composites*, (Ed. by E. P. Plueddemann), Volume 6 in *Composite Materials*, Academic Press, Chapt. V.
2. H. A. Freeman and E. P. Plueddemann, Proc. 31st Ann. EMSA Meeting (Ed. by C. J. Arceneaux), p. 130 (1973).
3. E. P. Plueddemann, Proc. Annu. Conf., Reinf. Plast. Compos. Inst., Soc. Plast. Ind. 24-A (1974).
4. E. P. Plueddemann, Fifth Akron Summit Conf. on Adhes., September (1974), to be published in *Adhesives Age*.

Catalysis of Silicone Elastomer Adhesion

Thomas W. Greenlee

Tremco, Incorporated

10701 Shaker Blvd., Cleveland, Ohio 44104

 Room-temperature curing silicone elastomers will adhere well to unprimed, unanodized aluminum and steel if they contain alkoxysilyl groups (at least initially) and if the substrate has been brought into contact with certain anions either before or during cure. Alkoxysilyl groups can be introduced into the polymer via mixed reactivity crosslinkers such as alkoxyacetoxysilanes and random $(MeHSiO)_x$ $(MeSi(OR)O)_y$ silicones. Methods for preparing such crosslinkers are given. Among good adhesion-promoting anions are nitrate, nitrite, bromide, iodide, tetraborate, phosphate, and carboxylate species including glycine. Fluoride, hydroxide, cyanide, and azide have no effect. Chloride, sulfate, perchlorate, and bicarbonate are of intermediate activity. Carboxylate anions can be applied either from salt solutions (rapid reaction) or as the free acids (slower reaction). In general, adhesion promoting ability is inversely related to anion affinity for alumina.

INTRODUCTION

 Silanol-terminated linear polydimethylsiloxane polymers can be crosslinked to elastomers by a variety of reactions. Among these is the reaction with acetoxysilanes, shown schematically in reaction sequence A. The first reaction takes place upon

A. $SiOH + RSi(OAc)_3 \rightarrow SiOSiR(OAc)_2 + HOAc$

  $2\ SiOSiR(OAc)_2 + H_2O \rightarrow SiOSiR(OAc)OSiR(Ac)OSi + 2\ HOAc$, etc.

compounding. Subsequent chain-extending and crosslinking reactions require water, which is normally provided by the atmosphere. This sequence provides a widely used one-part cure system for silicone RTV elastomers. A disadvantage is that, in general, polymers thus cured have poor adhesion to unprimed metals.

Another cure mechanism for silanol-terminated polymers is the catalyzed (usually by tin) reaction with alkoxysilyl species exemplified by alkyl ortho- and polysilicates. Schematically, it is

B. $4 \text{ SiOH} + \text{Si(OR)}_4 \rightarrow \text{Si(OSi)}_4 + 4 \text{ ROH}$

Reaction B is used to cure many two part silicone elastomers. These adhere even worse than those cured by Reaction A; in fact, some are used as mold making materials.

Thus, neither alkoxysilanes nor acetoxysilanes will cure silanol functional linear polymers to well-adhering elastomers.

MIXED REACTIVITY CROSSLINKERS

It was thus surprising to find that alkoxyacetoxysilane crosslinkers, e.g., diethoxydiacetoxysilane, give excellent unprimed adhesion when used in Reaction A.[2] We decided to ascertain the structural breadth of this process and, if possible, to determine the mechanism of improved adhesion.

A variety of reactions are available for the preparation of alkoxyacetoxysilane species. Among them are the reactions of tetraacetoxysilane with alcohols[2,3], tetraalkoxysilanes with anhydrides[4,5], of alkoxychlorosilanes with acetate salts[6,7], and of hydrosilanes with anhydrides as exemplified by

$(\text{RO})_3\text{SiH} + \text{Ac}_2\text{O} \rightarrow (\text{RO})_3\text{SiOAc} + \text{AcH}$ [8]

We used the very convenient acid-catalyzed equilibration shown here.

$\text{Si(OAc)}_4 + \text{Si(OR)}_4 \rightarrow (\text{RO})_n\text{Si(OAc)}_{4-n}$

Small amounts of acetate ester and disiloxane species were sometimes formed via

$\text{SiOAc} + \text{SiOR} \rightarrow \text{ROAc} + \text{SiOSi}$

(Such siloxane species are advantageous in RTV adhesives[9].) In each case we used an equilibrium mixture as crosslinker, making no attempt to isolate and maintain pure any single compound.

Table 1 shows the effect of these mixed reactivity crosslinkers on elastomer physical and adhesive properties. The runs reported are selected from among some fifty, all of which showed the same tendencies. Pure acetoxysilane crosslinkers (Runs 1-7) all give elastomers with poor adhesion to steel, all failing at ≤ 15 pli with visual clean peel. Runs 8-10, in which unequilibrated mixtures of alkoxysilane and acetoxysilane compounds were used, show modest improvement if any. (This improvement may be due to fortuitous equilibration during mixing.) As explained below, I believe tetrafunctional species such as diethoxydiacetoxysilane to be responsible for the adhesion increase. Such compounds cannot arise in Run 10, and in Run 9 require displacement of bulky 2-methoxyethoxyl groups. The compact ethoxyl groups in Run 8 are expected to equilibrate more easily. Clearly the mere presence of alkoxyl groups in the mix is insufficient for good bonding.

Runs 11-17 show the results when mixed reactivity crosslinkers are used. Runs 16 and 17, in which no tetrafunctional species can occur, show the lowest adhesion (which is, however, still better than that with any nonequilibrated crosslinker). Several different alkoxysilane compounds were equilibrated. Although typical elongation and tensile strength have been reduced slightly, adhesion is now greatly improved, with mixed clean-cohesive failure at 60-90 pli and 100% cohesive failure at > 90 pli. For attainment of good bond strength, crosslinker must contain SiOAc and SiOR groups on the same molecule.

THE MIXED REACTIVITY HYPOTHESIS

During cure acetoxyl groups in mixed reactivity crosslinkers react more rapidly with silanol than do alkoxyl groups, hence initially the latter are brought into the polymer network. This in itself does not insure good bonding, since elastomers cured by Reaction B initially contain such groups as well, and they adhere wretchedly to virtually any unprimed surface. We propose that polymer-bound alkoxyl groups can react with metal substrates to form direct metal oxide-to-polymer linkages (with attendant good adhesion) if the surface contains "adhesion promoting" anions. The "adhesion promoter" in elastomers derived from mixed reactivity crosslinkers is the acetic acid generated during cure. The crosslinker might react first (via its SiOR) with the substrate and then (via SiOAc) with the polymer--the overall effect would be the same.

Thus, "Reaction A" elastomers are not expected to adhere well in the absence of mixed reactivity crosslinkers since the polymer contains no SiOR groups. "Reaction B" elastomers do not adhere well because no promoting anions bring about reaction between SiOR and substrate.

Table 1

Mixed Reactivity Crosslinkers. Adhesion of Elastomer A to Steel.[a]

Run	Crosslinker[b] (parts by weight)	Adhes., (pli)	Tensile, (psi)	Modulus, (psi)	Elong. at Break (%)
1	MeSi(OAc)$_3$	2	905	110	630
2	EtSi(OAc)$_3$ [c]	15	925	60	800
3	Me/EtSi(OAc)$_3$ [d]	9	1018	77	765
4	CH$_2$=CHSi(OAc)$_3$	10	910	80	640
5	Me$_3$SiOSi(OAc)$_3$ [c]	8	650	30	1000
6	(AcO)$_2$SiO(Me$_2$SiO)$_3$Ac [c]	3	750	60	745
7	(AcO)$_2$SiMeCH$_2$CH$_2$SiMe(OAc)$_2$	10	435	110	310
8	Me/EtSi(OAc)$_3$[d](2) + Si(OEt)$_4$ (1) not reacted	40	1220	80	800
9	(5) + Si(OC$_2$H$_4$OCH$_3$)$_4$ (3) not reacted	6	750	110	560
10	(5) + MeSi(OEt)$_3$ (3) not "	15	990	70	787
11	MeSi(OAc)$_3$ (2) + Si(OEt)$_4$ (1) equilibrated [e]	115	608	100	505
12	Me/EtSi(OAc)$_3$ (5) + Si(OEt)$_4$ (3) "	96	580	158	480
13	(5) + Et Polysilicate[f] (3) "	100	607	70	587
14	(5) + Si(OC$_2$H$_4$OCH$_3$)$_4$ (3) "	95	830	95	610
15	(4) + Si(O-n-C$_3$H$_7$)$_4$ (3) "	80	615	90	560
16	(40) + MeSi(OEt)$_3$ (27) "	60	760	80	723
17	(40) + MeSi(OMe)$_3$ (21) "	50	573	100	503

a. Series 500 cold rolled steel, mill finish aluminum, 1/16 in. adhesive layer, 180° peel. Details in Experimental Section.
b. For adhesive formulation, see Experimental Section.
c. For preparation, see Experimental Section.
d. Equal masses MeSi(OAc)$_3$ and EtSi(OAc)$_3$.
e. 1% acid clay catalyst, 1-3 hours at 110°.
f. Union Carbide "Polysilicate 40", 1% HOAc catalyst, 19 hours at 110°.

ADHESION PROMOTION WITH SiOR-SiOH ELASTOMERS

There is no *a priori* reason why application of promoting anions could not be separated temporarally from interfacial bond formation, i.e. why substrates could not be treated with, for instance, acetic acid and subsequently with a "Reaction B" elastomer (which we designate Elastomer B) that would then be expected to adhere well. Results of such experiments are shown in Table 2. In order to simplify assumptions concerning substrate surface composition, aluminum alone was used. Only one adhesive formulation was employed; its constituents and properties are found in the Experimental Section.

Runs 19-23 show results of treating aluminum with acetic acid and then applying Elastomer B. A quick dip in glacial acetic acid gives no bond improvement, although a one hour treatment is almost as good as a treatment overnight. The acid apparently reacts slowly with the alumina substrate. Runs 22 and 23 show that the acid may be dissolved in toluene or even in air and still promote adhesion.

Other carboxylic acids and derivatives are seen in Runs 24-28. Even such powerful acetylating agents as acetyl chloride and acetic anhydride are without effect if contact time is short. Benzoic and propionic acids promote adhesion as well as acetic does, formic a bit less well. Contact times were prolonged, so the acids' effects are manifested.

Table 2
Adhesion of Elastomer B to Treated Aluminum

Run	Treatment[a]	Duration	Adhesion, (1 wk)	pli (1 m.)
18	None (control)		2-5	3-6
19	HOAc, glacial	ca. 1 min	---	5
20		1 hr	18	21
21		16 hr	22	26
22	5 wt % in toluene	"	14	19
23	Satd. vapor in air	64 hr	9	17
24	Acetyl chloride	ca. 1 min	2	2
25	Acetic anydride	"	5	9
26	Formic acid, pure	64 hr	5	17
27	Propionic acid, pure	16 hr	26	20
28	Benzoic acid, 5 wt. % in toluene	"	19	12
29	NaOAc, satd. in 2-propanol[c]	ca. 1 min	9	19
30		7 hr	10	16
31	NH_4OAc, 2 wt. % in 2-propanol	16 hr	18	22

Table 2 (Continued)

Run	Treatment[a]	Duration	Adhesion, (1 wk)	pli (1 m.)
32	Ca(OAc)$_2$, satd. in 2-propanol[d]	16 hr	9	17
33	NaOAc, 2 wt. % in water	"	10	18
34	Glycine, 5 wt. % in water	"	18	19
35	NaF 2 wt. % in water[e]	30 min	6	8
36	NaCl " (0.342 M)	16 hr	11	19
37	HCL " (e)	"	9	16
38	KBr "	"	16	17
39	KI "	"	18	22
40	KCN "(e)	30 min	9	9
41	NaN$_3$ "(e,f)	"	9	12[g]
42	Na$_2$B$_4$O$_7$·10 H$_2$O (2 wt. % in H$_2$O)	16 hr	21	25
43	H$_3$BO$_3$ "	"	5	13
44	NaHCO$_3$ "	"	7	14[h]
45	KNO$_3$ "	"	20	19
46	KNO$_2$ "	"	17	22
47	Na Phosphates, pH = 7, 2 wt. % in water	"	15	17
48	Na$_2$SO$_4$, 2 wt. % in water	"	9	22
49	KClO$_4$. 1 wt. % in water	"	9	15
50	Piperidine, pure	"	--	6
51	Methyl ethyl ketoxime, pure	"	--	7
52	Tributylphosphine, pure	"	4	5
53	NaOH, 6 M in water	ca.10 sec	10	9[j]
54	NaOH (1%, 3 min) then NaCl, 2% in water	16 hr	5	5
55	then KBr, 2% in water	"	5	5
56	then KI, 2% in water	"	6	6
57	NaCl and NaI (0.342 M each) in water	"	5	13
58	KNO$_3$ and Na$_2$SO$_4$ (0.198 M each) in water	"	5	10
59	Water	"	6	7
60	2-Propanol	64 hr	6	6
61	Dimethylsulfoxide	"	6	8
62	Acetonitrile	"	3	3
63	Nitrobenzene	"	2	2
64	Toluene	16 hr	3	6

Table 2 (Continued)

Note:
a. All at room temperature. Volatile materials let evaporate at room temperature. Solutions of nonvolatiles removed by rinsing with pure solvent before drying as above.
b. 1 in. by 4 mil Al strips bonded to Al panels by 5-8 mil elastomer. 180° peel. F< 10 pli is clean peel, 10-16 pli gives mixed failure, > 16 pli gives completely cohesive failure.
c. 0.54 wt. %.   d. 0.010 wt. %.
e. Light etching.   f. Vigorous gas evolution.
g. 13 pli after 8 months.   h. 19 pli after 8 months.
i. Vigorous etching.   j. 8 pli after 8 months.

Runs 29-34 show the effects of carboxylate anions as adhesion promoters. Simple dipping in sodium acetate solution is as effective as seven hours' immersion—hence the reaction here is rapid (Runs 29,30). Acetates of other cations are effective, as is glycine. Thus, carboxylate anions as well as the acids promote the SiOR-metal oxide reaction.

(In point of fact, acetic acid probably manifests itself ultimately as acetate ion in the case of mixed reactivity crosslinkers. A sample of hydrated alumina was exposed to glacial acetic acid overnight and then sucked dry. Before acid treatment, its infrared spectrum between 3800-1200 cm^{-1} consisted solely of broad hydrate water peaks (3450 and 1630 cm^{-1}). After treatment new peaks at 2920, 2850, 1575, 1465, 1415, and 1335 cm^{-1} appeared, all assignable to acetate ion. No signs of unionized acid carbonyl (1760 and 1710 cm^{-1} for monomer and dimer respectively) was seen. Mass gain and BET surface area measurements indicated a monolayer to be present (See Fig. 1).

Effects of halide and pseudo-halide ions on Elastomer B adhesion are seen in Runs 35-41. Fluoride (Run 35) gives little improvement even after a month, even though the surface was etched. Chloride solutions (Runs 36,37) give better bond strength and the effect is the same whether sodium or hydronium is the cation. Bromide and iodide (Runs 38,39) give even better performance, maximum bond strength being attained in a week rather than a month. This order of increasing adhesion exactly parallels the order of decreasing affinity of the anions for alumina[10]. Neither cyanide nor azide increases adhesion substantially (Runs 40, 41).

Effects of oxygenated salts are seen in Runs 42-49. Tetraborate (Run 42) gives excellent adhesion while boric acid (Run 43) does not. The former is known to absorb strongly to alumina[11] as does sulfate in acid solution[12]. The alumina affinities of sulfate and chloride are roughly equal, as are their effects on adhesion (Runs 36,48). Adhesion promotion by bicarbonate and perchlorate (Runs 44,49) is moderate, by nitrate, nitrite, and phosphate good (Runs 45-47).

Runs 50-53 show results (all negative) of various bases on adhesion. The slight increase shown by hydroxide may be due to surface roughening (cf. Runs 54-56). Of all anions, hydroxide is absorbed most tenaciously to alumina.[13]

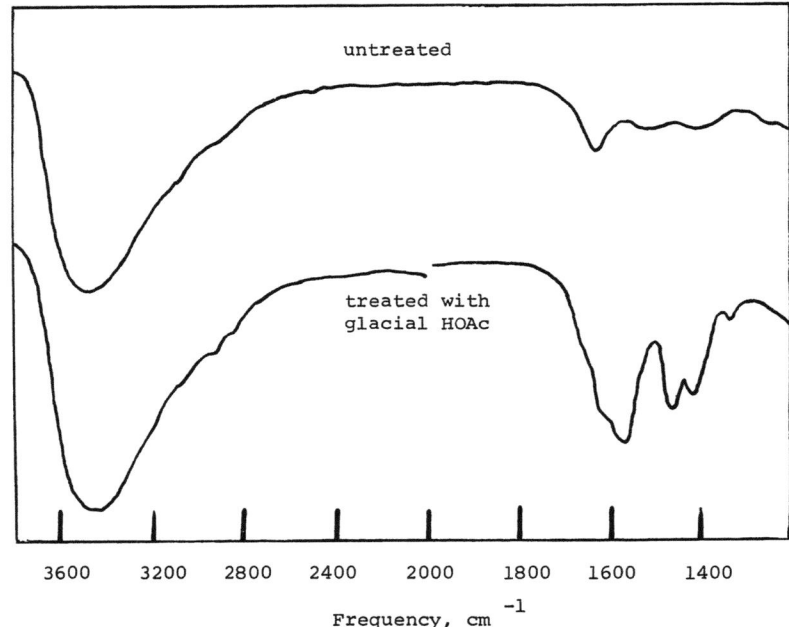

Figure 1. IR Spectra of Alumina.

It appears that alumina affinity and adhesion are inversely related. The obvious exception, tetraborate, is polyfunctional and may form a bridge between polymer and substrate. To test this hypothesis substrates were immersed in hydroxide, rinsed, and then treated with anions normally giving fair to good adhesion. Runs 54-56 show that, as expected, the halide ions cannot function. The surface is covered with hydroxide, which will give way to nothing else. In runs 57 and 58, aluminum is immersed in mixtures of anions, one absorbing more strongly than the other. In each case, as expected, adhesion is only mediocre. The more strongly absorbing species seems to prevail and substantial bond strength takes a month to develop.

Runs 59-64 show the results (none) of solvents alone. These serve in part as controls for the foregoing, and show clearly that the increased adhesion is not due simply to the removal of some weak boundary layer from the aluminum.

ADHESION PROMOTION WITH SiOH - SiH ELASTOMERS

Silanol-terminated silicone polymers can also be cured via Reaction C.

C. $\text{SiOH} + \text{SiH} \xrightarrow[\text{cat.}]{\text{Sn}} \text{SiOSi} + \text{H}_2$

A polyfunctional SiH compound is used as crosslinker. SiOR functionality can be introduced into such materials via Reaction D.

D. $\text{SiH} + \text{ROH} \xrightarrow[\text{cat.}]{\text{Pd}} \text{SiOR} + \text{H}_2$

Two different SiH crosslinkers, designated CX and CCX were used to cure Elastomers C and CC respectively. CX and CCX were partially akloxylated to crosslinkers DX and DDX respectively. From these latter crosslinkers, which are mixed reactivity crosslinkers in this cure system, were prepared Elastomers D and DD. In this way SiOR functionality can again be grafted into the polymer network, even though it is not needed for cure.

If the mixed reactivity hypothesis is correct, then substrates previously treated with "good" anions should adhere better to Elastomers D and DD than to C and CC. Table 3, shows that this is indeed the case. By far the strongest adhesion occurs when an SiOR functional material is cast against a substrate treated with acetic acid or sodium acetate.

RADIATION CURE STUDIES

If both SiOR groups in the polymer and "good" anions on the substrate surface are necessary for good bonding, absence of the former should preclude good adhesion even if the latter are present. This was shown in Runs 1-7. A more elegant demonstration is to radiation-cure filled silicone fluids in contact with treated and untreated aluminum.[14] This is a free radical reaction of the methyl groups and permits crosslinking without recourse to any silane chemistry whatsoever. If the adhesion-promoting agents work by somehow modifying surface topography, by making the surface more wettable, by removing weak boundary layers, or by increasing substrate affinity for silanol or siloxane groups, then they should work as well with radiation-cured polymers as with the others. If the agents

Table 3

Adhesion of SiOH - SiH Elastomers to Aluminum[a]

Run	Treatment[b]	Elastomer[c]	Pli (1 wk)	peel (1 m.)
65	None (control)	C	1	4
66		D	6	7
67	HOAc, glacial	C	10	44
68		D	48	>100
69	NaOAc, Satd. in 2-propanol[d]	C	6	10
70		D	9	40
71	None (control)	CC	14	10
72		DD	30	65
73	HOAc, glacial	CC	44	38
74		DD	95	>100
75	NaOAc, satd. in 2-Propanol[d]	CC	20	20
76		DD	95	>100

Note:
a. 1/4 in. by 4 mil Al foil bonded to Al panels by 1/16 in. elastomer. 180° peel. Details in Experimental Section.
b. All were 16 hr at room temperature.
c. Elastomers D and DD contain SiOR functionality, C and CC do not. See Experimental Section.
d. 0.54 wt. %.

do not work, then silane chemistry of some kind must be involved with their action.

Changes in wettability are unlikely, since the base polymer of Elastomers A, B, C, CC, D, DD, and E has a contact angle of <1.5° to untreated aluminum. Scanning electron micrograph (18,000 x) shows no surface change resulting from acetic acid treatment.

Table 4 shows results of such experiments. Neither silanol-functional nor non-functional polymers show any effect of substrate treatment. So the various physical explanations of their adhesion promoting action cited above are untenable. The "good" anions work by promoting reaction between polymer SiOR groups and the substrate.

EXPERIMENTAL

1. Elastomers

Formulations and mechanical properties are shown in Table 5.

Table 4
Adhesion of Radiation Cured Elastomers to Aluminum

Run	Treatment[a]	Elastomer[b]	Pli peel[c]
77	None (control)	E	2.4
78	HOAc, glacial		1.6
79	NaOAc, Satd. in 2-propanol[d]		2.2
80	None (control)	F	2.0
81	HOAc, glacial		2.7
82	NaOAc, satd. in 2-propanol[d]		2.0

Note:
a. All 16 hr at room temperature.
b. Elastomer E derived from silanol-terminated silicone fluid, Elastomer F from nonfunctional fluid.
c. 1 in. by 4 mil Al foil bonded to Al panels by 5-8 mil elastomer. 180° peel. All failure visually clean, 100% cohesive failure corresponds to 8.8 pli in this configuration.
d. 0.54 wt. %.

Table 5
Composition and Properties of Elastomers

	Wt. % in Elastomer							
	A	B	C	CC	D	DD	E	F
Hydrophobic silica	30	30	30	30	30	30	30	30
$HO(SiMe_2O)_xH$, M.W. ca 40,000	100	100	100	100	100	100	100	
$Me(SiMe_2O)_xSiMe_3$, M.W. ca 40,000								100
$Si(O\text{-}n\text{-}C_3H_7)_4$		4						
$(C_4H_9)_2Sn(OAc)_2$	0.08[a]	0.3	1	1	1	1		
$Me(HSiMeO)_xSiHMe_2$[b] (CX)			3					
CCX				4.5				
DX [c]					6			
DDX						6		
Various acetoxysilanes[d]	8-10							
Tensile strength, psi	d	946					360	340
Elongation at break, %	d	714	e	e	e	e	500	475
Modulus, psi	d	83					34	29

Note:
a. Not necessary for cure, gives quick skin.
b. Dow Corning 1107 Fluid.
c. See Experimental Section.
d. See Table 1.
e. Physical properties not measured.

Elastomers E and F were mixed on a three roll mill, the others in a Semco mixer after incorporation of the filler on the mill. Elastomers E and F were cured with 10.7 megarads cobalt-60-radiation at roughly 50°. The others were cured at ambient laboratory conditions.

2. Acetoxysilanes

$EtSi(OAc)_3$, bp:60-70°/0.08-0.13 torr, n_D^{25} = 1.4105 (lit.[15] values of 107.5-8.5°/8 torr, n_D^{20} = 1.4123,) was prepared from 1.00 mole $EtSiCl_3$, 3.25 moles Ac_2O, and 0.16 mole NaOAc at 100°.

$Me_3SiOSi(OAc)_3$ arose from 1.00 mole $Me_3SiOSiCl_3$ and 4.00 moles NaOAc in hexane, 2 hr at 60°. After filtration and evacuation, the product was used without further workup. $Me_3SiOSiCl_3$ arose from 12.0 moles $SiCl_4$, hexamethyldisiloxane (9.0 moles), Arquad 2 HT (15 g) (Armour and Co.) and 1,2-dimethoxyethane (300 cc) which were stirred seven days at room temperature, and stripped from the Arquad catalyst. Product (85%) distilled at 126° (lit. value of 127.9° [16]). (These disiloxane compounds were prepared by Paul Brown, whose synthetic methods are given above.)

$(AcO)_3Si(OSiMe_2)_3OAc$ was prepared from the tetrachloro- compound (1.00 mole) and NaOAc (5.4 moles) in cyclohexane, 19 hr at room temperature. After filtration and evacuation, the product was used without further workup. $Cl_3Si(OSiMe_2)_3Cl$ arose from $SiCl_4$ (1.00 mole), cyclic $(Me_2SiO)_3$ (1.00 mole), MeCN (50 cc), and DMF (5 cc), refluxing until the temperature began to rise above 57°. Product (84%) distilled at 42°/0.15 torr. Anal. calcd. for $C_6H_{18}Cl_4O_3Si_4$: Cl = 36.23%, Si = 28.62%. Anal. found: Cl = 36.08, 36.12%, Si = 28.35, 28.02%. (These compounds were also prepared by Paul Brown.)

$(AcO)_2SiMeC_2H_4SiMe(OAc)_2$ (49%) arose from the tetrachloro compound (1.00 mole) and Ac_2O (4.20 mole) at 80-105° until no more AcCl came over (this required several days). Excess Ac_2O was removed by evacuation and the product was crystallized twice from heptane, mp 73.5-75°. Anal. calcd. for $C_{12}H_{22}O_8Si_2$: C = 41.12%, H = 6.33%, Si = 16.03%. Anal. found: C = 40.5, 40.8%, H = 6.31, 6.38%, Si = 16.1, 16.0%. IR and NMR spectra are in accord with the expected structure. $Cl_2SiMeC_2H_4SiMeCl_2$ arose from Pt-catalyzed addition of equimolar $MeSiHCL_2$ to $CH_2=CHSiMeCl_2$ at 40-50°. Product distilled 99-104°/20-4 torr (lit. value of 109-11°/32.5 torr[17]). IR and NMR spectra were identical with those of authentic specimens. (Figure 2).

Other acetoxysilanes are products of the Dow Corning Corp.

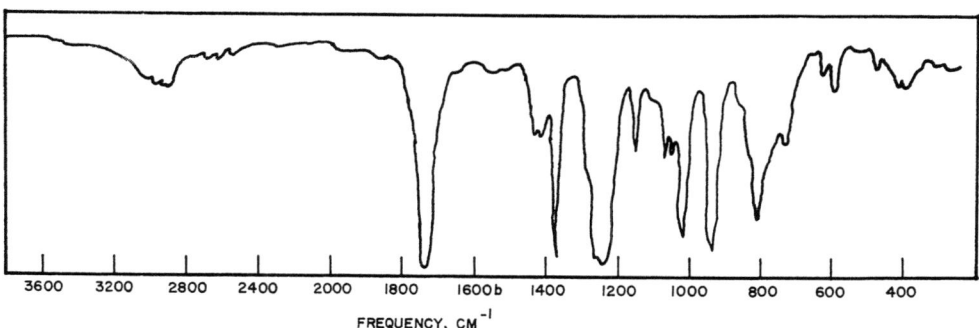

Figure 2. IR Spectrum of $(AcO)_2SiMeC_2H_4SiMe(OAc)_2$

3. Equilibration of SiOR and SiOAc Compounds

Starting materials (see Table 1) were stirred at 110° with an acid catalyst (acid clay, Amberlyst 16 resin (Rohm and Haas)) until no change occurred in the gas chromatogram during one half hour. Acid clay was used at 1 wt. % of the charge, Amberlyst resin at 7 wt. %.

4. Crosslinkers CCX, DX, and DDX

Crosslinker CCX is a proprietary Dow Corning product containing roughly five each Me_2SiO and $MeHSiO$ groups per molecule.

Crosslinker DX arose from treating Dow Corning 1107 Fluid (1.00 SiH equiv.) with methanol (0.33 mole) and 5% Pd-charcoal (1 wt. %) at 40-50°. Filtration and evacuation gave the desired substance. Analysis showed that 33% of the SiH groups had been removed. DX contained both SiH and SiOMe functionality.

Crosslinker DDX arose from reaction of CCX (1.00 SiH equiv.) with $CH_2= CHCH_2Si(OMe)_3$ catalyzed by a trace of H_2PtCl_6 at 90°. Analysis showed removal of four % of the SiH groups. DDX also contained both SiH and SiOMe functionality.

5. Prepolymers and Filler

Hydrophobic silica, $HO(Me_2SiO)_xH$, and $Me(Me_2SiO)_xSiMe_3$ are products of the Dow Corning Corporation.

6. Polymer Mechanical Properties

These were measured at ambient laboratory conditions on specimens pressed from 1/16 in. thick sheets of cured elastomer.

7. Adhesion Measurements

Substrates were cleaned by rubbing with a paper towel under trichloroethylene, rinsing with fresh solvent, and after solvent evaporation the process was repeated with MIBK. Test specimens for Elastomers A, C, CC, D and DD were prepared by extruding beads onto 4 x 6 x 1/16 unanodized aluminum panels and pressing 1/4 in. x 4 mil unanodized aluminum or 1/4 in. x 10 mil series 500 cold-rolled steel ribbons into the beads. Separation of 1/16 in. was maintained between ribbons and panels with a chase. After 1 wk. cure, the elastomer bead was cut through to the panel along each side of the ribbon. Adhesion was measured by 180° peel, 2 in/min jaw separation rate, at ambient laboratory conditions.

Test specimens for Elastomers B, E, and F were prepared by spreading uncured polymer on 1 in. x 4 mil unanodized aluminum ribbons and applying these to unanodized aluminum panels with a 10 lb hand roller. After cure these specimens were tested as above.

CONCLUSIONS

1. Dimethylsilicone compositions have potential for good adhesion to aluminum and steel if they contain SiOR groups.
2. This potential is realized if the substrate is exposed, either before or during silicone application, to sources of certain anions.
3. Anions strongly absorbed to the substrate (e.g. hydroxide and fluoride) do not promote adhesion via SiOR groups.
4. Anions less strongly absorbed (e.g. chloride and sulfate) promote adhesion slowly.
5. Bromide, iodide, nitrate, nitrite, tetraborate and carboxylate species give good adhesion rapidly.
6. The process is due to interfacial chemical reaction.

REFERENCES

1. This work was performed at the Dow Corning Corp., Midland, Michigan. Portions have been patented: T. W. Greenlee, U.S. 3,754,967; 3,754,969; 3,769,064; 3,772,240 (1973).
2. J. C. Goossens, U.S. 3,296,195 (1967); General Electric Corp., Brit. 1,064,930 (1967).
3. B. N. Dolgov, V. P. Davydova, and M. G. Voronkov, J. Gen. Chem. USSR $\underline{27}$, 1002 (1957).
4. H. W. Post and C. H. Hofrichter, Jr., J. Org. Chem. $\underline{5}$, 443 (1940).
5. R. P. Narain and R. C. Mehrota, J. Indian Chem. Soc. $\underline{39}$, (12), 855 (1962).
6. A. J. Barry, U.S. 2,405,988 (1946).
7. H. A. Schuyten, J. W. Weaver and J. D. Reid, J. Am. Chem. Soc. $\underline{69}$, 2110 (1947).
8. A. F. Reilly and H. W. Post, J. Org. Chem. $\underline{16}$, 387 (1951).
9. J. R. Schulz and W. H. Clark, U.S. 3,647,917 (1972).
10. S. Tustanowski, J. Chromatography $\underline{31}$, (1), 270 (1967).
11. T. W. Greenlee and O. Flaningam, unpublished results.
12. J. S. Fritz, S. S. Yamamura and M. J. Richard, Anal. Chem. $\underline{29}$, 158 (1957).
13. G. M. Schwab and G. Dottler, Angew. Chem. $\underline{50}$, (33), 691 (1937).
14. D. J. Fisher, R. G. Chaffee and E. L. Warrick, Rubber Age $\underline{88}$, (1), 77 (1960). E. L. Warrick, Ind. Eng. Chem. $\underline{47}$, 2388 (1955).
15. B. N. Dolgov, V. P. Davydova and M. G. Voronkov, J. Gen. Chem. USSR $\underline{27}$, 1593 (1957).
16. E. P. Mikheev and N. P. Filimonova, Soviet Plastics $\underline{1961}$, (8), 19.
17. V. M. Vdovin and A. D. Petrov, J. Gen. Chem. USSR $\underline{30}$, 838 (1960).

DISCUSSION

On the Paper by A. N. Gent and E. C. Hsu

R. J. Good (SUNY at Buffalo): (1) The energy of a dispersion force bond is believed to be nearer 0.5kcal than 2kcal. See the review paper by R. J. Good, in "Treatise on Adhesion & Adhesives", R. L. Patrick, ed., Vol. 1, Marcel Dekker, N. Y. (2) Have you considered the transition from locally adiabatic separation processes to isothermal processes, with decreasing speed of pulling? In your low-speed limit, the system is in thermal equilibrium with its environment and the processes resemble creep, rather than fracture.

A. N. Gent (Institute of Polymer Science): Thank you for your comment on dispersion bond energies. In answer to your second comment, the success of the WLF rate - temperature equivalence implies that the temperature itself is not important, but only the rate of segmental motion relative to the rate of peeling. Thermal conduction processes would not be expected to depend strongly upon the rate of segmental motion, i.e., upon temperature, and if they played a dominant role one would not expect agreement with the WLF temperature dependence. We conclude that adiabatic heating is not a major factor in these experiments.

S. E. Cantor (Uniroyal): Have you varied the microstructure of the elastomer?

A. N. Gent: No, although we have examined an EPR elastomer, with generally similar results.

F. M. Fowkes (Lehigh University): Perhaps the enhanced adhesion of the polybutadiene to "clean" glass might result from an acid-base interaction, for the olefin groups are weak bases and certain glass-cleaning methods leave the glass surface acidic.

A. N. Gent: Thank you for this suggestion.

K. L. Mittal (IBM Corp., Poughkeepsie): You have shown a linear increase in peel energy versus the amount of vinyl silane. I expected peel energy to become independent of amount of silane after a critical amount of vinyl silane. Would you please comment on this?

A. N. Gent: Apparently the amount of interfacial bonding did not reach an optimum level in our experiments. We estimate that there was at most about one vinyl group attached to the glass per 100 $\overset{\circ}{A}^2$ of surface, and that only a small fraction of these groups were employed in interfacial coupling to the rubber layer.

O. C. Elmer (*General Tire*): The cleaning of the glass surface may introduce microscopic imperfections and weakness or porosity that makes mechanical rubber/glass interaction possible. Microscopic glass removal has been reported in the literature. The fact that repeated cleaning increases the adhesion would agree with this hypothesis. I should think freshly fused glass may have fewest imperfections.

A. N. Gent: Thanks for the suggestion.

S. C. Sharma (*General Tire and Rubber Co.*): This question concerns the observed good adhesion between "clean" glass surface and polybutadiene. Is it possible that an "ultra'clean" glass surface (devoid of low mol. wt. adsorbed impurities) is really a higher energy surface than a glass surface coated with ethyl triethoxy silane and therefore bonds better to polybutadiene than the latter? Perhaps contact angle measurements on "ultra clean" glass surface would help resolve this issue.

A. N. Gent: Glass surfaces are generally regarded as high-energy surfaces in comparison with non-polar polymers. However, adhesion of a non-polar material to glass is attributed to the non-polar component only of the glass surface energy, and this component is not at all large. One would therefore expect polybutadiene to adhere to glass to the same degree as to a non-polar substrate, i.e. rather weakly.

On the Paper by R. W. Smith

S. D. Steen (*Dentsply International Inc.*): Would you elaborate on your observations with optical and SEM microscopy of surfaces obtained by sectioning with a "dull" knife or an abrasive wheel?

R. W. Smith (*Goodrich R. & D. Center*): This is strictly dependent upon the types of rubber involved. In some cases, significant detail is exposed with a dull knife and this has to do with differences in hardnesses of the rubbers involved in the construction. Generally, soft rubber surfaces exhibit coarse knife markings and harder rubbers show finer markings. The surface roughness exposed after sanding with a fine abrasive wheel varies in the same manner.

DISCUSSION

L. H. Lee (*Xerox Corp.*): How do you account for the presence of chlorine only on one side of one of your figures? Is it due to diffusion?

R. W. Smith: No, it is due to noise.

On the Paper by D. A. Given and R. W. Downey

A. I. Medalia (*Cabot Corp.*): The failure in your samples, which were all compounded with SRF black, was generally in the rubber rather than in the adhesive. Have you looked at more highly reinforcing blacks, also high-modulus vs. low-modulus blacks, to achieve higher strength and in the rubber and to promote the development of stronger adhesives?

D. A. Given (*Goodyear Tire and Rubber Co.*): Our compounding section investigated the use of other types of carbon black in the rubber used for the energy absorbing bumper. They found that the SRF black gave the best overall physical properties for this application. Some attention was given to the type of carbon black used in the adhesives, but no definite conclusions have been drawn at this time.

F. T. Parr (*Westinghouse*): What type of primer is type "B"?

D. A. Given: Chemlock 205.

On the Paper by E. P. Plueddemann and W. T. Collins

R. Iyengar (*DuPont Co.*): We are interested in your data on dependence of adhesion and modulus of stocks. In our work with textile fibers/adhesive/rubber system we find adhesion is inversely proportional to modulus, in contradiction to your observation that adhesion is directly proportional to modulus. Your comments?

E. P. Plueddemann (*Dow Corning*): Apparently the textile/adhesive/rubber system is quite different from the silane reinforced systems dealt in the paper. Consequently the effect of adhesion vs. modulus is not the same. Unmodified silica does not bond to rubber at all.

S. Schmukler (*Chemplex Co.*): 1) What is the mechanism of EPDM adhesion when using cationic styryl silane as opposed to the reaction SO_2Cl with amine in the Hypalon case and peroxide reaction with EPDM and the styryl group in the EPDM (peroxide) case? 2) Do you consider low density polyethylene like a rubbery elastomer?

E. P. Plueddemann: 1) The adhesion of Hypalon to aminosilane-treated fillers is, as suggested by Mr. Schmukler, most probably through reaction of $-SO_2Cl$ groups with the amine. Simple amine-

functional silanes as well as complex styryl-functional silanes
are effective adhesion promoters (Table II). Simple amine-functional
silanes, however, are not effective like the styryl silanes with
either S-vulcanized, or peroxide-vulcanized EPDM (Table I). The
styryl double bond must be very important with both rubbers, since
a cationic benzyl-functional silane (similar structure without the
double bond) is as poor as the aminosilanes with both rubbers. It
seems reasonable that styryl-functional silanes should co-react
with peroxide-cured EPDM, but I have no clear explanation why they
are so much more effective than vinyl- or methacrylate-functional
silanes. The styryl groups must also participate in S-vulcanization.
Their greater effectiveness may be due to their multifunctionalities
which provide higher cross-link density in the interphase region.
I don't understand, however, why this effect is less evident with
other S-vulcanized rubbers like SBR and NR. 2) Yes. I classify
LDPE as a flexible polymer with adhesion requirements similar to
those of elastomers.

On the Paper by T. W. Greenlee

E. P. Plueddemann (*Dow Corning*): I agree that covalent bonds are
almost certainly formed by reaction of alkoxy silanes with aluminum
acetate surface with elimination of alkyl acetate.

The Si-O-Al bond, however, is not stable against water, but sets
up an equilibrium in the presence of water to form Al OH + Si OH.
The silicone RTV mixtures invariably contain excess crosslinker
which can form a resinous layer at the interface with aluminum.
Thus, conditions are favorable (as described in the previous paper)
for water-resistant bonds to aluminum, even though the bulk of the
polymer is rubbery. The initial reaction of alkoxy silane with
aluminum salts appears from this paper to be critical in setting
up the initial adhesive contact.

PART FOUR

Natural Products and Structural Adhesives

Introductory Remarks

H. G. Arlt, Jr.

Arizona Chemical Company

Stamford, Connecticut   06904

From the earliest times, the ability to fasten objects together by innovative means has resulted in development of machines and structures. When simple mechanical fastening techniques were inadequate, they were supplemented by embryonic adhesive technology. Glue formulations were generally derived from plant gums and resins or cooking residues. Geographical availability of materials, rudimentary technology, and the inability to scientifically study adhesive bonds limited progress.

Over the past thirty years, new analytical techniques have been developed which allow the scientific study of adhesives. Our Symposium this week concerns the science of adhesives and this session focuses on Natural and Structural Adhesives. The papers today exemplify recently developed techniques and show the high science approach being applied to the adhesive bond. These studies replace historically empirical ones and begin to elucidate and separate the adhesive and cohesive factors.

The multi-disciplinary nature of this subject is evident from the titles of the papers. In this session alone, the chemistry and structure of tackifying terpene resins, the adhesion of pigments to textiles, the analysis of fracture surfaces by electron microscopy, the permanent bonding of aircraft structures and development of anerobic adhesives are elaborated.

The paper by C. E. Warburton concerning color printing on cloth binders examines cohesive and adhesive properties of pigment produced from the closely related copolymers of ethyl or butyl acrylate with acrylonitrile. Modern tensile, peel and shear tests are used, and surface energies calculated. The unique system chosen allows a

definitive separation of cohesive and adhesive effects, development
of a mechanism supported by surface energy calculations, and excellent correlation with the pragmatic crockfastness test.

The chemistry and structure of the tackifying terpene resins
are developed by E. Ruckel et al. These resins, produced commercially
from pine turpentine, since the mid-Thirties, are formulated with
natural rubber to produce pressure sensitive adhesives. More recently, the scope of their use has been broadened by formulation
with elastomers and waxes for hot melt applications. Empirical
application tests have developed a broad knowledge of utility but
little science or predictability. By use of sophisticated high
polymer techniques, polymerization mechanisms are used to explain
how the minor structural differences between the beta-pinene and
dipentene resins suit these resins respectively for pressure sensitive and hot melt adhesive usage. Again for use application the
critical aspects of the formulation are its adhesive and cohesive
properties as demonstrated by tack, shear and peel properties.

Permanent adhesive bonding of aircraft structures as elaborated
by J. Noland centers on why a specific 120° structural adhesive,
closely related to satisfactory 175° adhesives, failed and methods
of solving the problem. The failure was determined to be interfacial, progressive and to occur under low stress. Attenuated total
reflectance infrared spectroscopy indicated that surface chemical
instability was not the problem. Numerous tensile tests negated
cohesive failure of the formulation again pointing to adhesive failure at or near the metal surface. Examination of aluminum surface
preparation methods rapidly demonstrated the sensitivity of the
etched aluminum surface boundary layer to water. Once defined, this
problem was solved by careful surface preparation and choice of
surface primer. The interdisciplinary nature of adhesive investigations and required use of varied high science techniques necessary
to solve adhesive problems is well demonstrated.

The paper presented by J. P. Wightman et al, examines metal
surface before and after bonding with polyimide resins. Again, as
in the previous paper, the critical nature of the boundary layer
between the adhesive and the metal surface becomes evident. By use
of scanning electron microscopy, electron spectroscopy for chemical
analyses, and reflectance infrared spectroscopy, the actual surface
and boundary layers are examined. This paper develops experimental
and theoretical criteria for evaluating new cured adhesives.

This author does an excellent job interpreting data obtained
by application of modern experimental techniques and indicating
the value of the techniques applied for study and understanding
of the factors which affect bond-strength at the adhesive-substrate
boundary layer. The enumerated techniques correlate well with ten-

sile test results and in this system provide a good method for predicting resin bonding efficiency.

The review of anerobic adhesives presented by M. Hauser depicts a new type of adhesive. The materials described, cure (polymerize) only under anerobic conditions. In presence of air, curing of the adhesive is inhibited. The chemistry involved was first reported in 1940. Over the next twenty years, the state of the art progressed to the point where stable, commercially useful adhesives which cured only in the absence of air could be produced. The application was mainly for locking of threaded fasteners.

The major focus of this paper is on the chemistry of curing, formulation modifications, and properties of various anerobic adhesives developed since that time.

The session covers a broad field of science. A multiplicity of techniques and disciplines are utilized to approach specific commercial adhesive problems or to develop mechanisms and theories. The results discussed show the rapid progress of adhesive technology. Although the separate papers will have specific theoretical or practical impact, all are related by their dependence on science. The authors of these papers should be congratulated on their exceptional use of modern analytical methods and interdisciplinary techniques required to produce these valuable papers.

Editor's Note:

Both Drs. Warburton and Hauser chose to publish their papers elsewhere. A contribution by Dr. Griffith et al. is included in this Part.

The Use of SEM, ESCA and Specular Reflectance IR in the Analysis of Fracture Surfaces in Several Polyimide/Titanium 6-4 Systems

Thurman A. Bush, Mary Ellen Counts and J. P. Wightman

Virginia Polytechnic Institute and State University

Chemistry Department, Blacksburg, Virginia   24061

Scanning electron microscopy, electron spectroscopy for chemical analysis (ESCA) and specular reflectance infra-red spectroscopy were employed to characterize titanium alloy (Ti-6Al-4V) surfaces before and after bonding with polyimide resins. Water contact angles on the titanium alloy surface were shown to correlate with surface contamination. Diglyme and DMAC contact angles correlated with fracture strength of the completed adhesive joints formed by the condensation polymerization of benzophenone tetracarboxylic acid dianhydride (BTDA) and m,m'-diaminobenzophenone (m,m' DABP). Octane/water interfacial contact angles were used to show the presence of polar forces at the adhesive/adherend interface. Variations in adhesive strength were noted for condensation polymers formed in diglyme solutions of (i) BTDA and m,m' DABP, (ii) BTDA and m,p' DABP and (iii) m,m' DABP and pyromellitic dianhydride (PMDA). Scanning electron microscopy was used to observe the titanium alloy surfaces after various pretreatments and the surfaces of fractured joints. ESCA spectra were obtained for the cleaned alloy surface and for fracture surfaces. The intensity of the titanium peak in the ESCA spectra was related to the presence of thin polyimide films. Specular reflectance infrared spectroscopy was also used in the analysis of the fracture surfaces.

I. INTRODUCTION

A number of organic polymer resins which were discovered in the 1960's have shown promise as candidates for formulation as thermally stable adhesives (1). However, the adaptation of such novel polymers as practical adhesives has been hampered by a lack of sufficient experimental and theoretical criteria for evaluating new resins and predicting their suitability for adhesive purposes.

The processes for forming adhesive bonds between materials have been developed empirically. Current theories of adhesion remain controversial (2-4). The development of a general theory of adhesion has been deterred in part due to the experimental inaccessibility of interfacial interactions between solids and the difficulty in establishing the nature of the interface (3).

The objective of this work was the utilization of some recently developed techniques that may be of value in the characterization of the adhesive process between a titanium alloy and a variety of polyimide resin systems. The techniques utilized were electron spectroscopy for chemical analysis (ESCA), specular reflectance infrared spectroscopy, and scanning electron microscopy. Contact angles of various liquids on the titanium alloy were also measured. Specifically, the question arises to what extent are any of these techniques of value in the characterization of the interface and in the determination of interactions for the titanium 6-4/polyimide resin systems. Dwight and Riggs (5) successfully used ESCA, soft X-ray spectroscopy, contact angle hysteresis and electron microscopy to examine fluoropolymer surfaces.

II. EXPERIMENTAL

A. Samples

Panels of Ti-6-4 alloy adherend were obtained from the NASA-Langley Research Center. The panels were either used in the as-received condition or cleaned by the Pasa-Jell 107 method, a commercial process (American Cyanamid) for cleaning titanium alloy surfaces. The primary steps in this cleaning process are, briefly: sample immersion in degreasing 1,2-dichloroethane; immersion in an alkaline cleaner, SPREX AN 9 solution; pickling in an HNO_3/HF solution; and treatment with Pasa-Jell 107 (a chromate based acid paste).

Two sets of fractured lap-joint samples were obtained from the

NASA-Langley Research Center. The characteristics of the samples in the two sets are given in Table I. The first set of samples were lap-joints of Pasa-Jell cleaned Ti-6-4 panels bonded with one polyimide resin adhesive. The resin adhesive was prepared from benzophenone tetracarboxylic acid dianhydride (BTDA) and m,m'-diaminobenzophenone (m,m'DABP). The structures of these compounds are given in Table II. The uncured adhesive was applied on the adherend in the polyamic acid stage from either diglyme or DMAC solution and then heat cured to the polyimide resin form. This condensation polymerization reaction is shown below.

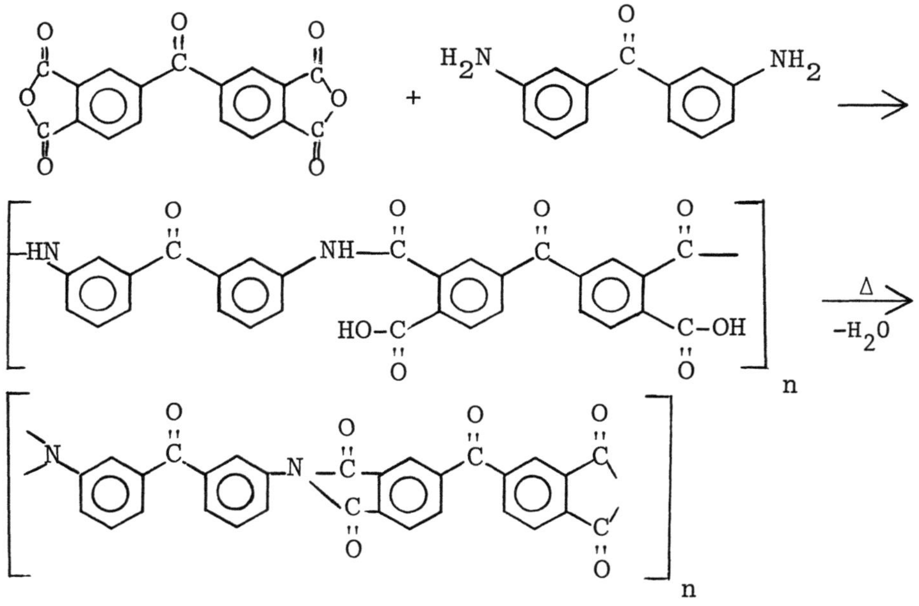

The second set of samples were lap-joints of Pasa-Jell cleaned Ti-6-4 panels bonded with various polyimide resin adhesives. The resin adhesives were prepared from BTDA or pyromellitic dianhydride (PMDA) and m,m'DAPB or m,p'-diaminobenzophenone (m,p'DABP). The structures of (PMDA) and (m,p'DABP) are also given in Table II. The uncured adhesive was applied on the adherend in the polyamic acid stage from the solvent diglyme and then heat-cured to the polyimide resin form. Tensile lap shear sandwich specimens were prepared by bonding 13 x 2.5 x 0.1 cm Ti-6-4 coupons with a 1.3 cm overlap. Typically, the coated coupons were air dried for 30 min at room temperature and then for 30 min at 60°C. Five successive coats were applied. The panels were overlapped at room temperature, placed under a constant pressure of 50 psi, and heated to 300°C at a rate of 5°C min^{-1}. The specimen was held at 300°C for

TABLE I

FRACTURE SAMPLE CHARACTERISTICS

Set	Code	Adhesive Sample	Solvent	Average Strength(psi)	$T_G(°C)$*
I	219D2	BTDA + m,m'DABP	diglyme	5280	
	220D3	BTDA + m,m'DABP	DMAC	2510	
II	1m2-517D(1,2,3,4)	BTDA + m,m'DABP	diglyme	3860	240-250
	1mp2-516D(1,2,3,4)	BTDA + m,p'DABP	diglyme	2073	260-270
	2m2-515D(1,2,3,4)	PMDA + m,m'DABP	diglyme	0	325-330

*maximum processing temperature – 300°C

TABLE II

STRUCTURES OF RESIN STARTING MATERIALS

BTDA = benzophenone tetracarboxylic acid dianhydride

PMDA = pyromellitic dianhydride

m,m'DABP = m,m'diaminobenzophenone

m,p'DABP = m,p'diaminobenzophenone

50 min. The lap-shear strength of each sample in both sets was determined at room temperature on a tensile tester (Cal-Tester Model TH-5).

B. Scanning Electron Microscopy

Representative samples were cut from both sets of the fractured lap-joint specimens. The samples were gold-coated and photomicrographs at various magnifications were obtained on an AMR scanning electron microscope (Advanced Metals Research Corporation: Model 900). Each sample surface was scanned totally to insure that the photographs were representative.

Five samples were cut from a bare Ti-6-4 panel. One sample, which served as a control, received no pretreatment and was placed into a vial. The Pasa-Jell cleaning process was applied to the four remaining samples. A sample after each step of the cleaning process was blown dry in a nitrogen stream and placed into a vial. Exposure time of the freshly cleaned material to the lab atmosphere was kept to a minimum. These five samples were examined in the scanning electron microscope.

C. Contact Angles

1. Materials. Distilled water was obtained from a Barnstead metal still. Mercury was obtained from the Glass Shop at the Virginia Polytechnic Institute and State University. Aldrich (99%) bis(2-methoxy ethyl) ether and octane were used. Dimethylacetamide (DMAC) was obtained from Burdick & Jackson (technical grade) and was distilled from calcium hydride. The degreaser used was 1,2-dichloroethane obtained from Fisher (ACS Certified). Metal coupons of the Ti-6-4 alloy (2.5 x 12 cm), and solutions of Pasa-Jell 107, HNO_3/HF, and SPREX AN 9 were furnished by the NASA-Langley Research Center. The polyamic acid (BTDA + m,m'DABP) was supplied by the NASA-Langley Research Center as a 20% solution of the polymer dissolved in diglyme [bis(2-methoxy ethyl) ether]. The solution was refrigerated to minimize degradation of the polymer.

2. Apparatus and Procedure. Contact angles of water, mercury, octane, DMAC, diglyme, and polymer resin were measured on prepared surfaces of Ti-6-4 samples with a Gartener Scientific microscope goniometer. Contact angle measurements were made on alloy surfaces cleaned by the Pasa-Jell method, and on the material as-received except for degreasing in 1,2-dichloroethane. Each liquid was introduced as drops delivered from a syringe inserted through the septum

of a custom optical cell. Saturation of the vapor phase within the
cell was insured by placing a small container of water within the
cell or, in the case of other liquids, placing several drops over
the surface in addition to the one being measured. Equilibrium
angles were recorded when the last two measurements separated by at
least fifteen minutes agreed to within ± 1°. Octane/water and
water/octane interfacial contact angles against the Ti-6-4 surface
were measured in a custom optical cell. The temperature for all
contact angle measurements was 25 ± 2°C.

D. Specular Reflectance Infrared Spectroscopy

A Unicam attachment was used with a Beckman IR-20A infrared
spectrophotometer in the specular reflectance studies. The spectrophotometer was operated in both the single and double beam modes.
IR reflectance spectra were obtained for the fractured samples of
Set II. The samples were placed in the reflectance attachment so
that the lapped portion of the panels covered the entire sample
window. This method allowed spectra of the samples to be obtained
in situ. Comparisons of intensities of individual peaks in the
different sample spectra proved to be unsatisfactory in evaluating
the amount of adhesive present on a panel. For this reason,
reflectivity as measured by percent transmission in a non-absorbing
region (2600 cm^{-1}) was considered as a possible characterization of
the fracture surface. The percent transmission in this region was
measured for each sample and also for a polished Ti-6-4 surface
and a Pasa-Jell cleaned Ti-6-4 surface.

E. Electron Spectroscopy for Chemical Analysis (ESCA)

The ESCA studies of the fractured samples from Sets I and II
were done with an AEI ES 100 photoelectron spectrometer using Al
Kα radiation (1486.6 ev.). Data acquisition was accomplished using
a AEI DS 100 Data System and a Digital PDP-8/e computer. Specific
spectrometer conditions are noted on the spectra which follow.
The cut samples were secured to the ESCA probe with double-sided
tape. ESCA spectra also were obtained for Ti-6-4 samples in the
as-received condition and cleaned by the Pasa-Jell method, and
coated with the polymer resin. Two polymer coated samples were
prepared by placing a 0.01 ml of 14% and 25% dilutions in
diglyme of the stock resin solution of BTDA and m,m'DABP on
cleaned Ti-6-4 panels. Each drop spread spontaneously over the
entire sample surface. The samples were dried at room temperature
for at least 24 hours prior to the ESCA and contact angle runs.

III. RESULTS AND DISCUSSION

A. Scanning Electron Microscopy (SEM)

1. <u>Adherend Surfaces</u>. The most striking feature in the scanning electron photomicrographs of the untreated metal surface is the amount of debris (large white particles) typically observed as shown in Figure 1. At 100 x the surface is noticeably fine

Fig. 1. Photomicrograph of Untreated Ti-6-4 Sample (100 X)

grained, whereas at the higher magnification (x1000) shown in Figure 2 the microscopic roughness readily becomes apparent. In addition, many smaller white presumably crystalline particles are contained in and projecting from a matrix of greyish material.

A photomicrograph (X1000) of the degreased sample in Figure 3 has the same surface features as the untreated sample except that the amount of debris is significantly reduced. The photomicrograph (X1000) in Figure 4 shows that the alkaline step of the cleaning

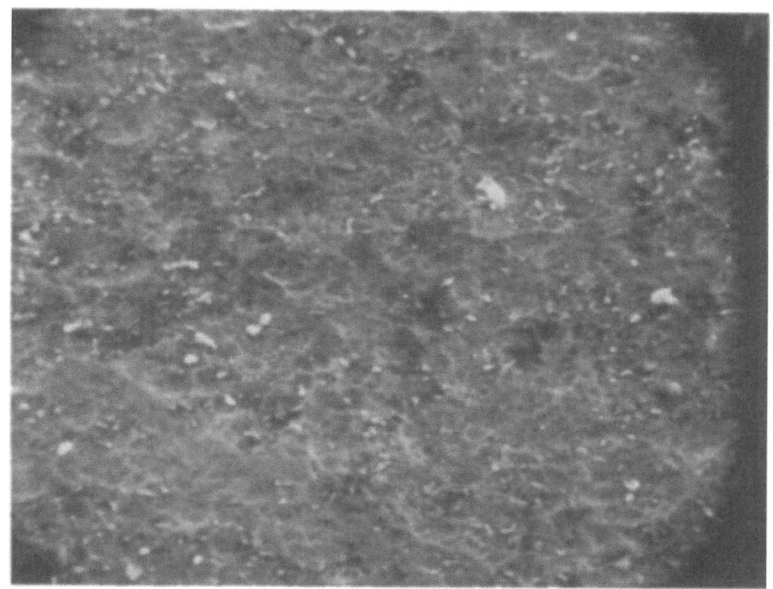

Fig. 2. Photomicrograph of Untreated Ti-6-4 Sample (900 X)

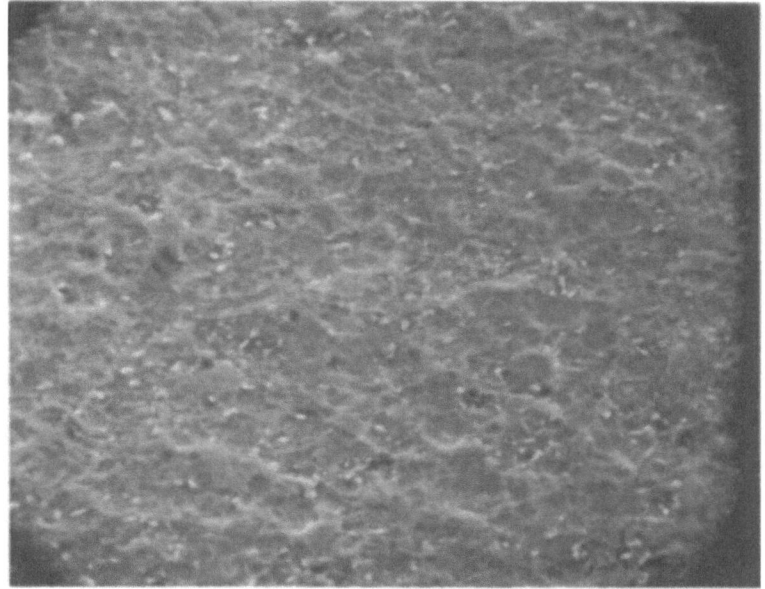

Fig. 3. Photomicrograph of Degreased Ti-6-4 Sample (900 X)

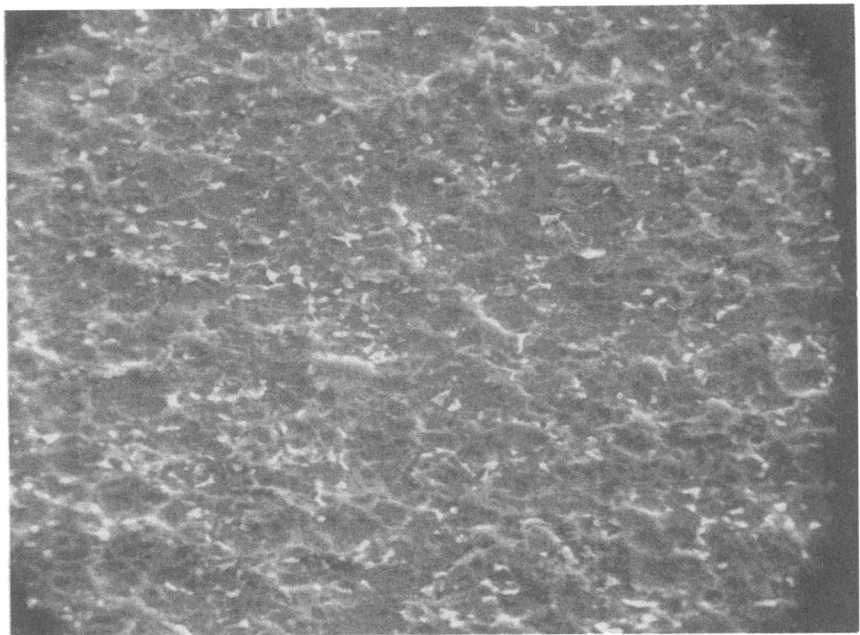

Fig. 4. Photomicrograph of Alkaline Cleaned Ti-6-4 Sample (1000 X)

process selectively etches the grey material, thus exposing more of the small white particles. Acid cleaning continues the selective etching of the grey material as indicated by the photomicrograph (X1000) in Figure 5. No discernibly different features were noted in the photomicrographs after the Pasa-Jell treatment. Thus, each step of the cleaning process produces distinct changes in surface features except the final Pasa-Jell treatment.

Ti-6-4 is an alpha-beta titanium alloy (6% Al, 4% V) readily available commercially (6,7). The alpha phase crystallizes in a hexagonally close packed array and the beta phase in a body centered cubic array. The beta phase is the high temperature form and exists in equilibrium with the alpha phase at room temperature. The photomicrographs in Figures 1 through 5 were compared with those of the ASM's Atlas of Microstructures of Industrial Alloys (6). On the basis of similarities in the photographs, the white particles in the photomicrographs are identified as the beta phase of the titanium alloy and the grey material as the alpha phase.

More of the beta phase particles of one to five microns in length are exposed in the cleaning process. The beta phase particles, occupying predominantly ridges and high points on the surface, would be expected to be the first points of contact for an adhesive

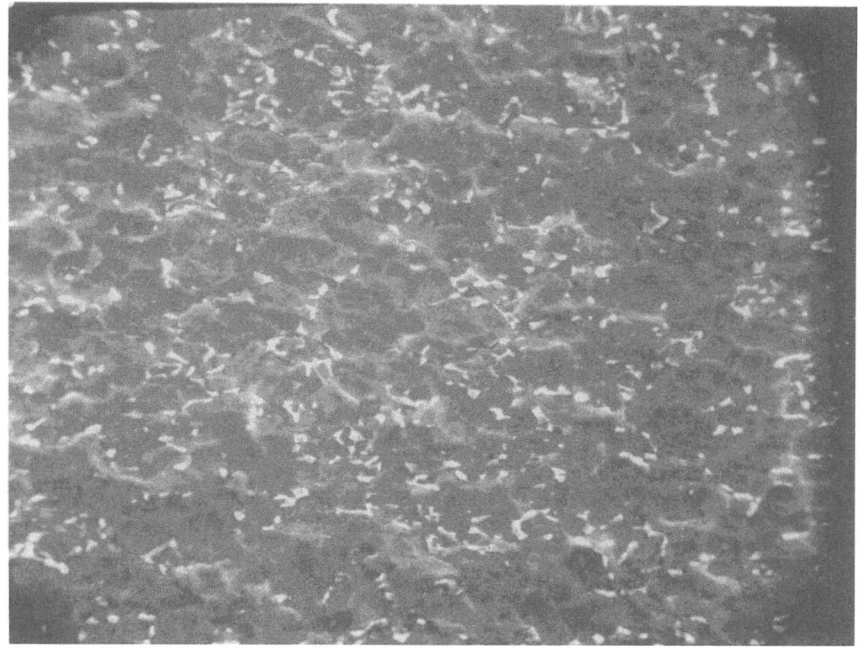

Fig. 5. Photomicrograph of Acid Cleaned Ti-6-4 (100 X)

material spread on the surface. Steps and ridges of the alpha phase become important features as etching takes place. Etching along planes of the crystal faces oriented oblique to the surface produces the steps and ridges as seen in Figure 5. Similar SEM results for Ti-6-4 have been reported by Hamilton (8).

Comparison of photomicrographs of three separate cleaned Ti-6-4 samples over a 12 month period shows the same details, thus indicating the reproducibility of the cleaning process in producing the surface effects noted.

2. Fracture Surfaces - Set I. Photomicrographs of the fractured samples from Sets I and II are shown in Figures 6 to 16. Diglyme was used as a solvent in Sample 21902 whereas DMAC was used as a solvent in Sample 220D3 in Set I. The photomicrographs of Sample 219D2 in Figure 6 and of Sample 220D3 in Figure 7 indicate dramatically the difference in the extent of surface coverage of the adhesive in these two systems. Sample 219D2 (diglyme) exhibits an almost complete coverage of the metal surface by the adhesive with only small patches of metal exposed indicative of cohesive failure. On the other hand, the surface of sample 220D3 (DMAC) has large areas of metal exposed as seen in Figure 7 indica-

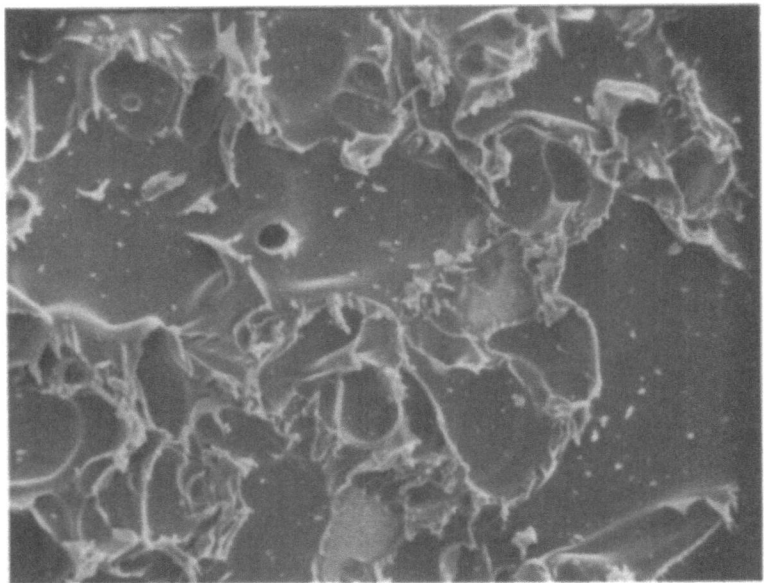

Fig. 6. Photomicrograph of Fracture Sample 219D2 (90 X)

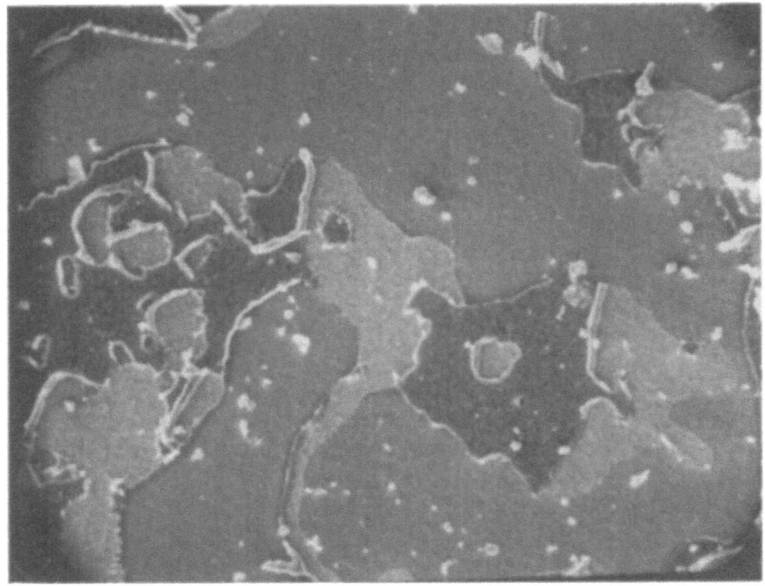

Fig. 7. Photomicrograph of Fracture Sample 220D3 (90 X)

tive of adhesive failure. The adhesive apparently did not wet the substrate in this system. Additional evidence for non-wetting of the substrate by the adhesive in the DMAC system compared to the diglyme system is seen on comparison of the photomicrographs in Figures 8 and 9. The adhesive/substrate interface in Figure 8 is

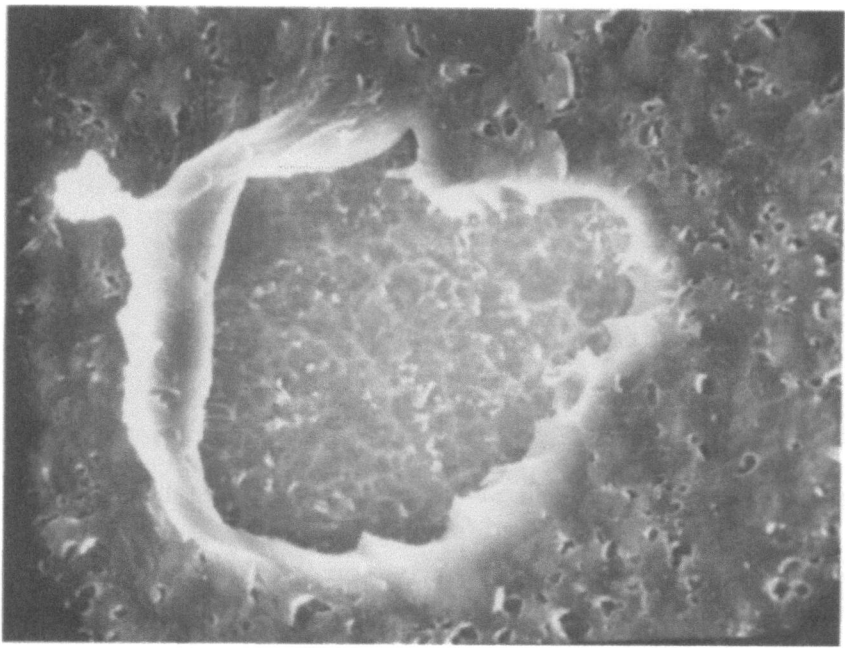

Fig. 8. Photomicrograph of Fracture Sample 220D3 (1000 X)

characterized by a sharp break whereas the same interface in Figure 9 is continuous. The better bonding in Figure 9 is obvious. The substrate surface of Sample 200D3 (Figure 8) appears to contain particles identified previously as the beta phase of the alloy whereas Sample 219D2 (Figure 9) shows less of this particular feature. Since the adhesive of Sample 219D2 has wet the surface, perhaps the fewer number of these particles observed is additional evidence for the presence of a film of adhesive on the surface.

The scanning electron microscope results described above correlate well with the breaking stress data of Table I. That is, the fracture strength of Samples 219D2 and 220D3 decreases as the extent of wetting or surface coverage decreases as seen in Figures 6 and 7.

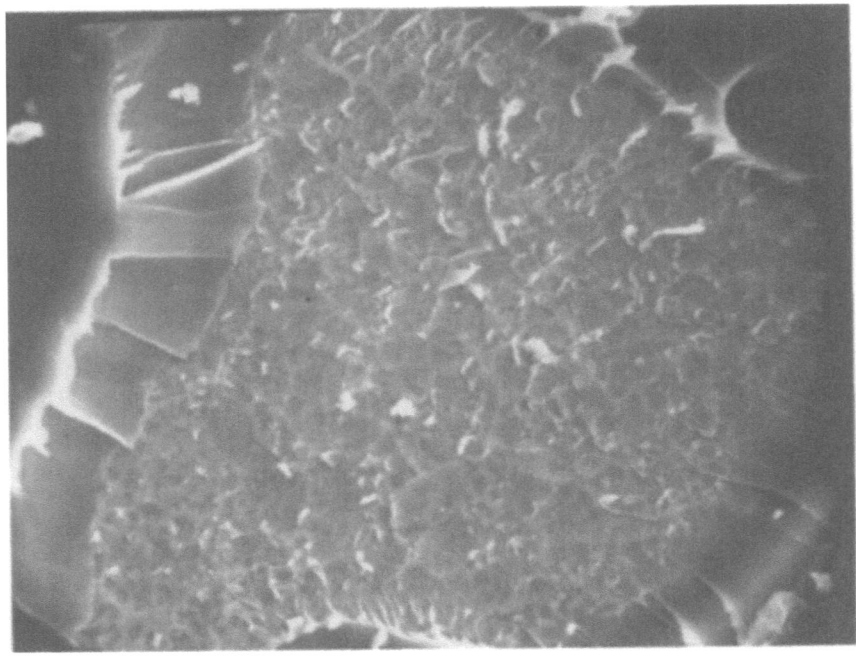

Fig. 9. Photomicrograph of Fracture Sample 219D2 (1000 X)

3. **Fracture Surfaces - Set II.** The samples in Set II were prepared twelve months after the samples in Set I. As noted in Table I, the average lap-shear strength of the samples in the 1m2-517 series (Set II) was 3860 psi. This series has the greatest lap-shear strength of the different series in Set II. The 1m2-517 series is the same adhesive-solvent system as the sample 219D2 in Set I. The difference in the absolute value of the lap-shear strength is not considered significant. In Figure 10 is seen a 100X photomicrograph of the sample 1m2-517D1. The features of this sample were representative of the other samples of this series. The significance of this photomicrograph was the apparent absence of the metal substrate structure. The excellent reproducibility of the SEM analysis of the fracture surface is demonstrated by the similarity of the features in Figure 10 and Figure 6 for Set I for the same BTDA + m,m'DABP/diglyme system. A closer examination of this sample at 500X can be seen in Figure 11. The smoothness of the pockets relative to the jagged areas is apparent. The jagged regions are believed to result from the fracture of contact areas between the two adhesive-coated panels when the samples were lap-shear tested. Adhesive strength might be substantially increased if more contact with the resin were possible. The pockets in Figure 11 represent non-bonding areas.

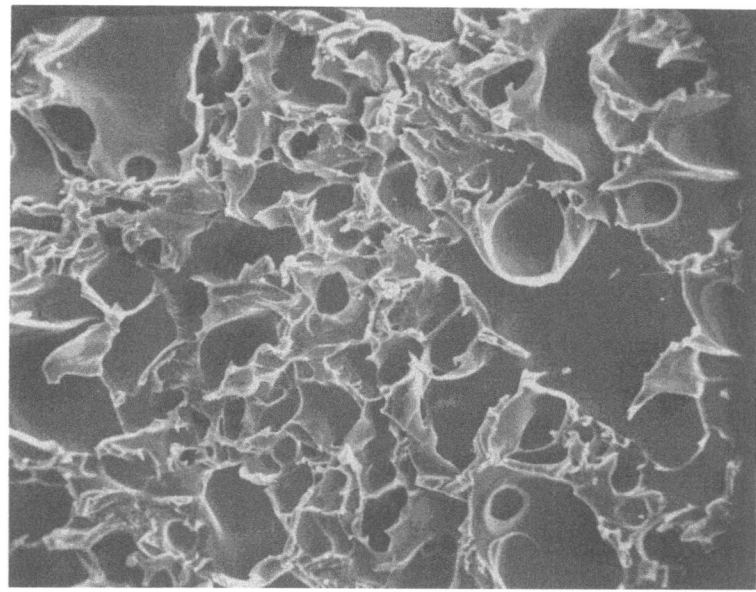

Fig. 10. Photomicrograph of Fracture Sample 1m2-517D1 (90 X)

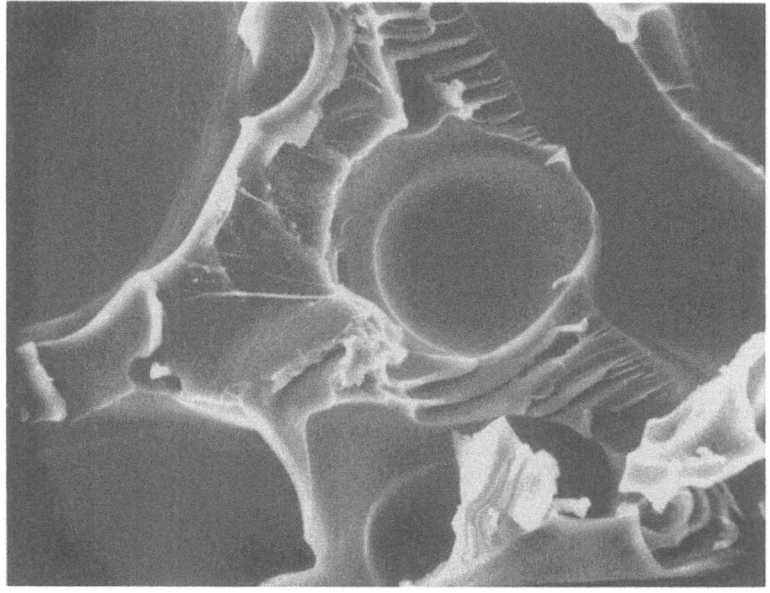

Fig. 11. Photomicrograph of Fracture Sample 1m2-517D1 (450 X)

The samples in the lmp2-516 series had an intermediate average lap-shear strength of 2073 psi (Table I). A 20X photomicrograph, Figure 12, of the sample lmp2-516D1 illustrates a structure very different from that observed for the lm2-517 series of fracture samples. The dissimilarity of the lm2-517 and the lmp2-516 samples is more apparent in Figure 13 (500X) which is a photomicrograph of one of the lighter regions seen in Figure 12. Whereas in the lm2-517 sample the adhesive appeared smooth, the adhesive in this photomicrograph appears porous and brittle-like. A closer examination of a darker region noted in Figure 12 is seen in Figure 14 (1000X). This region appears smooth with no evidence of metal substrate. The important feature to note in Figures 12-14 is again the lack of the metal substrate structure.

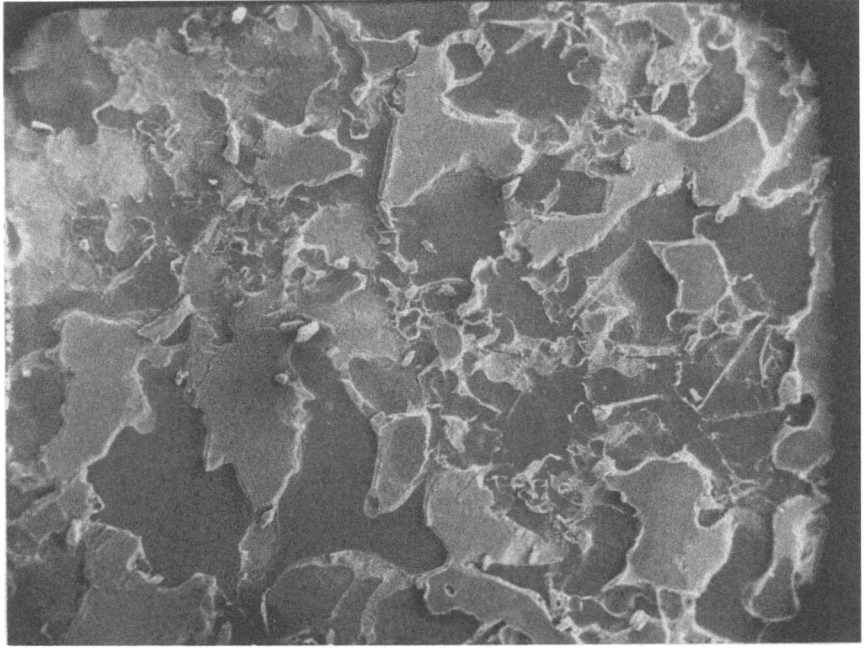

Fig. 12. Photomicrograph of Fracture Sample lmp2-516D1 (20 X)

A 20X photomicrograph of sample 2m2-515D1, which is representative of the samples of zero strength, is seen in Figure 15. This sample has a jig-saw puzzle appearance in that the adhesive is cracked and broken. This feature is more clearly seen in Figure 16 (500X). The metal substrate structure is apparent and

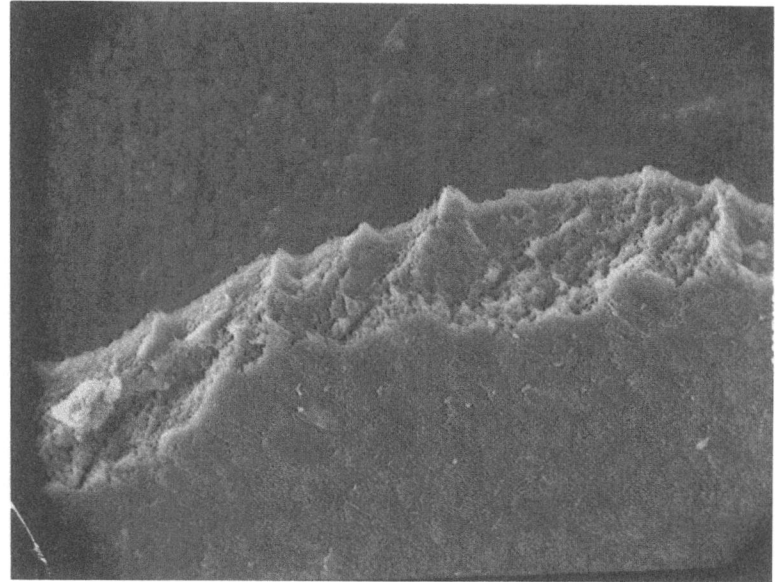

Fig. 13. Photomicrograph of Fracture Sample 1mp2-516D1 (450 X)

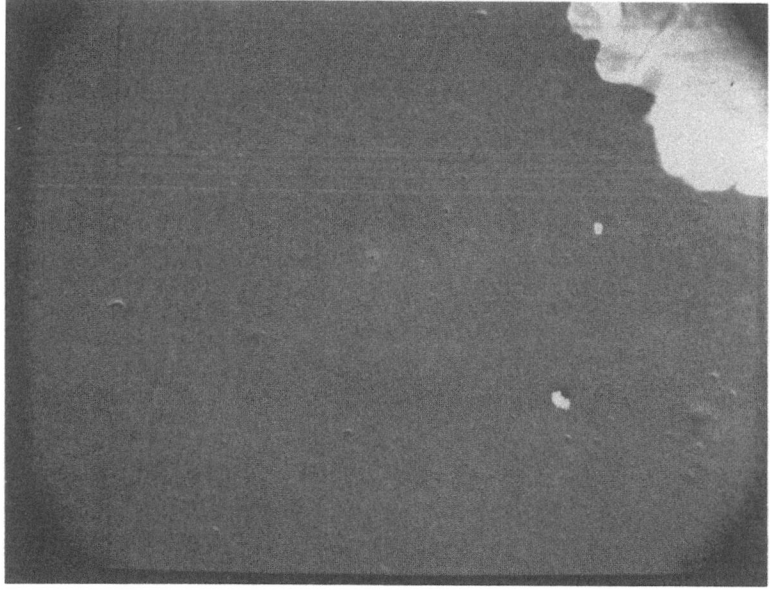

Fig. 14. Photomicrograph of Fracture Sample 1mp2-516D1 (900 X)

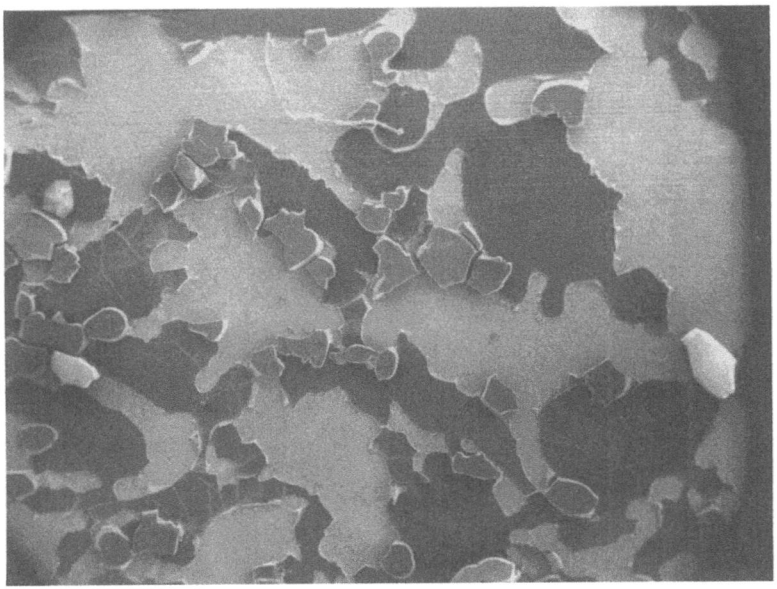

Fig. 15. Photomicrograph of Fracture Sample 2m2-515D1 (18 X)

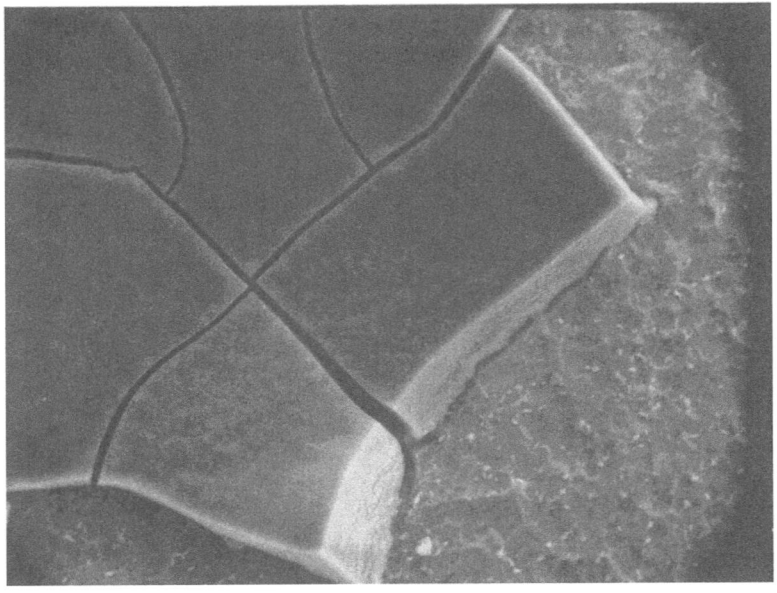

Fig. 16. Photomicrograph of Fracture Sample 2m2-515D1 (450 X)

there appears to be little wetting between the adhesive and the adherend.

The SEM results suggest cohesive failure in the 1m2-517 and 1mp2-516 series because of the absence of metal substrate structure. The lower lap-shear strength of the 1mp2-516 series compared to the 1m2-517 series is attributed to a difference in the cohesive strength of the two polyimide resins. An adhesive failure mode is suggested for the 2m2-515 series of fracture samples from the appearance of metal substrate.

B. Contact Angles

1. Contact Angles for Various Liquids With Ti-6-4. The advancing contact angles are given in Table III for various liquids with the titanium alloy. Each value represents the average of at least three independent measurements. The use of distilled water in the Pasa-Jell process produced a surface that gave a water contact angle between 5 and 15°, whereas the use of deionized water gave an angle some ten degrees higher. It was further noted that if the drying was done in a nitrogen stream and the water drop introduced while the sample was still in the nitrogen atmosphere, the drop would spread as observed by Harkins and Grafton (9).

The effect of laboratory air present in the drying step of the alloy cleaning process was examined by measuring the water contact angle at various times of exposure to laboratory air for the alloy

TABLE III

CONTACT ANGLES OF VARIOUS LIQUIDS ON PREPARED SURFACES OF TI-6-Al-4-V SAMPLES AT 25°C

Liquid	Preparation	
	Cleaned	As-Received
Water	0-25°	54°
Mercury	Not Measured	160°
Octane	0°	0°
Diglyme	0°	5°
DMAC	8°	23°
Octane/Water	175°-180°	140°
Polyamic acid solution	12°	28°

surface after the nitrogen drying step of the Pasa-Jell method. Typical results are given in Figure 17. After each measurement, the drop was evaporated under a nitrogen stream followed by exposure to lab air for the indicated time and application of a new drop for measurement. The water contact angle (Figure 17)

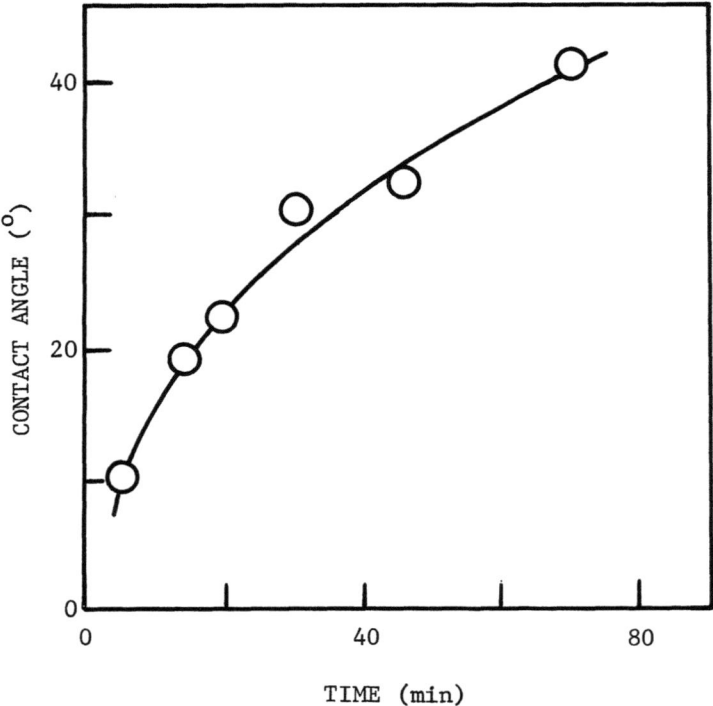

Fig. 17. Water Contact Angle Versus Lab Air Exposure Time for Cleaned Ti-6-4 Samples at 25°

after some four days of exposure to lab air was 60°. Koranyi and Acs (10) also reported an increase in the contact angle of water on glass from 4° to 23° within 4 hours after heating. The noted increase in contact angle (Figure 17) is taken to be indicative of contamination by adsorption of contaminants from laboratory air. This interpretation is in general agreement with the work of Bartell and Smith (11) for water contact angles on gold surfaces. They reported the following values for the conditions indicated: water vapor + pure air : 6°, water vapor + benzene vapor : 84°, and water vapor + lab air : 65°.

2. Octane/Water Interfacial Contact Angles. Measurement of the octane/water/titanium alloy interfacial contact angle verified the increasing contact angle noted in Figure 17. A freshly cleaned surface gave an interfacial contact angle of approximately 175° (Table III). Exposure to lab air for thirty minutes decreased the measured interfacial angle to 152°, close to the value observed for the untreated surface. Thus octane/water/solid contact angle decreased with increasing contamination of the Ti-6-4 surface.

The octane/water interfacial contact angle, according to Hamilton, (12), indicates the hydrophilicity of the cleaned metal surface and provides an estimation of the polar forces (non-dispersion forces) acting across the interface. Hamilton's equation is

$$\cos \theta = (\gamma_w - \gamma_o - I_{sw}) / 48.3$$

where γ_w is the surface tension of water (72.0 dynes/cm), γ_o is the surface tension of octane (21.8 dynes/cm), and I_{sw} is the interfacial free energy contribution from hydrophilic (polar) interactions at the solid/water interface. The value of 48.3 is the experimentally determined water/octane interfacial energy (dynes/cm). According to Hamilton (12), solids capable of dispersion interactions only have a octane/water/solid contact angle of 50°. For surfaces with polar sites, the contact angle is > 50° due to interaction of the polar sites with water. The smaller octane/water contact angles for the cleaned surface and the surface exposed to lab air shows the effect of contamination on contact angle.

Differences in the contact angles for diglyme and DMAC (Table III) on the titanium alloy make an interesting comparison in view of the scanning electron microscopy results. Diglyme wets the cleaned surface whereas DMAC has a finite contact angle. The as-received material which has a higher level of contamination still exhibits a smaller diglyme contact angle than does DMAC. The correlation between fracture strength and wetting as observed in the scanning electron microscope has been discussed above. The correlation is now further documented by the measured contact angles. The smaller contact angle for diglyme compared to DMAC may be indicative of the better wettability of diglyme for the titanium alloy.

Water and octane/water contact angles were measured on separate Ti-6-4 surfaces coated with 3% and 5% polyamic acid (BTDA + m,m' DABP) solutions in diglyme. The average values of the water and octane/water contact angles were $54 \pm 1°$ and $114 \pm 8°$, respectively. The high and constant water contact angle indicates a compact partly hydrophobic surface film (13). The value of 114° for the octane/water contact angle is greater than the 50° contact angle for dispersion forces only and implies that the polyamic acid film is capable of polar interaction.

Conflicting evidence on the role of contact angles in adhesion is found in the literature. Sharpe and Schonhorn (14) cite a zero degree contact angle for adhesive on substrate as a valid criterion for selection of a good adhesive. On the other hand, Muchnick (15) found a poor correlation between joint strengths and contact angles. At least for the titanium alloy/polyimide system, contact angles are significant as adhesion criteria for the following reasons: (1) in a qualitative way (ascertaining the wettability), contact angles of diglyme and DMAC correlated with the fracture strength of two samples as also shown by electron microscopy, (2) water contact angles were indicative of the level of contamination for the alloy surface, and (3) octane/water contact angles provided an insight into the nature of the forces capable of interacting at the alloy/adhesive interface.

C. Specular Reflectance Infrared Spectroscopy (SRIS)

The SRIS study was undertaken in an attempt to correlate the intensity of absorption peaks on the different samples to the amount of adhesive remaining on the panels. Reflectance spectra of the adhesive were obtained for all samples in Set II as expected because the scanning electron photomicrographs showed significant amounts of adhesive present on all samples. It should be emphasized that this is an in-situ method for the infrared analysis of fracture surfaces. The following assignments were made based on the major absorption peaks for sample 1mp2-516: 700, 840, 920, 970, and 1080 cm^{-1} ($\delta\phi$), 1200 and 1370 cm^{-1} (ν C-N), 1270 and 1730 cm^{-1} (ν C=O).

It proved impossible to make any definite correlations between peak intensities and the quantity of adhesive present on a panel because of reflectivity differences of the samples. These differences prevented peak height comparisons from a common base line.

For this reason, reflectivity as measured by percent transmission was used to characterize the fracture surfaces. The percent transmission was measured at 2600 cm^{-1} where no absorption occurred. The percent transmission of a sample from each series in Set II and also for a polished Ti-6-4 panel and an as-received cleaned Ti-6-4 panel as determined in both single beam (SB) and double beam (DB) modes are listed in Table IV.

The percent transmission value obtained in the single beam mode for the fracture surfaces of Set II are plotted against the respective lap-shear strength in Figure 18. There appears to be a somewhat linear relationship between the percent transmission and the shear strength of the samples. This may be of no more than

TABLE IV

SRIS PERCENT TRANSMISSION OF FRACTURE SURFACES

Sample	Spectrum #		
	#18-SB	#19-DB	#24-SB
1m2-517	21 ± 2	19 ± 1	11
1mp2-516	39 ± 4	44 ± 4	19
2m2-515	56 ± 2	63 ± 3	29
Ti-6-4, cleaned			45
Ti-6-4, polished			97

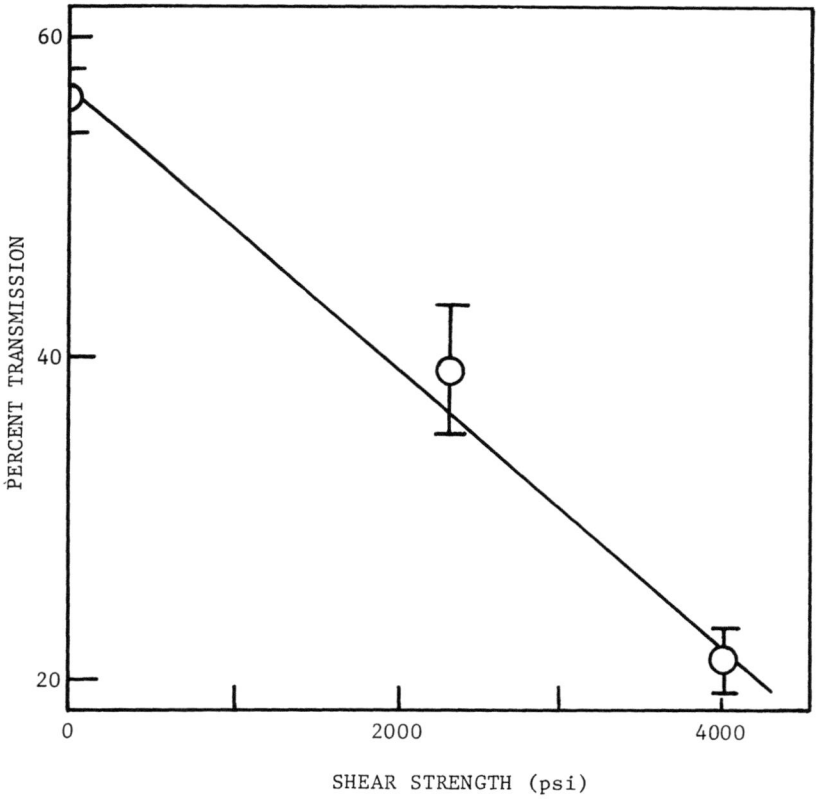

Fig. 18. Percent Transmission vs. Shear Strength for Set II Sample

qualitative significance but it is interesting to note that the samples of zero shear strength (series 2m2-515) in which bare metal was seen in the SEM micrographs have the highest reflectivity.

D. Electron Spectroscopy for Chemical Analysis (ESCA)

Figure 19 shows typical ESCA spectra for the titanium alloy selected from 23 separate ESCA runs on samples that were either in the as-received state or cleaned. Table V is a listing of the binding energies for titanium, oxygen, nitrogen, and carbon corrected for the work function of the spectrometer. The uncertainty in assignment of the cited binding energies is $\pm$ 0.3 ev at a 95% confidence level determined for five independent sample runs. The literature values and assignments in Table V are from Siegbahn (16). The binding energies for titanium and oxygen are in excellent agreement with the values of 457.9 and 529.6 ev reported by Hamilton (8) for Ti-6-4 samples. The observed shift in the binding energies of the titanium doublet from the literature values is taken to indicate the presence of an oxide film on the surface of the Ti-6-4 sample. The oxygen peak was typically broad as shown in Figure 19b with the main shoulder being of lower energy. The carbon peak in some spectra exhibited shoulders indicating different types of carbon contamination in contrast to the sharp peak exhibited in Figure 19c.

TABLE V
BINDING ENERGIES IN THE ESCA SPECTRA OF THE TI-6-AL-4-V SAMPLES

| Element | Binding Energy (ev) | | | Assignment |
	Cleaned	As-received	Lit. Values	
Ti	463.4	462.9	461	$2p_{1/2}$
	457.7	457.1	455	$2p_{3/2}$
O	529.3	528.9	532	$1s_{1/2}$
C	(284)	(284)	(284)	$1s_{1/2}$
N	ND	398.7	399	$1s_{1/2}$

ND - not determined

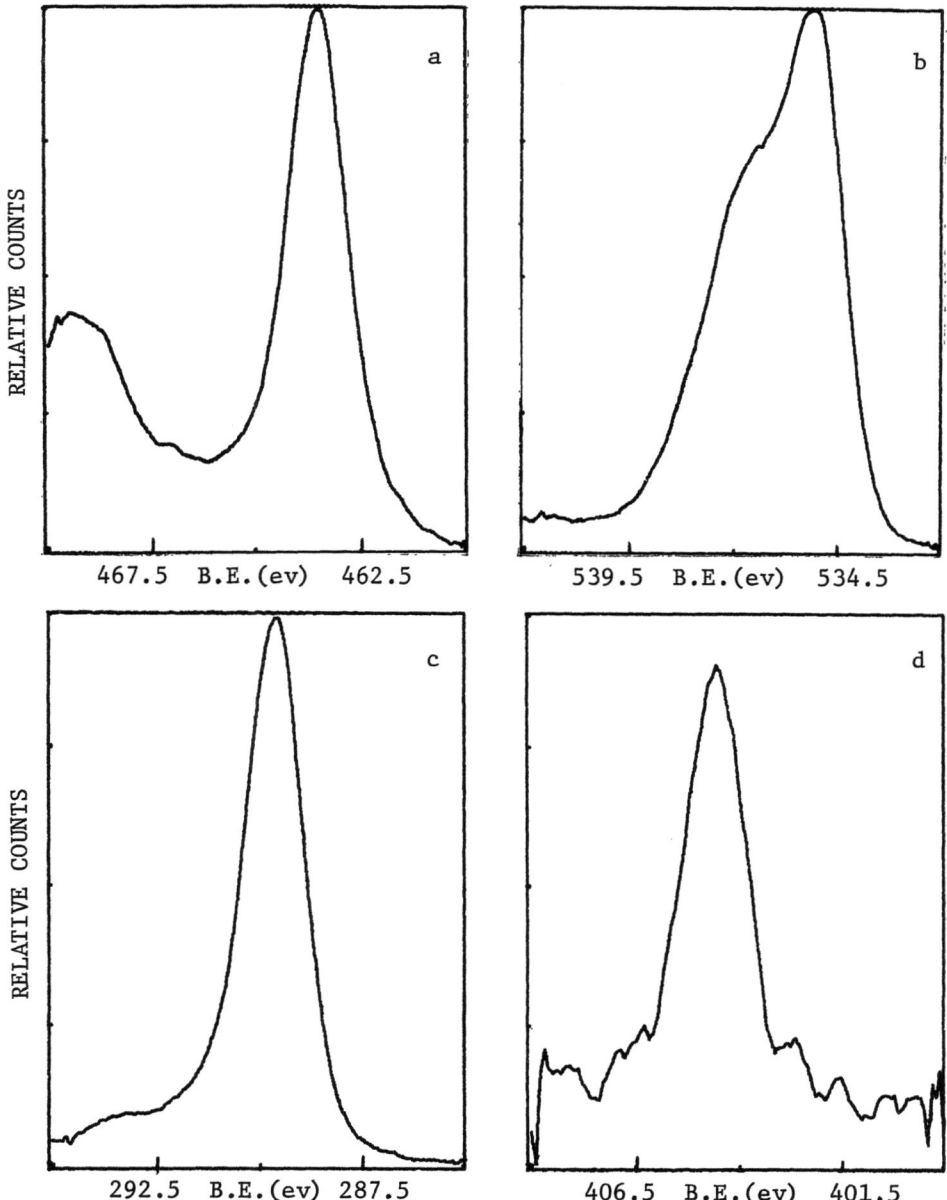

Fig. 19. a. Typical ESCA Spectrum of Titanium $2p_{1/2}$ and $2p_{3/2}$ Electron
b. Typical ESCA Spectrum of Oxygen $1s_{1/2}$ Electron
c. Typical ESCA Spectrum of Carbon $1s_{1/2}$ Electron
d. Typical ESCA Spectrum of Nitrogen $1s_{1/2}$ Electron

No Ti signal was observed in the ESCA spectra of separate Ti-6-4 surfaces coated with 3% and 5% polyamic acid (BTDA + m,m' DABP) solutions in diglyme. Since the escape depth of secondary electrons is 50-100 Å, the absence of a titanium signal is due to the presence of a uniform film thicker than 50-100 Å. The conclusion of a uniform film rather than a patch film is consistent with the measured contact angles.

ESCA studies proved to be of great value in assessing the failure mode of the lap joints. The escape depth of the photo ejected electrons from the Al Kα x-rays is 50 to 100 Å. It would be expected then that, if the samples were covered with a film thicker than 50-100 Å, the ESCA spectra of the elements of the adherend would not be observed. The lack of a Ti signal is seen in Figure 20b for sample 1m2-517D3 and a strong Ti signal is seen in Figure 20a for sample 2m2-515D4. The results of the ESCA studies on the samples in Set II are summarized in Table VI. Carbon, oxygen, and nitrogen were found on all fractured samples and on the as-received Pasa-Jell cleaned Ti-6-4 panel. The binding energies, BE_ϕ, have been corrected for the work function of the spectrophotometer using the 2s electron of carbon at 284 ev. The intensity of the peak as measured by the ratio of the difference (Δ) between the maximum and the minimum counts peak width (ω) at half-height are also noted.

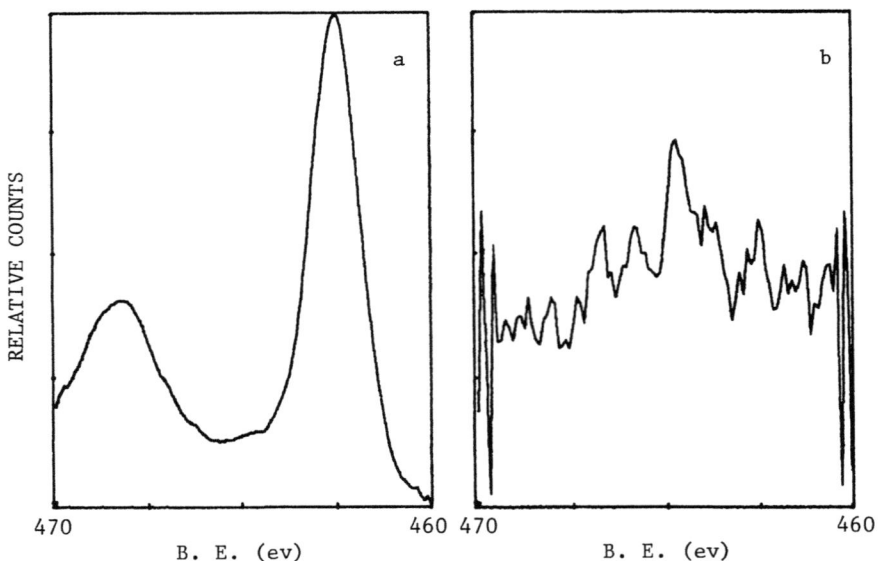

Fig. 20. a. ESCA Spectrum of Ti in Fracture Sample 2m2-515D4
          Scanning: 100 channels, 3 sec/channel, 18 scans.
       b. ESCA Spectrum of Ti in Fracture Sample 1m2-517D3
          Scanning: 100 channels, 3 sec/channel, 18 scans.

TABLE VI

ESCA ANALYSIS OF FRACTURE SAMPLES

Sample Type		Ti-6-4 as rec'd		1m2-517D3	1mp2-516D4	2m2-515D4
Sample #		04230	05270	04180	05041	05043
Ti 2p3/2	BE_ϕ(ev)	457.0	458.0	NP*	NP	457.8
	Δ/MIN	5.70	4.70	0.14	0.14	2.82
	ω(ev)	1.8	2.0	NC**	NC	1.7
C 1s$_{1/2}$	BE_ϕ(ev)	(284)	(284)	(284)	(284)	(284)
	Δ/MIN	6.87	6.75	24.7	34.0	10.6
	ω(ev)	1.6	1.7	2.1	1.7	2.2
O 1s$_{1/2}$	BE_ϕ(ev)	529.3***	529.5***	531.0	532.4,531.2	531.8,529.4
	Δ/MIN	2.33	2.07	3.22	2.00	2.19
	ω(ev)	NC	NC	NC	NC	NC
N 1s$_{1/2}$	BE_ϕ(ev)	399.1	399.6	399.4	399.5	399.2
	Δ/MIN	0.41	0.41	0.89	0.39	0.87
	ω(ev)	2.1	2.5	2.0	2.1	2.3

*NP - no peak   **NC - not calculable   *** High energy shoulder

The most significant feature to note is that for the fracture samples (Table VI) no Ti signal was found for samples 1m2-517D3 or 1mp2-516D4 indicating the presence of a film at least as thick as 50-100 Å on the panels. This film is seen for the 1m2-517 and 1mp2-516 series in the SEM photomicrographs of Figures 10 and 12. The SEM results showed that bare metal was present on the 2m2-515 samples and the strong ESCA Ti signal for sample 2m2-515D4 clearly supports this finding. The ESCA results demonstrate the utility of the ESCA technique in establishing unambiguously adhesive or cohesive failure at the molecular level of 50-100 Å.

The present work demonstrates at least 3 factors contributing to adhesive strength. Differences in <u>wettability</u> were observed for the BTDA + m,m'DABP resin in diglyme compared to the same resin in DMAC (Set I). A higher adhesive strength was noted in the more wettable system. Alternately, <u>relative solubility</u> could be another factor to account for differences in adhesive strengths for BTDA + m,m'DABP and BTDA + m,m'DABP, BTDA + m,p'DABP and PMDA + m,m'DABP all in diglyme (Sets I and II). Decreasing adhesive strengths were noted with expected decreases in relative solubility. Finally the three resins (Sets I and II) have different <u>glass transition temperatures</u>. Again, the increasing adhesive strength parallels decreasing T_G values (see Table I). Work is in progress to delineate between these several factors.

IV. CONCLUSIONS

The techniques of (1) contact angle measurement, (2) electron spectroscopy for chemical analysis, (3) specular reflectance infrared spectroscopy, and (4) scanning electron microscopy are all of value in the characterization of the titanium alloy/polyimide resin adhesive system. The titanium alloy was identified as being composed of an α and a β-phase based on scanning electron microscopy. Scanning electron photomicrographs revealed definite changes in surface topography of the titanium alloy after the alkaline cleaning and the acid pickling steps of the cleaning process. A correlation of wettability to fracture strength for the DMAC and diglyme solvent systems was made by use of the scanning electron microscope. Failures in both the adhesive and cohesive mode were noted in the scanning electron photomicrographs for fracture surfaces. The DMAC and diglyme contact angles on the titanium alloy correlated with fracture strength. Octane/water-titanium alloy interfacial contact angles indicated that both the polyamic acid film and the alloy surface can interact by non-dispersion forces. Atmospheric contamination reduces the octane/water/solid contact angle. The infrared spectrum of fracture surface can be obtained in situ by specular reflectance infrared spectroscopy. The reflectivity of fracture surfaces is directly related to fracture strengths. Analysis of ESCA spectra based on binding energies and peak intensities can be used to detect the presence of ultra-thin adhesive surface layers.

Further, the techniques of Scanning Electron Microscopy (SEM), Specular Reflectance Infrared Spectroscopy (SRIS), and Electron Spectroscopy for Chemical Analysis (ESCA) have proved complementary in this investigation of the relationships between adherend surfaces and adhesive properties. As SRIS results have shown the presence of adhesive on all fracture samples, ESCA and SEM results further clarified the nature of the fracture surface through the presence or absence of a Ti ESCA spectrum and the observation or lack of observation of the substrate structure in the SEM photomicrographs. It is concluded from the results of the three techniques that for the Set I samples, cohesive failure was noted for 219D2 whereas adhesive failure was noted for 220D3. Cohesive failure was noted for samples 1m2-517 and 1mp2-516 and adhesive failure was noted for 2m2-515 in Set II.

ACKNOWLEDGEMENT

Financial support for this work under NASA Contracts NAS1-10646-14 and NAS1-10646-25 including graduate research assistantships for two of us (TAB, MEC) is acknowledged gratefully. The very capable experimental assistance of Dr. James S. Jen, Frank Mitsianis, Charles Potter, and Dr. T. L. St. Clair is acknowledged.

REFERENCES

1. T. L. St. Clair and D. J. Progar, presented in part at the ACS Macromolecular Secretariat Symposium on Science and Technology of Adhesion, Philadelphia, Pa., April 1975.
2. J. J. Bikerman, *The Science of Adhesive Joints*, 2nd Ed., Academic Press, New York, 1968.
3. J. R. Huntsberger in *Treatise on Adhesion and Adhesives*, R. L. Patrick, Ed., Marcel Dekker, Inc., New York, New York, 1967, Chapter 4.
4. S. Wu, J. Phys. Chem., $\underline{74}$, 632 (1970).
5. D. W. Dwight and W. M. Riggs, J. Colloid Interface Sci., $\underline{47}$, 650 (1974).
6. L. Taylor, Ed., *Metals Handbook, Atlas of Microstructure of Industrial Alloys*, Vol. 7, American Society for Metals, Metals Park, Ohio.
7. ASTM Spec. Tech Pub. No. 204, Symp. on Titanium, American Society for Testing Materials, Philadelphia, Pa.
8. W. C. Hamilton, Applied Polymer Symp., No. 19, 105 (1972).

9. W. D. Harkins and E. H. Grafton, J. Amer. Chem. Soc., $\underline{44}$, 2665 (1922).
10. G. Koranyi and M. Acs, Acta. Chem. Hung., $\underline{24}$, 333 (1960).
11. F. E. Bartell and J. T. Smith, J. Phys. Chem., $\underline{57}$, 165 (1953).
12. W. C. Hamilton, J. Colloid Interface Sci., $\underline{40}$, 219 (1972).
13. L. S. Bartell and R. J. Ruch, J. Phys. Chem., $\underline{63}$, 1045 (1959).
14. L. H. Sharpe and H. Schonhorn, Adv. Chem. Ser., $\underline{43}$, 189 (1964).
15. S. N. Muchnick, WADC Tech. Rept. 55-87 Part II, Feb. 1958.
16. K. Siegbahn et al., *ESCA-Atomic, Molecular and Solid State Structure Studied by Means of Electron Spectroscopy*, Almquist and Wiksells, Uppsala, 1967.

The Chemistry of Tackifying Terpene Resins

E. R. Ruckel, H. G. Arlt, Jr. and R. T. Wojcik

Arizona Chemical Company

Stamford, Connecticut 06904

Terpene Resins are low molecular weight hydrocarbon resins prepared by cationic polymerization of certain terpenes. They are used as tackifiers in pressure-sensitive tapes, masking tapes, hot melt coatings and adhesives, laminating adhesives and rubber solution adhesives. Terpene resins can be considered as solid solvents for a rubber which function by solubilizing the smaller tack-bestowing molecules from their dispersion in the mass of a rubber.

This paper will discuss the cationic isomerization polymerization of beta-pinene, limonene and alpha-pinene as regards polymer structure, molecular weight, molecular weight distribution, spectral data and reactivity ratios. These monomers are activated by a cationogen derived from a Lewis Acid catalyst/co-catalyst system. The propagating species derived from the above terpenes rearranges to energetically preferred structures prior to propagation so that the repeat unit of the macromolecule does not possess the structure of the original monomer. Termination, at least in the case of alpha- or beta-pinene, appears to occur by rearrangement of the active end to a non-propagating bicyclic camphenic type moiety.

INTRODUCTION AND HISTORICAL

A pressure sensitive adhesive is one which is permanently tacky, requires no activation by heat, solvent, or moisture, and which will adhere strongly to most surfaces upon application with a minimum of pressure. Tack is defined as instant low order adhesion developed by mere contact with a variety of dissimilar surfaces. The major components of a pressure sensitive formulation are elastomer and tackifier. Performance depends primarily on three factors: tack, adhesion, and cohesion. The chemistry and physical characteristics of the terpene resins combine to produce materials uniquely fitted for use as tackifying resins. This paper is concerned with pressure sensitive adhesives formulated with tackifying resins. It will not include other adhesive systems such as acrylate copolymers and vinyl ester copolymers which use no tackifying resin.

Terpene resins are old, in fact, the oldest reference to polymerization was recorded in 1789 wherein turpentine was treated with sulfuric acid.[1] More modern milestones are a U.S. patent issued to Emile Rouxeville in 1909 for subjecting hydrocarbons such as turpentine to sulfuric acid to produce a resin which was said to resemble various India rubbers.[2] Then in 1933 aluminum chloride catalysis for terpene polymerization was patented by the Gulf Refining Company.[3] Later in 1950, a fundamental publication by Roberts and Day[4] appeared in the non-patent literature and anticipated much of the later work. Commercial terpene resins produced for adhesive applications resulted from modification of disclosed processes, catalysts and terpene feedstocks.

The most important single property of a resin is its molecular weight. It may be correlated to physical properties and utility. If a polymer property is plotted versus molecular weight, there occurs a "leveling off" in the property at a particular molecular weight which varies for each resin or polymer. In most instances, it is necessary to attain this minimum molecular weight range to get the desired physical properties, e.g., polypropylene has to have a molecular weight of 50,000 and acrylonitrile 35,000 to be useful in common polymer applications. By contrast, the relatively low molecular weights of terpene resins at which properties plateau, coupled with their narrow molecular weight distribution make them unique and useful for adhesives. High polymers, in general, attain their strength and hence utility from entanglements of extremely long chains and reinforcement with secondary valence bonds, whereas terpene resins attain their utility from low molecular weight, rapid change in viscosity with temperature, newtonian liquid behavior and good solubility.

This paper discusses properties and uses of tackifier resins and concerns our studies of the cationic polymerization of terpenes, the resultant structures, and physical properties of the terpene resins.

PROPERTIES AND USES OF TACKIFIER RESINS

A typical plot of the softening point-molecular weight curve for a beta-pinene resin, popular as a tackifier for pressure sensitive tapes, is shown in Figure 1. These resins are prepared from a commercial β-pinene feedstream containing approximately 80% β-pinene.

It is generally recognized that the preferred beta-pinene resins are those that have ring and ball softening points from 115 to 135°C. You will notice that maximum utility appears on the bend or levelling off of the softening point-molecular weight curve. At this bend, the terpene resins are transparent amber glasses.

The tackifying resin can be thought of as a solid solvent for the rubber elastomer.[10] Usually, solubility is affected by molecular weight; the smaller the molecule the higher the solubility. While low molecular weight in a tackifying resin is desirable, there is a practical limit to this feature. A low molecular weight semisolid resin will impart tack but adhesives made with such a resin fail cohesively. Conversely, adhesives made with resins having a softening point beyond 135°C lack in tack. Empirically, beta-pinene resins of 115° softening point impart the best balance of tack and cohesive strength.

The tackifying resin appears to operate by bringing out the smaller, tack-bestowing molecules from their dispersion in the mass of the rubber. The solubility of the longest chains of a rubber is at best limited, so we can speculate that the tackifying resin exhibits a gradient solvent effect--totally solubilizing the shortest chains, partially solubilizing those of intermediate size (i.e. operating as a plasticizer), and effectively not operating on those of highest molecular weight.

The processes milled rubbers of commercial sources have molecular weights between 60,000 and 350,000 which correspond to degrees of polymerization of 1,000 to 5,000. These elastomers, because of their high molecular weight fractions, provide the cohesive strength to a formulation and, in addition, possess a latent tackiness. The high molecular weight of rubber and other elastomers allows modification with large amounts of other substances without serious loss of cohesive strength.

The use of terpene resins has been expanded recently with the rapid growth of hot melt adhesives. Many of the properties required of a pressure sensitive adhesive are valuable assets for hot melts. These materials are usually composed of a synthetic elastomer, a tackifier and a wax extender. The chemical and structural studies following were designed to define terpene resin structure and provide analytical data on these resins which could be used to predict the specific utility of a resin in a pressure sensitive or hot melt formulation.

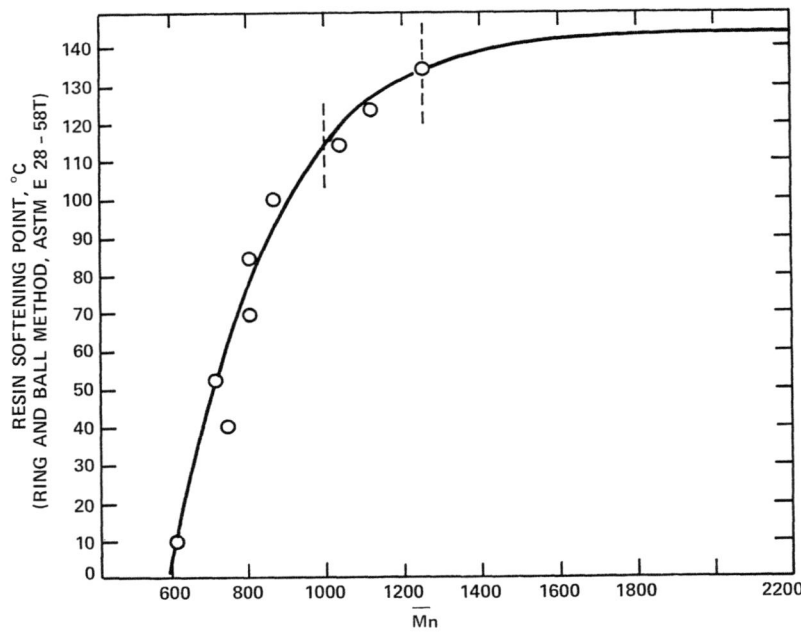

Figure 1. Effect of Molecular Weight on Softening Point of Beta-Pinene Resins.

TERPENE POLYMERIZATION

Terpene resins are prepared by a solution polymerization process.[5] Because of the large amount of heat generated during the polymerization, monomer is metered into a slurry of catalyst in solvent at such a rate (with appropriate cooling) so as to maintain the temperature at a set interval, for example 40-50°C. The catalyst is inactivated by immersing the resin solution in water, the inorganic catalyst remnants removed by washing, and the isolation of the resin by distilling off the solvent. Commercial resins are prepared from the following terpene monomers.

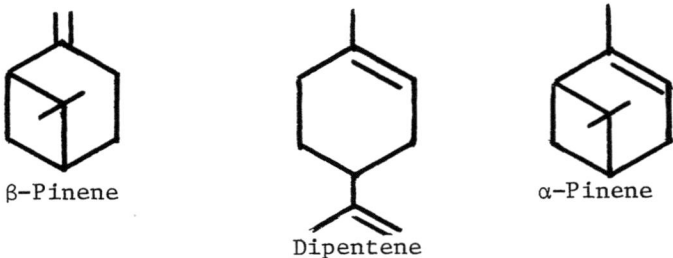

A. Beta-Pinene Resins

Figure 2 will illustrate the initiation and propagation steps which occur in beta-pinene polymerization. The main repeating unit in a beta-pinene resin is a ring having 1-4 disubstitution which results in an extended chain in solution. i.e., such a chain has a large hydrodynamic volume. This polymer can be visualized as a perfectly alternating copolymer of isobutylene and cyclohexene. Peracid oxidation indicates approximately one olefinic group per mer unit.

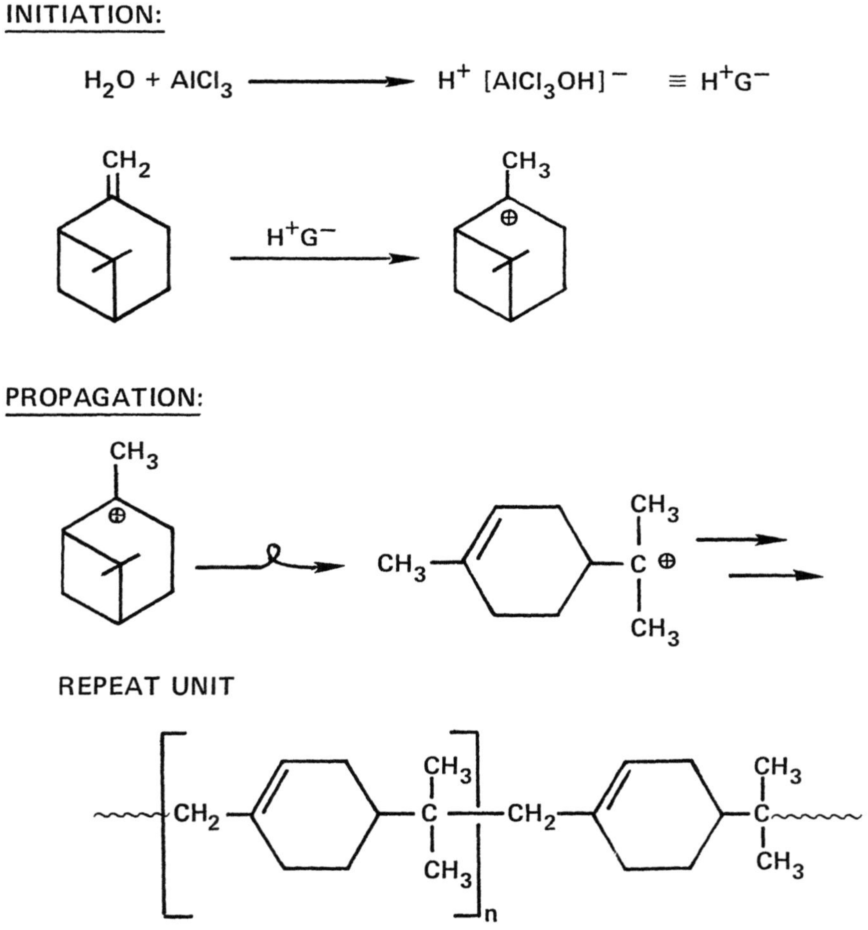

INITIATION:

$H_2O + AlCl_3 \longrightarrow H^+ [AlCl_3OH]^- \equiv H^+G^-$

PROPAGATION:

REPEAT UNIT

Figure 2. Beta-Pinene Polymerization.

Figures 3 and 4 will illustrate several chain transfer steps which occur. A review of the recent literature revealed that the molecular weights obtained from the polymerization of beta-pinene, d-limonene or alpha-pinene were of a low order, about 1000-2000. Our initial effort was to determine if impurities in these systems caused the low molecular weight or whether this feature was an intrinsic property of the terpenes. Polymerizations on the high vacuum rack with highly purified monomer indicated that the terpene monomer possessed a high chain transfer constant which was responsible for the low molecular weights. No conditions were found to produce resins with molecular weights in excess of 3500. The camphenic carbenium ion depicted in Figure 4 will be non-progressive[6] for steric reasons and its presence is probably the reason for the limited molecular weight.

Good analogy for the rearrangement mechanism to a non-propagative camphenic end cited above is seen in the acid-catalyzed Wagner-Meerwein isomerization of alpha- or beta-pinene which expands the four membered ring to the [2:2:1] bicylco ring system,[7] (Figure 5).

It is noteworthy to mention an interesting but puzzling effect. If a beta resin is hydrogenated for saturation, it loses its tack inducing capabilities. The softening point is lowered and the resin becomes more compatible with elastomers such as ethylene-vinyl acetate. Thus, it is not simply the ring, but a combination of ring plus unsaturation that produces tack.

B. Dipentene Resins

The initiation step is similar to that described for beta-pinene as illustrated in Figure 6. Propagation through the terminal methylene group would be predicted. However, the determination of olefin content by NMR, ozonolysis and perbenzoic acid oxidation indicates that approximately one-half of the mer units have unsaturation. The endocyclic or ring double bond is thus involved in the polymerization and is consumed in some manner. To substantiate this theory, the structurally similar model compound 8,9-p-menthene was polymerized under identical conditions. Only dimer was obtained. The double bond in the ring thus permits successful polymerization of dipentene.

Alternatively, we might speculate that the polymerization of dipentene proceeds by initiation at the tri-substituted olefinic ring position and the carbenium ion undergoes cyclic polymerization to yield a structural unit as shown in Figure 7. Butler[8] has shown that the polymerization of the related 1-methylene-4-vinylcyclohexane proceeds partially by cyclization.

TACKIFYING TERPENE RESINS TO AROMATIC SOLVENTS

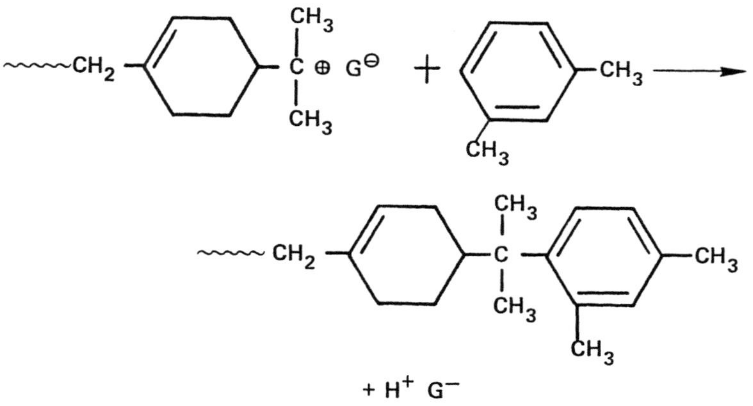

Figure 3. Chain Transfer.

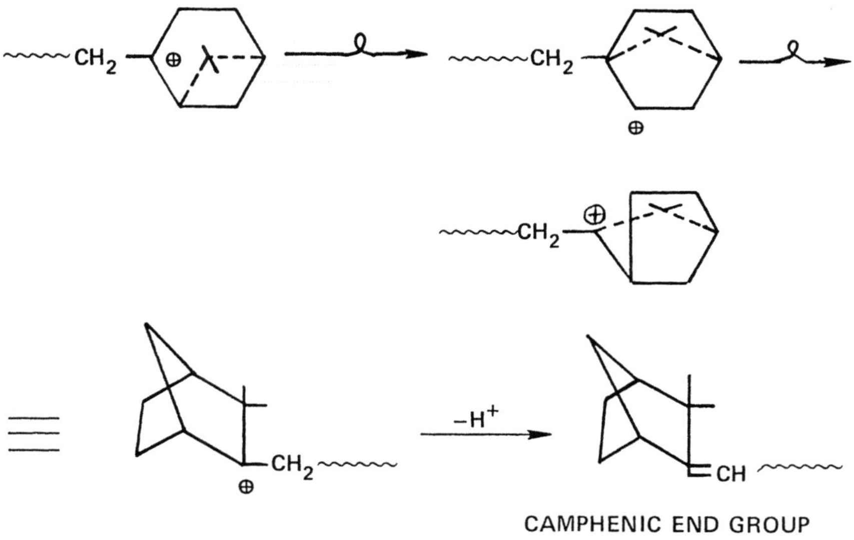

Figure 4. Rearrangement to Non-Propagating Living End Forces Chain Transfer to Monomer.

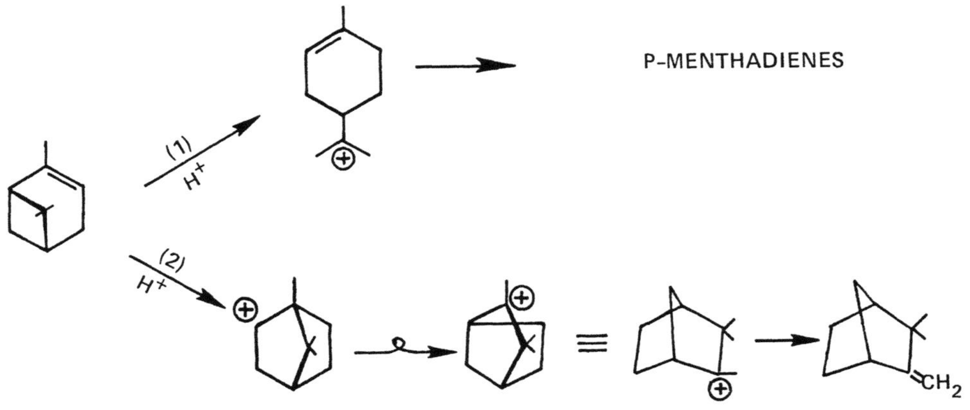

Figure 5. Possible Rearrangement Pathways.

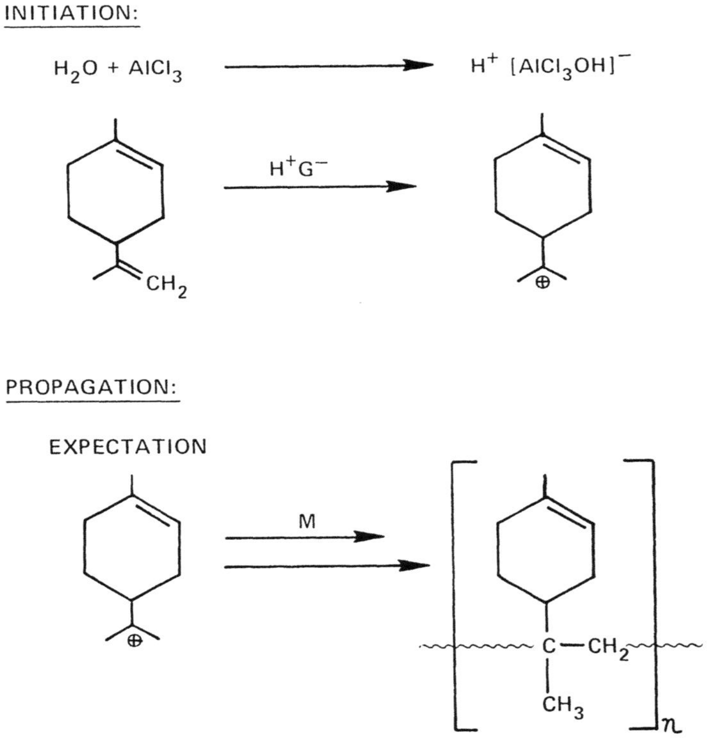

INITIATION:

$$H_2O + AlCl_3 \longrightarrow H^+ [AlCl_3OH]^-$$

PROPAGATION:

EXPECTATION

Figure 6. Dipentene Polymerization.

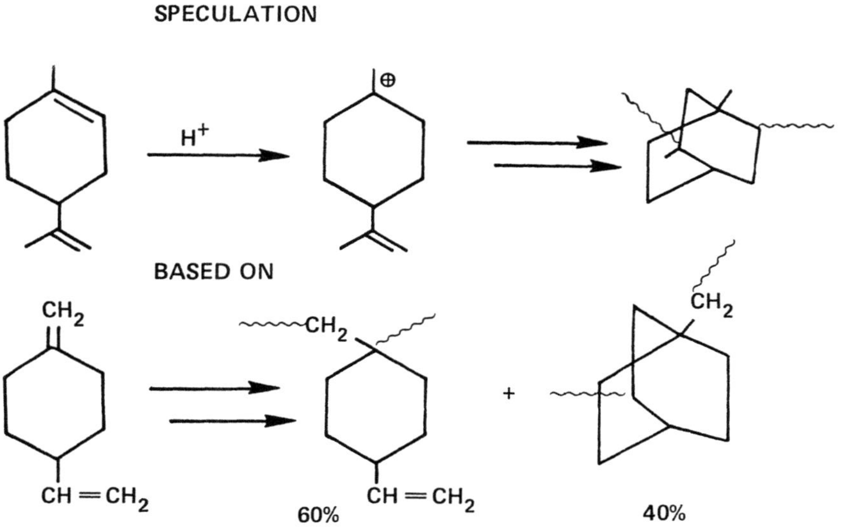

SPECULATION

BASED ON

(REF: G.B. Butler et al, J. Pol. Sci., Part A, Vol. 3, 723-732 (1965))

Figure 7. Propagation

More probably, the pendant isopropyl carbenium ion attacks the residual double bond of the penultimate mer unit and thus forms a ring with subsequent polymerization proceeding from the penultimate mer unit.

Schematically illustrated:

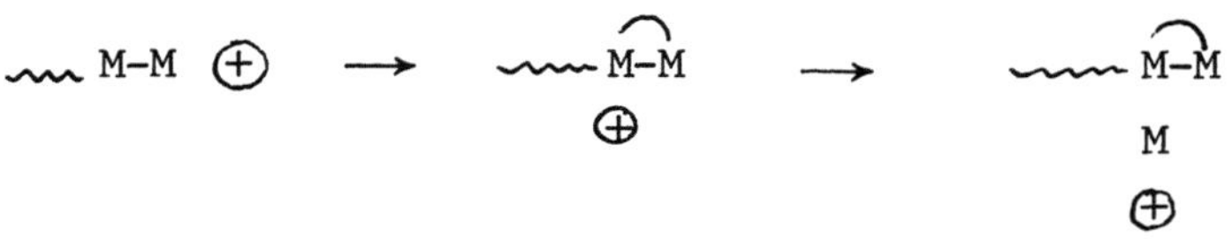

~~~ M-M ⊕ → ~~~ M-M ⊕ → ~~~ M-M
                                M
                                ⊕

A structural representation based on this postulation is presented in Figure 8. By invoking either of the above mechanisms, we can satisfactorily explain the presence of only one double bond per every two to three mer units.

C. Alpha-Pinene Resins

This monomer is the most difficult of the common terpenes to polymerize since it does not possess an exocyclic methylene group. Although alpha-pinene forms the same initial carbenium ion as beta-pinene (Figure 9), the propagation step is difficult for steric

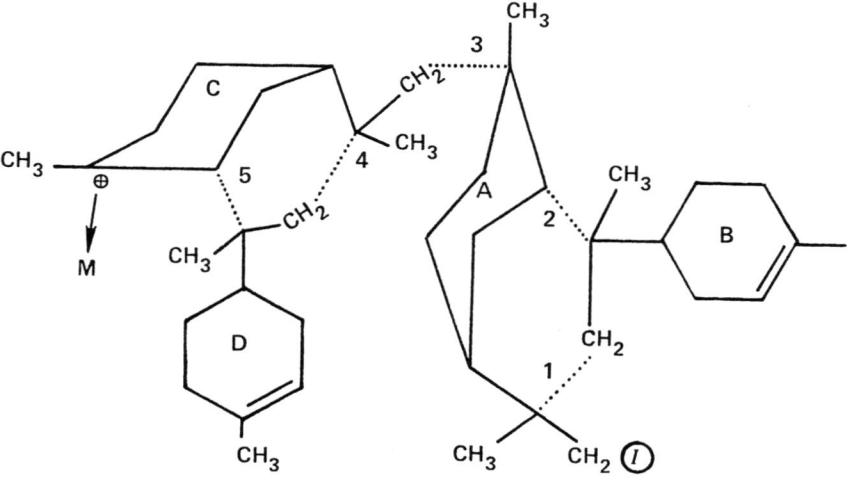

Figure 8. Proposed Partial Structure for Dipentene Resin.

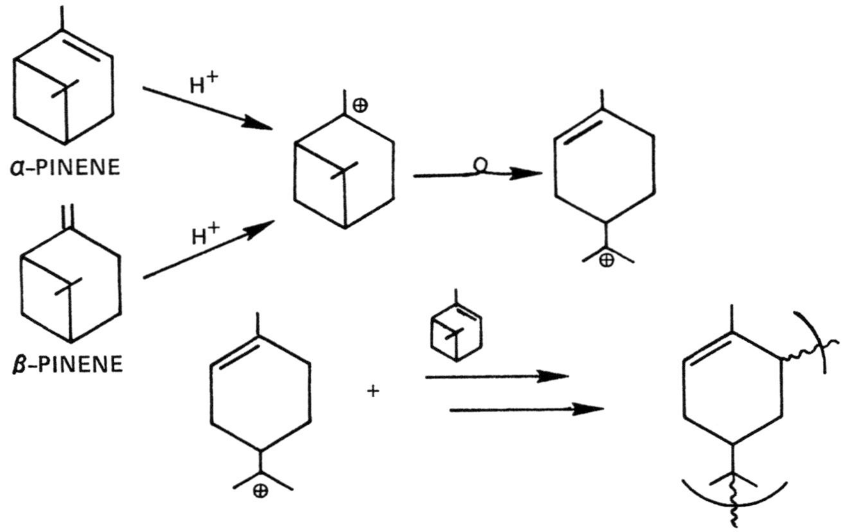

Figure 9. α-Pinene Polymerization.

reasons. Substantial amounts of dimer form during the polymerization of alpha-pinene indicating that propagation from dimer to trimer is a difficult step.

The peracid oxidation of alpha-pinene resin showed that approximately two-thirds of the mer units contain an olefin indicating that in the remaining one-third, the four membered ring expands and results in a saturated mer unit possessing the [2:2:1] bicyclic system.

Alpha-pinene dimer was found to contain one olefinic group per dimer suggesting the latter to possess a double bicyclic placement.

All other reasonable dimer structures would contain more than one olefinic group.

ANALYTICAL METHODS

Two spectroscopic techniques were used to gain insight into the partial structure of the polymer chain. Infrared absorption indicated the presence of two types of methyl groupings, gem dimethyl and single methyl.

$$\sim\!\!\sim\!\!\underset{\underset{CH_3}{|}}{\overset{\overset{CH_3}{|}}{C}}\!\!\sim\!\!\sim \quad \text{doublet at 1370, 1388 cm}^{-1}$$

$$\sim\!\!\sim CH_3 \quad \text{singlet at 1380 cm}^{-1}$$

The former arises from a beta-pinene chain or from a dipentene initiating mer unit. Single methyl arises from a beta-pinene initiating mer unit or a dipentene chain. Typically, the gem dimethyl/single methyl ratio is 85/15 for a beta-pinene and 15/85 for a dipentene resin.

NMR spectroscopy shows the presence of aromatic, olefinic and aliphatic hydrogens. The aromatic hydrogens are from the incorporated solvent while the olefinic and aliphatic hydrogens are from the terpene. Higher polymerization temperatures reduce the amount of olefinic absorption of beta resins since rearrangements presumably occur by hydride transfer to produce non-absorbing tetra-substituted olefin. Figure 10 presents the aliphatic region of the NMR spectra. The comparative spectral data shows a substantially reduced amount of inter ring CH_2 present in the alpha-pinene resin.

The number average molecular weights of representative commercial terpene resins, determined by vapor pressure osmometry, are presented in Table 1. Feedstreams for the dipentene and α-pinene resins were 95-99% pure. The feedstream for the β-pinene was already described.

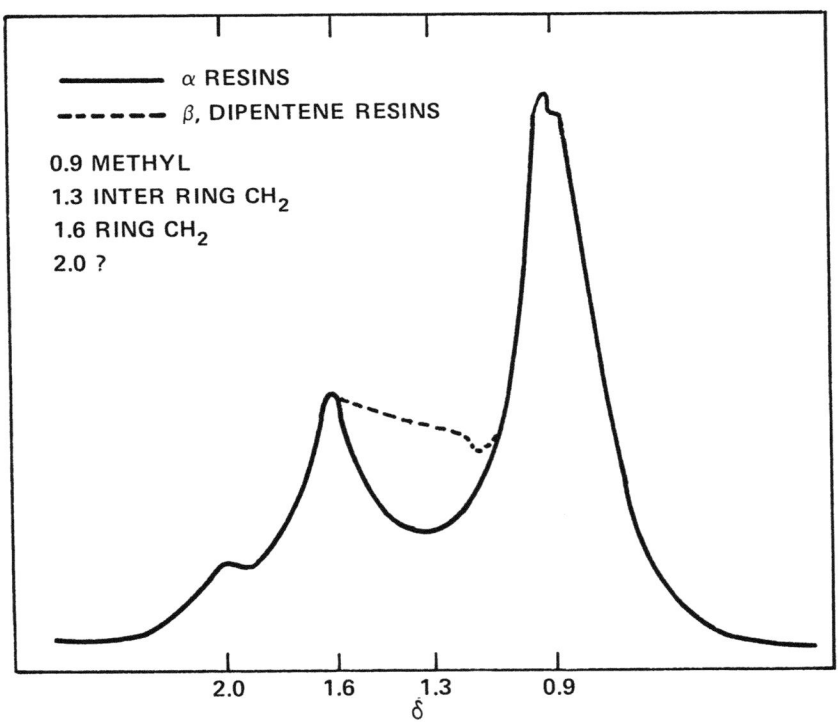

Figure 10. NMR Spectral Data.

Table 1

Mn of Typical Terpene Resins

R & B, °C	β-Resin	Dipentene Resin	α-Resin
85	815	570	725
100	870	675	775
115	1030	720	815
125	1110	760	830
135	1230	810	870
Mw/Mn (115°)	1.9	1.4	

For a specified softening point, a dipentene resin has a lower molecular weight than a beta-pinene resin indicating that the dipentene polymer structure is more rigid and more compact than that of a beta-resin. The density of dipentene resins is in fact higher than that of beta-resins, 0.995 to 0.980. Although dipentene resins have a higher softening point/mer unit, they have a smaller hydrodynamic volume and hence form solutions of lower viscosity.

The molecular weight distribution of a beta-pinene resin was monitored during polymerization by gel permeation chromatography. Since initiation, propagation and termination processes are all proceeding simultaneously, we obviously observe a distribution of molecular weights. It is not surprising that the molecular weight distribution was observed to change during the polymerization when one considers the multitude of physical changes which are occurring, e.g., the heterogeneous to near homogeneous catalysis, the increase in viscosity due to polymer formation and the change in dilution. Figure 11 illustrates the variation of the molecular weight distribution which occurs during this type of polymerization process.

The shape of the distribution curve can be meaningful as to various utilities. A very convenient method of qualitatively determining the width of the molecular weight distribution is by observing the cloud point of the 10:20:20 blend of resin:wax:ethylene-vinyl acetate mentioned earlier. Beta resins that do not exhibit a cloud point, i.e., do not reach a cloud-free state, typically have wide or even bimodal molecular weight distributions. The cloud points of the initial, intermediate and final resins are 155, >210 and 170°C respectively.

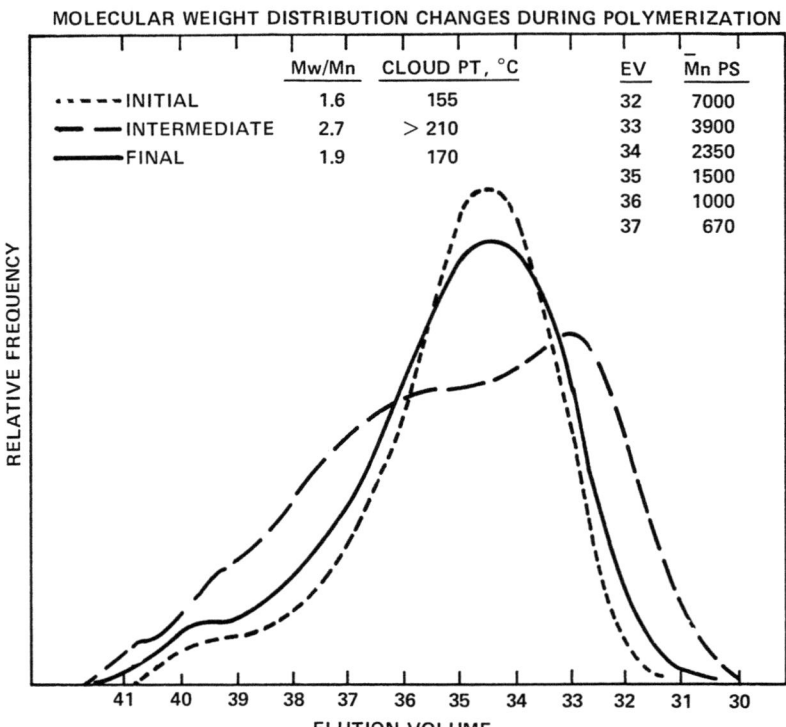

Figure 11. Molecular Weight Distribution Changes During Polymerization.

We can easily see that the structures of beta-pinene and dipentene resins are quite different. The polymer chain of a beta-resin is more extended and flexible than that of a dipentene resin. Thus, from a consideration of the structures of the polymer repeating units, dipentene resins should exhibit a lower viscosity than beta-pinene resins at equal degree of polymerization. Actually, the viscosity of dipentene resins also has a greater dependence on temperature; it is reduced to a greater extent than a comparable beta-resin. Dipentene resins are also more compatible with ethylene-vinyl acetate copolymers. The cloud point, obtained in a compatibility test using a 10:20:20 blend of resin:wax:ethylene-vinyl acetate, is about 90°C and can be compared with about 175°C for a beta-resin. Thus, these resins are preferred over beta-pinene resins for hot melt adhesives. Since formulations employing dipentene resins reach a compatible cloud-free liquid state at a lower temperature, less oxidation is likely to occur. Dipentene resins have also been found to be more color stable than beta-pinene resins, which probably reflects the presence of fewer olefinic sites where oxidation can occur. Dipentene resins also tend to be thermally stable because of their multiple strand structure.

It was clearly shown that beta-pinene resins were equivalent to an alternating copolymer of isobutylene and cyclohexene units as proposed originally by Roberts and Day.[4] Dipentene resin structure was shown to be compact and stiff by comparison of softening point and molecular weight. The determination of olefin content and model monomer studies showed a structure not of the type to be predicted by polymerization via the pendant olefin. The size difference and rigidity of these two types of terpene resins appear to be the outstanding properties which result in the predominant use of beta-pinene resins for pressure sensitive use and dipentene resins for hot melt use.

ADHESIVE CHARACTERISTICS

The objectives of the structural studies were to define terpene resin structures and provide analytical data which could be used to correlate structure and properties measured on the terpene resins with end use applications. Structures were well defined, but accurate prediction of adhesive formulation utility from resin properties has not been possible. Results of application studies are described in this section.

The following figures show how the change in chemical structure of the tackifying resin can markedly alter adhesive characteristics.[9] Figure 12 shows how the three adhesive characteristics of a beta-pinene resin compare with a dipentene based resin. The dipentene resin shows greater tackifying efficiency; however, its cohesion (or ability to hold) is markedly reduced.

Aging stability of a pressure sensitive adhesive depends to some extent on the stability of the tackifying resin; however, of equal or even greater importance are the aging characteristics of the base rubber in the adhesive. Natural rubber ages comparatively rapidly; the deterioration evidenced by a lengthy period of reversion, during which the elastomer becomes progressively softer and more gummy. Continued aging will cause the rubber to again change, this time becoming hard and brittle. Anti-oxidants are always used to prolong the usable life of the unvulcanized rubber. Styrene-butadiene rubber (SBR), however, deteriorates by becoming harder and more brittle without first going through a gummy stage. For this reason, many pressure sensitive adhesives utilize a mixture of natural and SBR rubbers; the increasing hardness of the SBR counteracts to some extent the increasing softness of the natural rubber and this also tends to prolong the usable life of the adhesive. Figure 13 compares the adhesive characteristics of dipentene and beta-pinene resins in a 50-50 mixture of natural and SBR rubber. In the mixed elastomer base, the adhesive characteristic for the dipentene resin more nearly approximate the results obtained with natural rubber base. The beta-pinene resin, on the other hand, shows reduced Hold and Peel Adhesion.

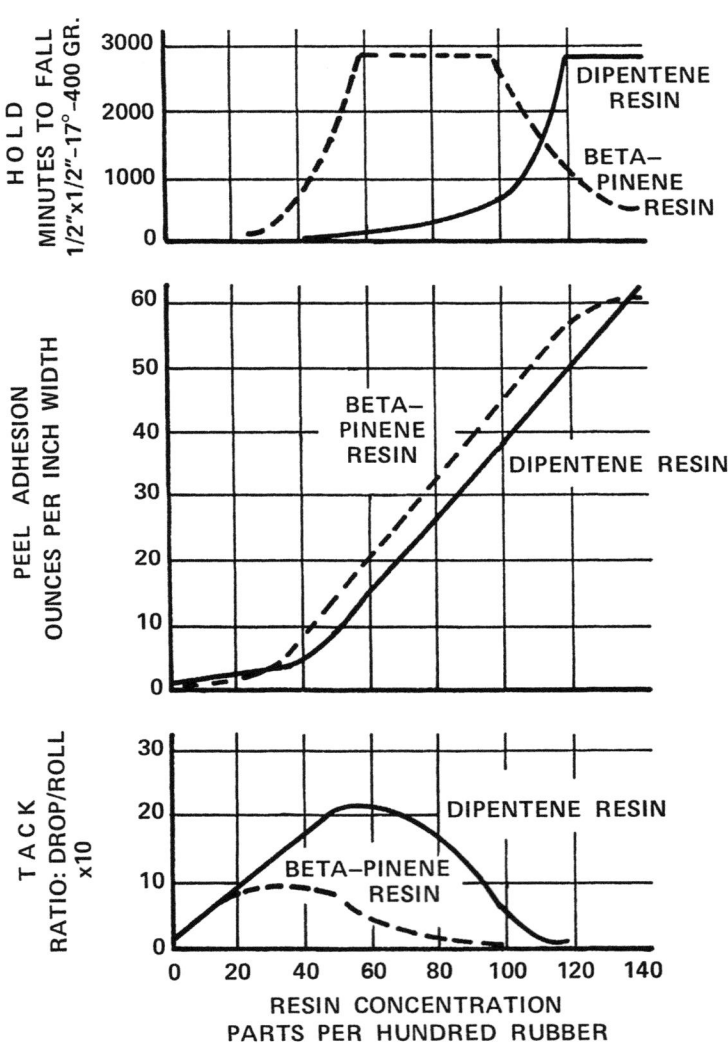

Figure 12. Adhesive Characteristics
Polyterpene Resin - Natural Rubber.

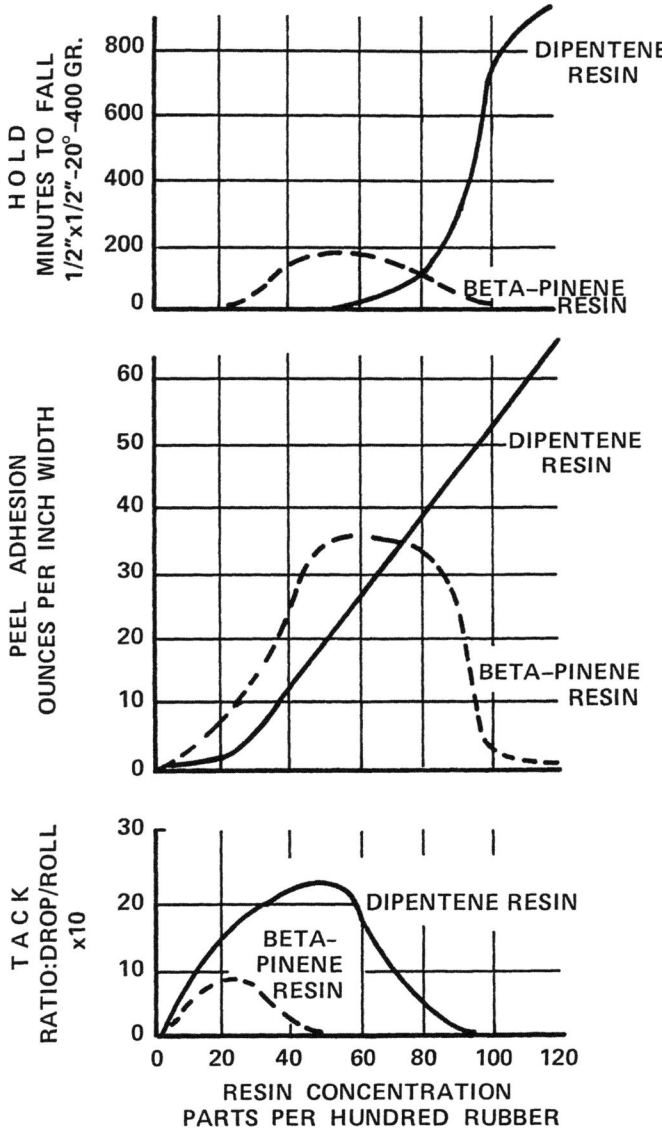

Figure 13. Adhesive Characteristics
Polyterpene Resin - 50:50 NAT:SBR

A hot-melt adhesive may be defined as a material containing 100% active components applied in a hot molten form which will form a bond between substrates upon cooling below its melting point. The advantages offered by hot-melt adhesives are many and vary in importance depending upon each individual application. No organic solvents are required. Hot-melt adhesives generally offer versatility not obtained with conventional aqueous adhesive systems in bonding to a wide range of materials under varied conditions. The following advantages are realized by the use of hot-melt adhesives: 100% active components, fast set, cleaner operation, less sensitive to change in surrounding conditions, reduction of fire hazard, easier storage, and better resistance of bonds to moisture. There are many variables which affect bonding with hot-melt adhesives. The most important are: temperature and viscosity of the adhesive melt, porosity of the substrate, adhesion to the substrate, cohesive strength of the adhesive, condition of the substrate, compression during bond formation, and storage conditions.

CONCLUSIONS

It is obvious that relatively small variations in the chemical structure of tackifying resins can exert appreciable effects on its physical properties and these in turn effect the adhesive efficiency of the resin in adhesive applications. Our studies focused on the chemistry of polymerization, have aided us in defining the resin structure and should aid us in the future in designing resins to meet the needs of the adhesives industry.

REFERENCES

1. R. Watson D.D., Chemical Assays, Vol. III, p. 5; MDCCLXXXIX.
2. Emile Rouxeville, U.S. Patent 919,248, 4-20-1909.
3. S. M. Cooper, Gulf Refining Company, U.S. Patent 1,938,320, 5-21-1933.
4. W. J. Roberts and A. R. Day, J. Am. Chem. Soc. 72, 1226 (1950).
5. E. R. Ruckel et al, Arizona Chemical Company, U.S. Patent 3,737,418.
6. A. Takata, T. Otsu and M. Imoto, Kogyo Kagaku Zasshi, J. Chem. Soc., Japan, Industrial Chemistry Section, 69(4):715-718, April 1966.
7. C. M. Williams and D. Whittaker, J. Chem. Soc. 668 (1971).
8. G. B. Butler, M. L. Miles and W. S. Bray, J. Pol. Sci., Part A, Vol. 3, p. 723 (1965).
9. H. P. Weyman, Naval Stores Review, pp. 6-11, February 1965.
10. C. DeWalt, Adhesive Age, pp. 38-45, March 1970.

Some Factors for Achieving Environmental Resistance in 120°C Structural Adhesives

James S. Noland

American Cyanamid Company

Stamford, Connecticut   06904

With the increased usage of 120°C cured, rubber modified epoxy structural adhesives for aluminum airframes, certain service problems have been observed which have been attributed to environmental factors. The problems associated with the combined effects of sustained load, elevated temperature and high humidity upon the aluminum substrate, corrosion inhibiting primers, and the structural epoxy adhesive matrix are discussed. A particular type adhesive matrix, based on acrylonitrile/butadiene rubber modified bisphenol type epoxy systems is discussed in detail, and important advances in the preparation of more moisture resistant aluminum (oxide) surfaces are reviewed.

INTRODUCTION

With the broad proliferation of low temperature (120°C) structural bonding, particularly into very expensive, large commercial aircraft, certain service problems have occurred manifested as corrosion and bond line deterioration. In this discussion, we should like to summarize the results of several years of a continuing program in our own organization and in the laboratories of certain aircraft manufacturers toward understanding and correcting the problem areas. Several investigators[1,2,3,4], have considered the degradation of epoxy adhesives by water; however, the previous investigators have generally considered theoretical aspects of the problem, and have used models which are quite different than the typical structural adhesives developed for the aerospace industry.

We wish to concern ourselves with the popular latent catalyzed, rubber modified, epoxy film adhesives developed for low temperature bonding of the main aircraft alloys, i.e., 2024 and 7075, bare and clad, aluminum. The maximum service environment we shall consider is 60°C, 100% relative humidity. Similarly, let us consider the problems with realistic structural loads, following the common adage that any material called "structural" should be capable of sustaining a constant stress which is about 25% of its failing stress.

The problem, of course, is to understand the factors which will provide reproducible, strong, and permanent joining for aircraft structure. There has been considerable concern because material specification tests in standards of even three or four years ago, failed to predict the service problems. This has resulted in major programs in the industry to evaluate these factors. We shall review some aspects of this work and present what we believe is a rational approach to the subject of obtaining durable bonds in hot, wet environments under sustained loading.

EXPERIMENTAL

Bulk Tensile Tests

The adhesive film, unsupported, is cured between shimmed Teflon sheets in the normal cure cycle to yield films .015 to .025 inches thick. Dogbone tensile specimens are cut from the film, measured, and conditioned at the desired temperature-humidity for a minimum of 72 hours. This duration was found to be adequate for the films to attain equilibrium water content, and samples showed no change upon further exposure.

Films were tested, at temperature, in an Instron Test Machine, with a 1/2 inch extensometer mounted within a one inch gauge length. A cross-head speed of 0.2 in./min. was used. Six replicates were used in each case.

The resulting curves were analyzed for:

1. Ultimate stress
2. Yield stress
3. 0.25% Offset stress*
4. Yield elongation
5. Tensile modulus

* Point of intersection of a line parallel to the tangent at 0.25% elongation.

ESCA

Samples of 0.030 in 2024 T3 bare aluminum were pre-cut, to suitable size for mounting in the spectrometer, with a tab to facilitate handling. The metal was subjected to 10 minute etch at 62-63°C in a sodium dichromate-sulfuric acid solution (per Section One, Handbook of Adhesives, Bloomingdale Aerospace Products, Havre de Grace, Maryland 21078), rinsed ten minutes in running, deionized water, and dried one hour at 63°C. One sample was then additionally subjected to the phosphoric acid anodization as outlined in BAC 5555[13].

The samples were characterized by routine examination with a Hewlett Packard ESCA Spectrometer, Model 5950A. After analysis, the samples were withdrawn and placed immediately in a humidity chamber at 60°C, 100% R.H. After one hour they were removed, air dried, and re-examined. Surface water was removed during evacuation in the sample chamber.

DISCUSSION

The problem area was described very succinctly by A. W. Bethune[5] of the Boeing Company. The 120°C curing modified epoxy adhesives "exhibited a sporadic disbonding within a few thousand service hours which could not be consistently related to model, location, or unique service conditions". Examination revealed that failures were: 1) interfacial, 2) progressive, and 3) occurred under low stress, i.e., insufficient to cause plastic deformation of thin gauge aluminum skins.

In the interest of duplicating these features, tests involving sustained load at elevated temperature and humidity were devised, and indeed, the performance of the (120°C) adhesives tested was particularly disappointing, even in such a "realistic" environment as 1000 psi, 60°C, and 100% R.H.

If we consider that these structural adhesive joints consist of three different parts, i.e., adhesive, primer, substrate, and that we must have satisfactory performance in all aspects, it is convenient to separate these factors in our consideration. The criteria for the long term performance of any material under load at any particular environment are that the material possess:

a) Adequate strength
b) Chemical stability to the environment
c) Resistance to damage under all the expected service conditions of stress, fatigue, etc.

Although the strength of an adhesive is a somewhat ambiguous term with the magnitude of the number dependent upon the type of test and the rate of loading, in general we can dismiss a mechanical strength requirement in terms of the present discussion of durability to sustained loads at 60°C and 100% R.H. The bulk tensile strength and the tensile shear strength of the 120°C systems in question far exceed the 900-1000 psi typical failure load values in the 1/2 inch lap specimens. Secondly, these 60°C tensile shear strength values of the present typical 120°C adhesives are in excess of those of many adhesives which have superior sustained load performance. (See Table 1)

Focusing our immediate attention on the adhesive matrix, the absorption of water in these systems does not produce significant chemical degradation, i.e., hydrolysis, at 60°C.

The infrared spectrum was determined by the attenuated total reflectance technique (ATR) on the fracture surface of a sample of a typical example of the adhesive in question, exposed in the 60°C, 100% R.H. cabinet for 40 days. Figure 1 shows absorption in the -OH region, 3400 cm^{-1}, attributable to water, relative to the unexposed control. These samples, dried (Figure 2), show no significant changes except removal of the water. The results are summarized in Table 2 as absorbance ratios, relative to an internal standard, the absorbance at 1180 cm^{-1}. The major change is a decrease in -OH absorption. Other changes in ratio are well within experimental error by the technique.

Similarly, the requirement of chemical resistance for these systems is not the limiting factor. The chemical nature of the various linkages in epoxy adhesives is generally similar, in both 120°C and older 175°C adhesives. The structural adhesives invariably involve the amine type curing systems and a large proportion of highly polar, hydrophilic groups, including quaternary nitrogen groups, -OH and > N-H functionality. In addition, these current 120°C systems employ carboxyl functional nitrile rubbers which are linked to the epoxy matrix through ester linkages. Therefore, all such systems absorb water in high humidity environment. We would expect that the rate of moisture permeation and its effect on physical properties would depend upon such factors as the exposure temperature relative to the glass transition temperature of the matrix, the actual concentration of polar groups, and the crosslink density. The primary effect of water is believed to be interference with the intermolecular hydrogen bonding. The effect has been described and we have effectively attacked the problem in the manner of Kwei[4]. However, it is not the effect on ultimate strength which is critical. For a sustained load capability, it is essentially a matter of whether the sample is damaged by the load.

TABLE 1

Sustained Loading (870 psi, 100% R.H., 60°C) of Various Systems

Adhesive	Average Tensile Shear Strength No Load Control	% UTS	Time to Failure Days
A – 120°C AN/BD Mod. Epoxy	3075	28	40
B – 120°C AN/BD Mod. Epoxy	3300	26	30
C – 120°C AN/BD Mod. Epoxy	3250	27	30
D – 175°C Al Filled Epoxy	3025*	29	No failures in 180 days (term)
E – 175°C Epoxy-phenolic	3430*	25	No failures in 180 days (term)
F – 175°C Vinyl Epoxy	2575*	34	No failures in 62 days (term)

* Initial L.S.S., 60°C, 50% R.H.

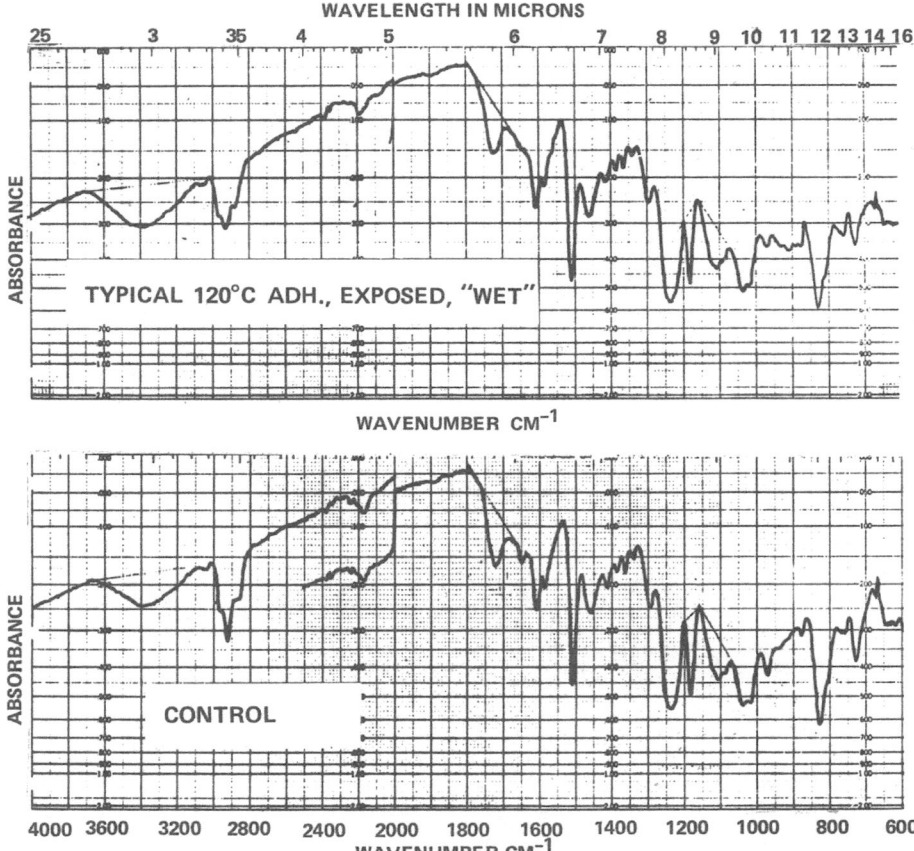

Figure 1. Infrared Study.

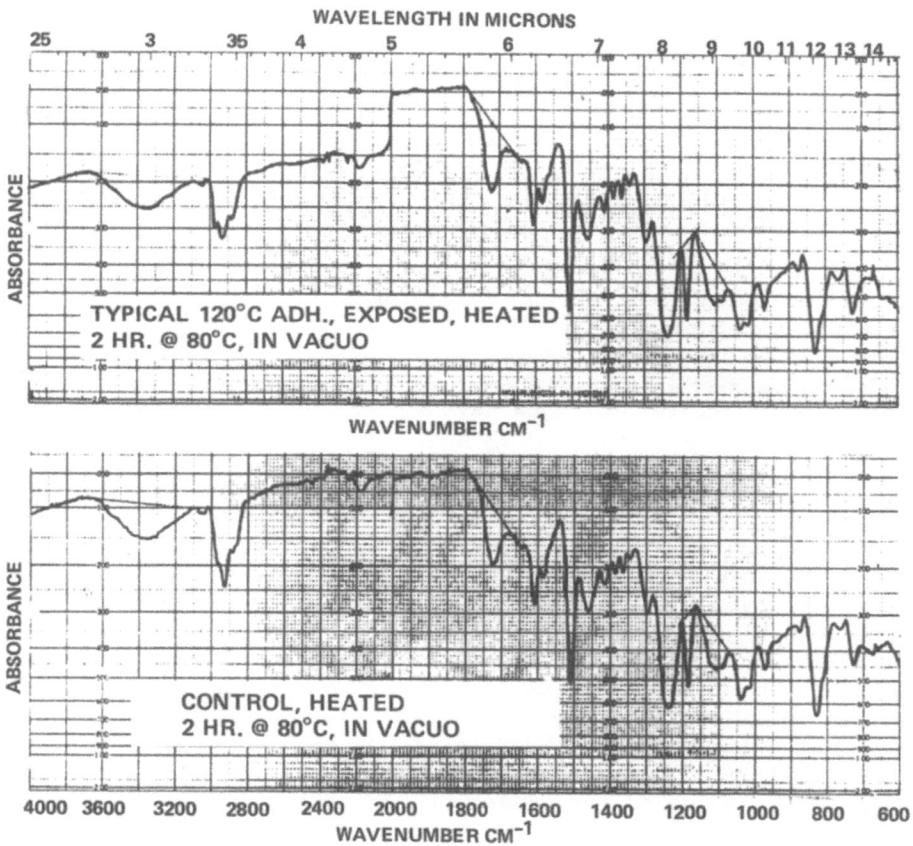

Figure 2. Infrared Study.

TABLE 2

Infrared Study of Humidity Exposed 120°C Epoxy Adhesives

	A 3400 cm^{-1} / A 1180 cm^{-1}	A 1720 / A 1180	A 1110 / A 1180
Control Sample	0.19	0.42	0.55
Exposed Sample	0.48	0.38	0.57
Control Dried	0.26	0.43	0.48
Exposed Dried	0.21	0.43	0.50

The phenomenon of rubber toughening of glassy materials has been widely investigated in recent years. A very excellent review of the subject has been provided by Bragaw[6] at an ACS Symposium in 1970. Professor F. J. McGarry[7] of the Massachusetts Institute of Technology and his students have shown that these same mechanisms are responsible for the toughening of thermosetting systems as well. That is, that incorporation of a precipitated rubber phase, which is "adequately" bonded at the matrix interface, raises the "fracture surface energy" of the material by 1-2 orders of magnitude. The peel toughening is achieved by dissipation of the tensile energy to fracture into myriad microscopic crazes, i.e., localized tensile deformations in the matrix, instead of a catastrophic crack in the neat glassy material. The rubber effectiveness depends upon the ability of the glassy resin to deform under stress. Therefore, the highest peel toughness is achieved with maximum deformation, i.e., the minimum number of crosslinks. This mechanism is manifested in the whitening which is observed if one applies a tensile stress to a cured film of these adhesives. The white regions represent scattered light due to the difference in refractive index between the crazed and uncrazed material. These crazes are partially void, and would be expected to accelerate the ingress of moisture into the matrix. Therefore, an adhesive system should not be loaded beyond its critical stress in normal service. We have considered this a major criterion in developing the new generation of adhesives.

For a screening test we chose to cure the adhesive material in film form and to test the material in simple "dogbone" tensile tests. The justification for using tensile tests, when structural designers inevitably seek to minimize the tensile stress in their applications of structural adhesives needs some explanation.

It has long been recognized that edge tensile stresses in structural adhesive applications may be very severe. At the same time, the advent of rubber modified epoxy adhesives with their high peel toughness, and resultant high tensile shear loads at failure, have perhaps prompted users to consider applications where significant tensile stresses are encountered. Certainly the tensile shear loading of one-half inch lap joints is well known to have large tensile stress concentrations at the edges.[8,9] The relatively recent applications of cleavage tests to adhesives[10] is clearly a tensile deformation. It was apparent that most of the tests for environmental durability were, in fact, largely tensile tests.

The justifications for screening systems by bulk tensile properties are: first, as previously indicated, the toughening mechanism for rubber modified epoxy systems involves stress crazing, and resulting voids, upon application of tensile stress beyond a certain critical value, as noted. Obviously, such a "stressed" specimen has reduced load bearing capability, and the voids will aid the damaging ingress of moisture. Secondly, it is helpful to know the tensile properties of the bulk adhesive in terms of judging its load bearing capability in any mode. Although calculation and measurement of stress distribution in bonded joints is very complex, it is obvious that knowing the general behavior of the bulk adhesive would be helpful in judging the likely performance of systems. Therefore, we have tested adhesive formulations for their initial tensile properties, and also for the retention of those properties when tested at equilibrium with the desired environment.

In order to predict the behavior of formulations, we have obtained the necessary data, in tension, to prepare master curves after Kwei[4]. One measures a parameter, say 0.25% offset stress, (see experimental section) at various temperatures under a variety of constant humidity conditions. (See Figure 3) When placed on a temperature scale, it is possible to shift the various isotherms to form a continuous curve. The amount of shift, in temperature, is basically the temperature equivalence of the humidity. The advantage of such a curve is that one can determine the performance for a wide variety of temperature and humidity conditions. We compare master curves of a former 120°C adhesive and one of our latest formulations in Figure 4. This improved performance is achieved with attention to the critical chemical factors of crosslink density of the resin, the backbone stiffness, and the effectiveness of the rubber precipitation during the curing reaction.

After development of adhesive systems which were theoretically designed to be capable of carrying sustained load at elevated temperature and humidity (as opposed to earlier systems where those conditions had not been criteria for acceptance), a realistic evaluation of the service performance of 120°C adhesives revealed the striking fact that these adhesives never exhibited cohesive (adhesive matrix)

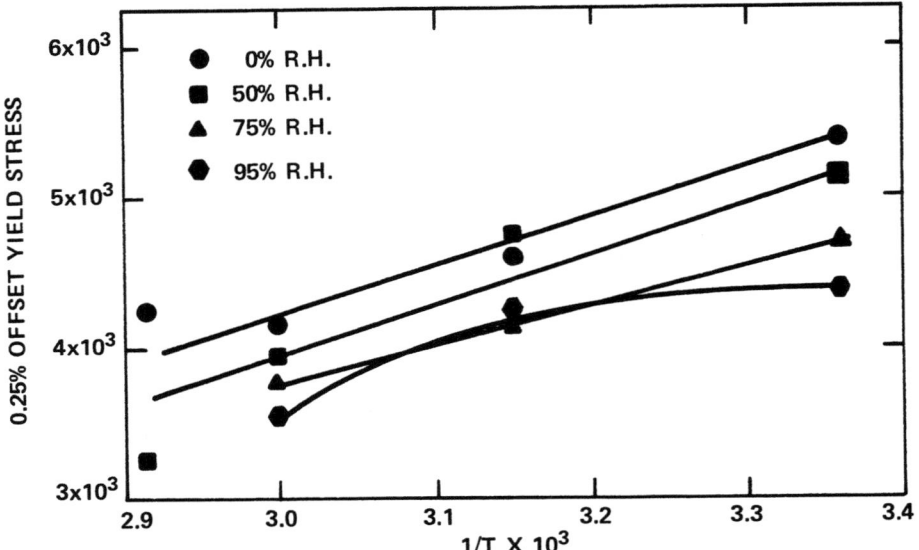

Figure 3. Temperature-Humidity Effects on Bulk Properties of Improved Epoxy Adhesive.

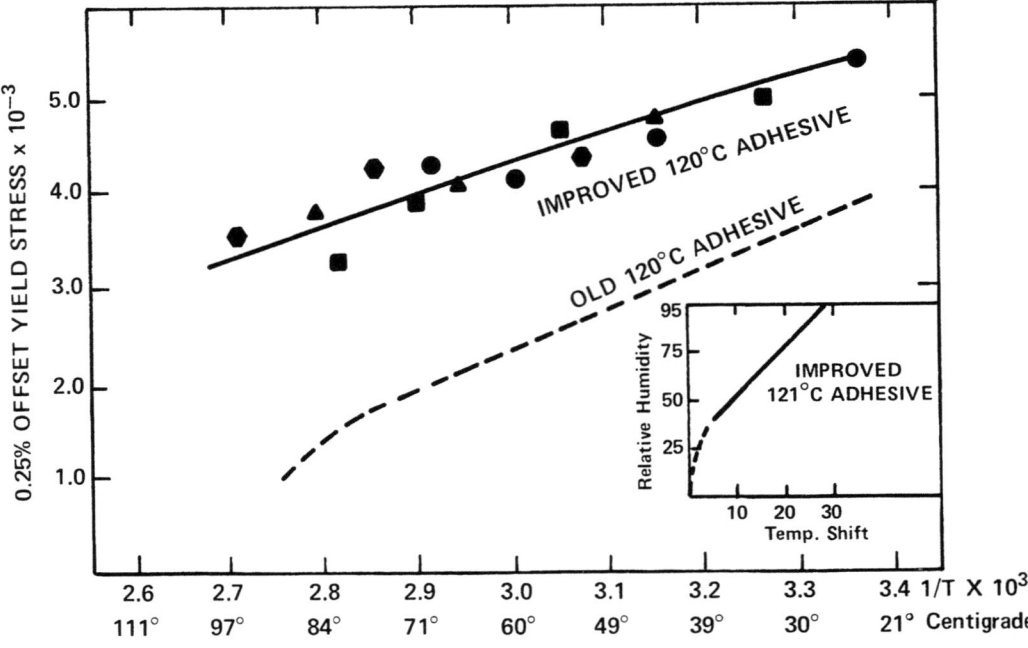

Figure 4. Master Curve, 0.25% Offset Yield Stress, 0% R.H. vs 1/T.

failure. Invariably, it was the bond interface which was degraded in some manner. It was obvious that this was a classical case of "weak boundary layer". The whole subject of interface preparation and stabilization needed to be examined, particularly for the 120°C bonding conditions.

The persistent interfacial failure with 120°C adhesives in actual service forces us to consider that the "standard" FPL etch for aluminum preparation may not be optimum for all bonding conditions. There are many years of outstanding service on this surface, but these surfaces had never before been evaluated with the particular combination of lower bonding temperatures and the unique testing program of combining heat, humidity, and sustained stress.

Fortunately, a great deal of work has been accomplished in a short time, and notably by aircraft manufacturers as well as adhesives suppliers. There are several important contributions in this area. First, in the area of FPL etch, the important consideration is what kind of bonding surface is provided by the preparation method. The chromic acid/sulfuric acid not only removes air oxide and leaves base metal; it also has a chemical potential which produces a very thin anodic type oxide layer of the surface. This oxide layer is porous, due to the dissolving action of the strong acid mixture, and thus the surface produced may be characterized as a thin, porous anodic oxide. (A. W. Smith[11] compared it to a 3V chromic acid anodize based on impedance measurements.) The optimum conditions for this etch as to time, temperature, and composition have been studied at Fokker[12] and by Smith[11] and generally a somewhat higher concentration of sodium dichromate or chromic acid was recommended than was commonly used.

In discussing the FPL etch with practitioners, the general opinion is that improper rinsing represents a major problem in the bonding industry. We have observed bond failures when one portion of a bonded area is completely degraded at the interface, and the remaining portion is intact. In certain instances, this has been correlated with the way the metal samples were hung when rinsed and dried. In general, it is important to cool and rinse the FPL etched metal immediately to obtain optimum surface properties. The continuance of the Cr^{+6} oxidation reaction, without maintenance of the optimum concentration, has been shown to be deleterious to the oxide stability. The improperly etched or rinsed surface is <u>very</u> sensitive to water. The quality of the etch is dramatically monitored by a wedge/cleavage type specimen such as that developed by Boeing Company[13]. However, it should be recognized that these surfaces are always degraded by water, and the ultimate stability of this system depends upon protecting that surface from moisture (upon the effectiveness of the primer or the adhesive matrix). In our laboratory, we have shown that only one hour at 60°C, 100% R.H. seriously changes the FPL etch surface revealed by ESCA examination. (Figure 5)

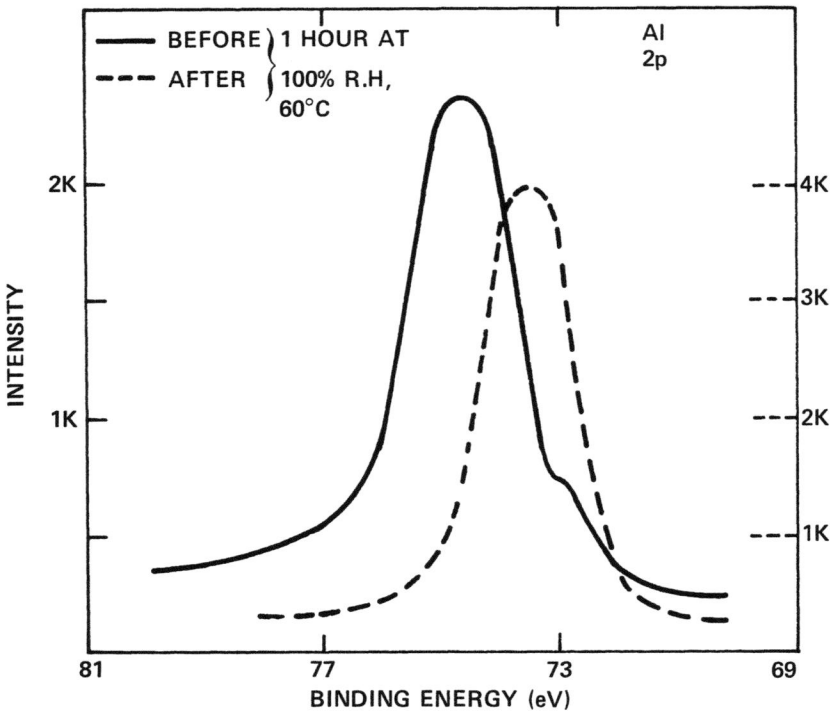

Figure 5. ESCA - 2024T3 Bare Aluminum FPL Etch.

We have established that the interfacial failures observed in the tests under load at elevated temperature and humidity are in fact failures in the aluminum oxide layer. Therefore, we postulate that these failures are due to a weaker, gelatinous type hydrated oxide layer. This "gelatinous bochmite" is known to form on the surface of "activated aluminum" in water at the boiling point.[14] We have shown that the oxide layer from the FPL etch exposed to 65°C, 100% R.H. (Figure 5) is indistinguishable from the same surface treated with boiling water in ESCA examination.

The most interesting development in recent bonding technology, in my opinion, is the work at Boeing on a new bonding surface disclosed in Bethune's[5] article. We are fortunate to have become aware of much of the literature work coincident to Boeing's development, and would like to emphasize that we have independently corroborated that the bonding surface is indeed much more resistant to hydration than the aluminum oxide from FPL etch. (Figure 6) I would draw attention to an excellent review article by Diggle[14] and the discussion of the chemical theories for this resistance to hydration in phosphoric acid anoidized aluminum. ESCA studies are in progress in hope of further defining this interesting system.

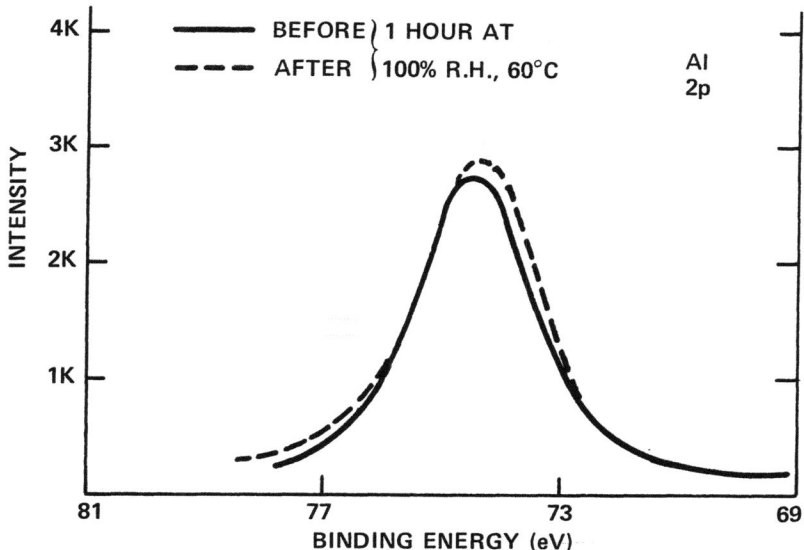

Figure 6. ESCA - 2024T3 Bare Aluminum BAC 5555 Anodize.

The final part of the adhesive system which will be considered is the corrosion inhibiting primer. In any discussion of primers for adhesives in a non-industry forum, it is important to consider that they serve a multiplicity of functions in the aircraft industry. First, and perhaps foremost, the primer, generally a dilute solution of resins compatible with the adhesive system, provides a means for achieving rapid wetting and spreading of the resin over the freshly prepared substrate (reactive metal oxide). The oxide layer is protected against ambient oxygen and moisture effects, extraneous organic molecules, and has a controlled surface for subsequent bonding operations. This function is particularly critical in the use of the film adhesives which have limited flow and wetting characteristics. The development of bondable, pre-cured, corrosion resistant primers has introduced several complications into this picture.

It is important to recognize that these corrosion inhibiting primers are fixed at the interface by the curing reaction, and that the primer forms a discrete layer which is the adhesive bond of fundamental importance for interface protection, and indeed for environmental resistance, since interfacial failure is the primary failure mode in the 120°C curing systems in question. Due to the balance of properties required of this primer, e.g., the spreading, wetting function, the protective function, subsequent bondability, durability under load, and fundamental environmental protection for the aluminum bonding surface, there are many limitations on the resin compositions which may be employed. It should be noted at the outset that the chromate pigment in these primers is primarily useful in the protective coating aspects of the primer's use. In the relatively protected bond line, the pigment particles are essentially encapsulated by the resin, and therefore can contribute little of the known, soluble chromate function of corrosion inhibitors.

Due to this balance of properties required of the primer, we can appreciate the restrictions placed upon the composition which may be employed for this purpose. For example, the flow characteristics needed for complete wetting and effective "sealing" of the porous aluminum oxide surface demand low melt viscosity during the curing reaction at 120°C. This involves the use of low molecular weight, highly functional resin types chosen for their bonding characteristics to the oxide. Such cured resin systems are inherently brittle, but soluble high polymers for toughening purposes would make objectionable viscosity and flow restriction, aside from other deleterious effects. The application temperature limitations (max. 120°C) are also involved. Higher temperature cures would permit better wetting and would also generally permit higher glass transition temperatures of the resin. It will be recalled that the older 175°C adhesives in Table 1 had better performance. Therefore, it is important to appreciate the various functions of the primer, and to give these functions their proper priority in terms of the kinds of requirements one actually has in bonding. Making use of these criteria and chemical species which are expected to give improved bonding to the aluminum oxide, we have now developed improved primers which have significantly better water resistance. (Figure 7) This durability is also corroborated in sustained load and fracture toughness testing.

In summary, we propose that the problem of sustained load capability, and general environmental durability of structural bonds with low temperature (120°C) adhesives can be corrected with particular attention to the factors outlined below:

1. The metal preparation method should be optimized and controlled to produce an aluminum oxide of maximum porosity, strength, and resistance to hydration.[5,13,14]

2. A corrosion inhibiting primer of maximum protection against moisture should be employed.

3. The adhesive system should be capable of sustaining the design load, without stress damage when plasticized by heat and moisture.

Finally, there should be a practical method for calculating stress levels in bonded structure so that we can more adequately match the material to the design requirement.

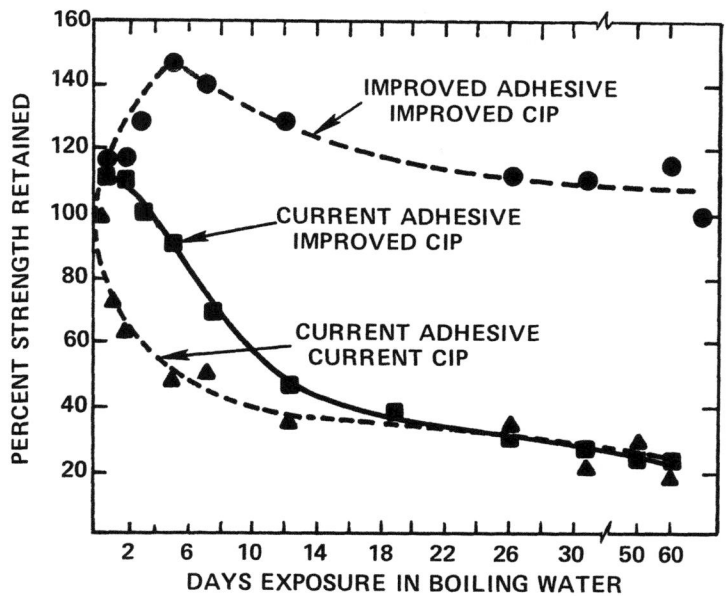

Figure 7. Percent Original Strength Retained, Bell Peel Test.

In conclusion, we are convinced that structural bonding in the aerospace field is on the threshold of a new scientific approach to bonding problems. We look forward to the day when we may see large scale bonding of primary structures, and the real potential of this fastening method can be realized.

ACKNOWLEDGEMENTS

Infrared spectroscopy work and ATR studies were carried out by Mr. John P. Falzone of these laboratories. ESCA experiments were carried out by Dr. J.S. Brinen and Mr. W. R. Doughman; anodization experiments by Dr. T. B. Reddy and Ms. L. Maxine Mull.

REFERENCES

1. C. Kerr and S. Orman, Brit. Poly. J. 2, 71 (1970); ibid, 2, 97 (1970).
2. H. Schonhorn and H. L. Frisch, J. Poly. Sci., Poly. Phy. 11, 1005-1011 (1973).
3. L. H. Sharpe, S.A.E. Tran. 700067 (1970).
4. T. K. Kwei, J. Appl. Poly. Sci. 10, 1647 (1966).

5. A. W. Bethune, Presentation, 19th Nat. SAMPE Sym. & Exh., Los Angeles, California, April 25, 1974.
6. C. G. Bragaw, *Adv. in Chem. Series* 99, 86 (1971).
7. F. J. McGarry and A. M. Willner, *Proceedings of ACS Symposium on Epoxy Resins* 28, #1, 512-526 (1968).
8. E. W. Kuenzi and G. H. Stevens, U.S. for Prd. Res. Note, FPL-011, September (1963).
9. M. H. Pahoja, *Stress Analysis of Adhesive Lap Joint Subjected to Tension, Shear Force, and Bending Moments*, T.N.A.M. Report #361, Univeristy of Illinois, Urbana (1972).
10. E. J. Ripling and S. Mostovoy, J. Adh. 3, 145-163 (1971).
11. A. W. Smith, Boeing Scientific Research Lab., Doc. D1.82.1003.
12. R. Exalto, P. F. A. Bijlmer, R. Schiekelman, *Pickling of 2024 T3 Clad as Pretreatment for Bonding*, Royal Neth. Aircraft Factories, Fokker, Report No. 1091, April 1970.
13. A. W. Bethune, Personal Communication, Boeing Materials and Technology, Boeing Commercial Aircraft Company.
14. K. Wefers and G. M. Bell, Alcoa Technical Paper No. 19, Alcoa Research Laboratories, 1972.
15. J. W. Diggle, et al., Chem. Reviews 69, 365 (1965).

A Fluoro-Anhydride Curing Agent for Heavily Fluorinated Epoxy Resins[*]

James R. Griffith, Jacques G. O'Rear and Joseph P. Reardon

Naval Research Laboratory

Washington, D.C. 20375

The cure of epoxy resins is commonly accomplished with amine or anhydride curing agents. The synthesis of suitable curing agents for fluorinated epoxy resins has been a problem since the fluorine contents in excess of 50% by weight are compromised by the curing agent unless it also contains fluorine. Fluoro-amines have several unfavorable characteristics including poor shelf life, poor color stability and unagressive cure function. Consequently, a new anhydride, 4-(2-hydroxy-hexafluoro-2-propyl) phthalic anhydride, has been synthesized. It is a colorless compound which is easily obtained as a supercooled liquid, is readily compatible with epoxy resins of the highest fluorine contents and which reacts readily to yield strong plastics of high fluorine content.

INTRODUCTION

The most common type of "epoxy resin" consists of an epoxy component and a curing agent which becomes as essential in the final molecular structure as the epoxy component. In particular, an unfluorinated curing agent dilutes the fluorine content of a fluorinated epoxy resin such that the final composition is not as highly fluorinated as the epoxy component. Since the progression toward more heavily fluorinated resins is a desirable trend in new materials

---

[*] This paper was presented to the 169th Meeting of the American Chemical Society, April 1975.

research, we undertook to synthesize an effective fluorine-containing anhydride for use with the fluorinated epoxy resins previously reported[1,2]. Fluorinated amines have been previously considered as epoxy curing agents[3], but they have several disadvantages which include poor long-term chemical stability (this is particularly true of aliphatic fluoroamines), unagressive reaction behavior with respect to epoxies, and excessive cost factors.

In adhesive applications, fluoroanhydride-cured fluoroepoxies should have a number of favorable properties including: (1) excellent wetting of substrates because of the low surface tension of the resin, (2) convenient processing characteristics, including, a liquid precured nature, a long pot-life, and reasonably short cure cycles at temperature, and (3) strong, tough adhesive bonds of exceptional resistance to environmental degradation. The hydrophobic-oleophobic qualities of the cured product should be ideal for critical bonding applications such as those of the aircraft industry. The resistance to ultraviolet light is also exceptional due to the large quantity of fluorocarbon within the network structure. For applications which require a lack of color in the adhesive (for example, lens cements), it is possible to produce compositions which yield clear, colorless plastics indistinguishable in appearance from poly(methyl methacrylate). As bonding agents or matrix resins for composite structures, the materials also hold considerable promise because of the processing ease of the precured resin and the inherent stability of the cured product.

Fluoro-Anhydrides

Most epoxy curing agents of the anhydride type are derivatives of phthalic anhydride or are structurally similar to it. On this anhydride there are four available ring positions and a possible means of introducing fluorine would be substitution of these aromatic positions. Since this would allow a maximum of only four fluorine atoms and since such a multiplicity of aromatic fluorines are easily displaced by nucleophiles, the substitution of fully fluorinated aliphatic units on one or more of these positions is preferable. A means of attaching substantial quantities of fluorocarbon is via an ester linkage, and some fluorinated aliphatic ester derivatives of trimellitic anhydride acid chloride have been previously reported.[3]

In the present work, a most efficient means of introducing a substantial quantity of fluorine into a phthalic anhydride type of curing agent is presented. Not only are the chemical yields high, but also, the process is attractive economically since all the fluorine is supplied by hexafluoroacetone, a relatively low cost component. The following reaction sequence illustrates the process:

EXPERIMENTAL

Preparation of (2-Hydroxy-2-propyl)-3,4-dimethylbenzene, (I)

This compound was prepared from o-xylene and hexafluoroacetone: bp 101-102°C/20.0 mm Hg; n_D^{25} 1.4334; lit. (4) bp 200-200.5°C/760 mm Hg.

Preparation of 4-(2-Hydroxyhexafluoro-2-propyl) phthalic acid, (II)

A mixture of (I) (80.0 g; 0.294 mole), potassium permanganate (196 g; 1.24 moles) and 0.15 N aqueous sodium hydroxide solution (3000 ml) was stirred and maintained at 90-93°C for 4 hours. The reaction mixture was cooled and filtered to remove manganese dioxide. The alkaline purple filtrate was acidified with 12 N hydrochloric acid (140 ml), decolorized with sodium sulfite, and the clear solution extracted with ether (1.35 kg). The ether extract was dried ($MgSO_4$), filtered, the filtrate diluted with toluene (200 ml) and the resulting mixture concentrated at reduced pressure to a mass of white crystals. Dispersal of the white crystals in boiling toluene, followed by filtration of the cooled dispersion led to analytical white crystals of (II): 86.5g, 88.4% yield; mp 183-185°C. Anal. calcd. for $C_{11}H_6F_6O_5$: C, 39.77; H, 1.82; F, 34.31. Found: C, 39.94; H, 1.80; F, 34.46.

Preparation of 4-(2-Hydroxyhexafluoro-2-propyl) phthalic anhydride, (III)

Compound II (43.0 g, 0.129 mole) was placed in a flask (300 ml) and heated in a silicone bath (200°C) for 15 minutes. The evolved water amounted to 2.30 g; theory, 2.33 g. A short path distillation of the viscous residue gave analytical (III) as a viscous, supercooled liquid which gradually crystallized: 37.7 g; 92.5% yield; bp 125°C/0.3 mm Hg; mp 75°C. Anal. calcd. for $C_{11}H_4F_6O_4$: C, 42.05; H, 1.28; F, 36.28. Found: C, 41.95; H, 1.22; F, 36.36.

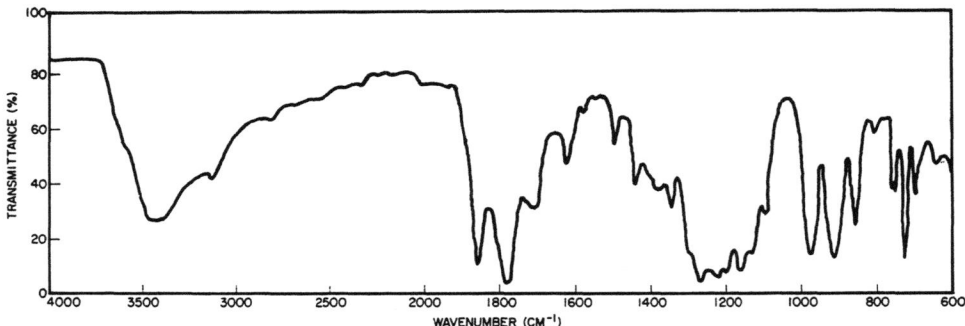

Figure 1. Infrared Spectrum of 4-(2-Hydroxyhexafluoro-2-propyl) phthalic anhydride.

Cure of Fluoroepoxy

The NRL fluorinated epoxies are sufficiently high in fluorine content that they are frequently incompatible with unfluorinated curing agents before the cure reactions are sufficiently advanced to induce compatibility. This anhydride is immediately compatible in the following system, for example:

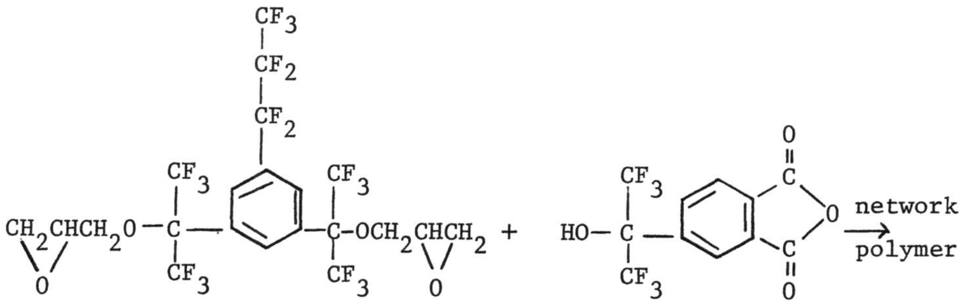

If one-to-one equivalence of epoxy and anhydride is used in the pre-cured solution, the fluorine content of the final polymer is nearly 48%. Although the melting point of the pure anhydride is 75°C, once melted it is easily retained in the liquid state until blending of the components to form a solution is accomplished. After the solution is formed, there is no tendency for precipitation of the anhydride to occur.

The cure reaction illustrated above is typical of epoxy-anhydride reactions. A catalytic amount of tertiary amine is an effective accelerator for the process, and the product is a tough, rigid solid which has the appearance of a typical cured epoxy.

Surface Tensions of Fluoroepoxies

The most common type of liquid epoxy resin, the diglycidyl ether of Bisphenol-A, has a surface tension of 47-49 dynes per centimeter[5]. Liquid fluoroepoxy resins of approximately 50% fluorine content by weight have about one-half this surface tension, as illustrated by the following Table.

A consequence of this low surface tension is that low energy surfaces such as that of Teflon (critical surface tension = 18.5 dynes/cm) are wetted rather well, and although a strong adhesive bond is not obtained to a Teflon surface, powders of Teflon are readily dispersed into fluoroepoxy liquid to form an extender pigment for the resin. More easily wetted surfaces, such as that of glass, are rapidly and thoroughly covered by fluoroepoxy.

Table 1

Surface Tensions of Liquid Fluoroepoxy Resins as Determined by the Cassel Tensiometer

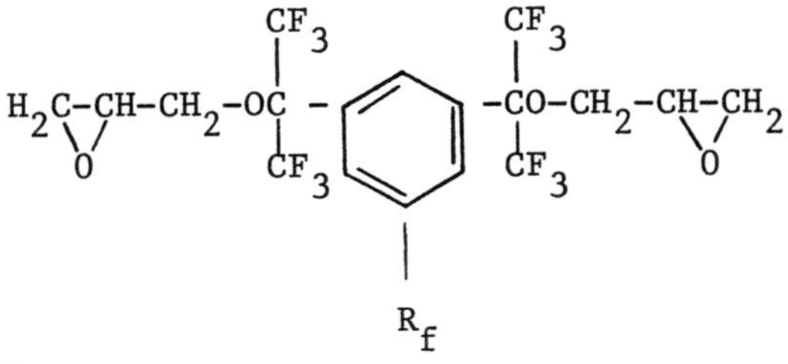

$$H_2C-CH-CH_2-O-\underset{CF_3}{\overset{CF_3}{C}}-\phi-\underset{CF_3}{\overset{CF_3}{C}}-O-CH_2-CH-CH_2$$
(with epoxide groups at each end and R_f on the phenyl ring)

$R_f = CF_2CF_2CF_3$

	Temp., °C	Surface Tension, dyne/cm
Nozzle A	20.0	24.8
	31.0	23.7
	41.5	22.8
Nozzle B	21.7	25.0
	32.5	24.2
	41.5	23.6

$R_f = CF_2(CF_2)_5CF_3$

	Temp., °C	Surface Tension, dyne/cm
Nozzle C	23.0	23.8
	32.0	22.8
	40.0	22.1
	50.2	21.3

REFERENCES

1. J. G. O'Rear and J. R. Griffith, 165th Meeting of American Chemical Society, Division of Organic Coatings and Plastics Chemistry, Preprints 33, No. 1, 657 (1973).
2. J. R. Griffith, J. G. O'Rear and S. A. Reines, CHEMTECH, 311 (1972).
3. J. R. Griffith and J. E. Quick, 155th Meeting of American Chemical Society, Division of Organic Coatings and Plastics Chemistry, Preprints 28, No. 1, 342 (1968).
4. B. S. Farah, E. E. Gilbert and J. P. Sibila, J. Org. Chem. 30, 998 (1965).
5. H. Lee and K. Neville, *Handbook of Epoxy Resins*, McGraw-Hill Book Company, New York, p. 21-10 (1967).

DISCUSSION

On the Paper by T. A. Bush, M. E. Counts and J. P. Wightman

H. Arlt (*Arizona Chemical Co.*): Appearance on photos of three types of failure cohesive, adhesive, and thin layer. Smooth failure - is this monomolecular?

J. P. Wightman (*Virginia Polytechnic Institute*): No. Recent results were obtained with energy diapersion x-ray analysis (EDAX) in the scanning electron microscope. No Ti signal was noted in the smooth regions (see Figure 14) which would indicate thicknesses considerably greater than monomolecular.

D. Dwight (*DuPont Company*): I have used the same techniques (ESCA, contact angles, electron microscopy, etc.) with similar success to understand bonding of "Teflon" ^{R}, copper, aluminum with epoxies and acrylics. Can you rationalize why your ESCA data on C/N intensities are so different on the two samples that show relatively good shear strength?

J. P. Wightman: Although a polyimide film is present on the fracture surfaces of samples 1m2-517D3 and 1mp2-516D4 (Table 6), the atomic percents of C and N are nominally 5.4% and 71%, respectively. Thus, the C/N intensities are reasonable.

A. J. Kinloch (*Ministry of Defense, U. K.*): It has been suggested that the type of oxide on the titanium alloy surface, for example whether it is rutile or anatase, can affect the water resistance of the subsequently made joint. Did you conduct any experiments, for example x-ray diffraction analysis, to identify the structural type of titanium oxide formed by your Pasa-Jell surface pretreatment?

J. P. Wightman: We have made no x-ray diffraction measurements. Hamilton and Lyerly (Picatinny Arsenal Tech. Rept. 4185, March, 1971) have shown that only the anatase structure of Ti 6-4 bonded durably.

On the Paper by E. R. Ruckel

P. O. Powers (*Consultant*): Why wasn't Mosher theory of phenyl indane structure of dimer mentioned?

E. R. Ruckel (*Arizona Chemical Co.*): I am unaware of this theory. Actually the study of the dimer structure was a very small facet of this work. The structure of the dimer illustrated during the talk was one obtained using the $AlCl_3/(CH_3)_3SnCl$ catalyst/cocatalyst system.

Author Index

(The numbers in parentheses are references for discussion participants. Pp. 1-438 are found in Volume 9A, pp. 439-860 in Volume 9B.)

A

Abrikossova, I. I., 828
Acs, M., 393
Aggarwal, S. L., 24
Aggias, Z., 186
Ahagon, A., 281, 288
Akhiezer, A. I., 632
Akopyan, L. A., 16
Albrecht, K.-D., 278, 327
Aleinikova, I. N., 827, 828
Allan, A. J. G., 165
Allen, K. W., 165, 514
Allen, R. P., 314
Alsalim, H. S., 514
Alter, H., 165
Andrews, E. H., 15, 17, 53, 127, 185, 186
Andries, J. C., 314
Andrus, P. G., 827
Aoki, Y., 185
Arlt, Jr., H., 361, 395, (437)
Arnold, V. W., 493
Arnold, W., 828
Aronson, M. P., 723, 725, 734
Ashkin, A., 829
Athey, R. J., 466
Atlas. S. M.. 29
Azrak, R. G., 233, 249, (253), (254)

B

Badeseu, Th., 761
Bailey, A. I., 828
Bair, H. E., 186
Baker, C. S. L., 278
Banks, W. H., 722
Barbaris, M. J., 165
Barnard, D., 278
Barr, D. B., 217
Barry, A. J., 253
Bartell, F. E., 90, 393
Bartenev, G. M., 16
Barth, B. P., 249
Bartz, K. W., 185
Bascom, W. D., 15, 167, 215, 288, 501, 510, 511, 613, 614
Bassemir, R. W., 722
Bates, Jr., T. R., 677
Bauer, R. F., 186
Bean, A. J., 722
Becker, J. H., 722
Beecher, J. F., 215
Behrndt, K. H., 15
Bell, G. M., 428
Bell, J. P., 185
Bell, V. L., 198
Bennett, E. C., (496)
Bennett, S. J., 704
Berger, S. S., 338
Bernett, M. K., 90, 91
Berry, J. P., 127, 614
Bersch, C. F., 562, 613
Bethune, A. W., 428
Beyer, D. J., 677
Bewig, K. W., 90
Bickley, W. G., 705
Bijlmer, P. F. A., 428
Bikerman, J. J., 127, 166, 184, 392, 515, 647

Bishop, E. T., 214
Black, A. L., 314
Black, W., 828
Blount, J., (495), (496)
Böhme, G., 828
Bohme, R. D., 185
Boielle, J. H., 723
Bolger, J. C., 185
Bolz, L. H., 515
Borchert, S. J., 214
Bottrell, N. L., 165
Boucher, E. A., 165
Boyd, G. E., 166, 613
Boyer, R. F., 232
Bradford, R. D., 215
Bradley, R. S., 828
Bragaw, C. G., 428
Bragole, R. A., 17
Bramit, T. M., 16
Bray, W. S., 412
Bright, K., 15, 16, 91
Brockman, W., 166
Brown, J., 614
Brown, R. J., 278
Bruno, M. H., 721, 761
Bryant, P. J., 90
Buchan, S., 278, 327
Buck, W. H., 215
Buff, F. P., 827
Bullett, T. R., 15, 185
Burdon, R. S., 632
Burks, H. D., 198
Burta-Gaponovitch, L. N., 828
Burton, J. D., 704
Bush, T. A., 365
Butler, G. B., 412

C

Cagle, C. V., 483, 676
Caldwell, J. R., 185
Campion, R. P., 15
Cantor, S. E., 17, (355)
Carlick, D. J., 761
Carlton, H. E., 827
Carrier, G. B., 90
Carter, C. P., 633
Casimir, H. B. G., 827
Caspar, D. L. D., 827
Cassidy, P. E., 186, 288

Cella, R. J., Jr., 215
Chang, M. D., 510
Chang, V. F., 723
Charlesworth, D. H., 722
Chatterjee, P. K., (253)
Chenevey, E. C., (252)
Cherry, B. W., 165
Chow, T. S., 687, 704
Christiansen, A. W., (170)
Churaev, N. V., 623, 827
Cirlin, E. H., 215, 663, 761
Clark, D. T., 15
Clark, H. E., 826, 852
Clark, W. H., 353
Clarke, J. A., 186
Clay, J. R., 827, 828
Clayfield, E. J., 734
Clayton, D. W., (169)
Clinton, W. C., 90
Collins, W. T., 329
Colson, J. P., 53
Conley, R. T., 288
Cooper, S. L., 214
Cooper, S. M., 412
Corn, M., 828
Corten, H. T., 510, 562, 613
Cottington, R. L., 215, 510, 511
Counts, M. E., 365
Craig, R. G., 647
Craig, T. O., 215
Crane, L. W., 215
Crocker, G. J., 166
Crosley, P. B., 562, 613
Cross, N. L., 632, 647
Cumming, A. P. C., 483
Curtin, J. L., 721

D

Dahlquist, C. A., 165, 167
Daniels, H., 828
Dann, J. R., 67, 185, 761
Davis, J. K., 90
Davis, M. H., 827
Dauksys, R. J., 17, 467, 472
Day, A. R., 412
Dean, R. B., 166
Dear, D. J., 734
DeBruyne, N. A., 165
deDiesbach, J., 198

AUTHOR INDEX

deJongh, J. G. V., 828
Dekker, H., 16
De Lollis, N. J., 165, 613
Derenzo, D., 721
Derjaguin, B. V., 16, 166, 633, 827, 828
Depew, H. A., 314
Desper, C. R., 232
Dessauer, J. H., 826, 852
Dettre, R. H., 90, 91, 127
De Vries, K. L., 510
Dewalt, C., 412
Dexheimer, R. D., 185
Dickie, R. A., 215
Dietz, E., 105
Diggle, J. W., 428
DiMarzio, E. A., 53, 54
Dimick, B., (252)
Dine-Hart, R. A., 198
Doggett, T., 722
Dolgov, B. N., 353
Domine, J., 493
Donald, D. K., 827
Doss, R. C., 16
Dottler, G., 353
Downey, R. E., 315
Drake, R. S., 511
Dreidger, O., 167
Drobek, J., 90
Drude, P., 53
Dukes, W. A., 597, 613
Dunaburg, M. S., 721
Dunbar, L., 614
Duncalf, B., 721
Dunn, A. S., 721
Dupre, A., 632
Durelli, A. J., 705
Dwight, D. W., (169), 392, (437)
Dyckerhoff, G. A., 165, 185
Dynes, P. J., 215, 663, 723, 735, 761

E

Eberhart, J. G., 90
Edeleaner, C., 705
Edwards, M. E., 473
Eichinger, B. E., 126
Eichorn, R. H., 827

Eick, J. D., 127
Eihl, T. R., 493
Eirich, F. R., 54
Elbing, E., 126, 167
Eley, D. D., 514
Elliott, J. R., 467, (496), 853
Ellison, A. H., 91
Elmer, O. D., (356)
Emerald, R. L., 827
Ennis, J. L., 721
Erb, R. A., 90
Erf, R. P., 16
Erickson, D. E., 165, 185
Erickson, P. W., 288, 613
Ershova, I. G., 633
Estes, G. M., 214
Evans, B., 288
Exalto, R., 428
Ezekiel, H. M., 29

F

Fan, P. L., 621, 633, 635, (707)
Farah, B. S., 435
Fedotova, O., 198
Fedyabin, N. N., 633
Feldman, A., 90
Fery, N., 105
Filimonova, N. P., 353
Fisher, D. J., 353
Fisher, R. A., 632, 647
FitzSimmons, V. G., 91
Flanigam, O., 353
Fleer, G. J., 53
Fleischmann, R. M., 288
Flory, P. J., 126
Folt, V. L., 314
Folman, M., 633
Fontana, B. J., 53
Forbes, W. G., 514
Ford, J. E., 466
Forrestal, L. J., 514
Forshirm, A., (252)
Fox, H. W., 90, 126, 166
Fox, P. G., 614
Fowkes, F. M., 90, 166, 185, (355), 613, 663, 761
Fowler, P. R., 16
Frank, C., 278

Frederick, J. R., 677
Freeman, H. A., 338
Freese, P. V., 53
Frenkel, J., 852
Frisch, H. L., 16, 185, 215, 428, 514, 613, 723, 852
Frisch, K. C., 215, 483
Frischkorn, L. H., 722
Fritz, J. S., 353
Fuller, K. N. G., 614

G

Gagosz, R. M., 16
Gaik, U., 186
Gardon, J. L., 167
Gati, Gy., 761
Gattermann, L., 198
Gaudioso, S. L., 722
Gehman, W. G., 614
Geisselmann, H., 828
Gent, A. N., 165, 185, 186, 281, 288, (355), (356), 514, 614
Gertzmann, A. A., (250)
Gibbs, J. W., 126
Gilbert, E. E., 435
Gillis, P. P., 704
Gilman, B., (616)
Gilman, J. J., 704
Gingell, D., 828
Gipe, H. F., 722
Girifalco, L. A., 126, 167
Given, D. A., 315 (357)
Givens, W. G., 90
Gladding, E. K., 215
Gledhill, R. A., 591, 613
Goddu, R. F., 288
Goel, N. S., 763, 827
Good, A., (617)
Good, R. J., 37, 90, 107, 126, 127, 166, 167, (170), 186, (355), (707), 761
Goodier, J. N., 704
Goodman, M., 31
Goodrich, F. C., 632
Goossens, J. C., 353

Gracia, R. F., 722
Grafton, E. H., 393
Grant, W. H., 43, 53, (169)
Gray, Jr., T. F., 185
Gray, V. R., 166, 167
Greenlee, T. W., 339, 353, (358)
Greenwood, L., 613
Gregg, S. J., 633
Griffith, A. A., 127, 168, 613
Griffith, J. R., 429, 434, 435, 665, 676
Gronskaya, E. V., 16
Gross, M. E., 16, 186, 257
Grosskreatz, J. C., 705
Gruntfest, I. J., 127
Gundlach, R. W., 826
Gunther, W. H. H., 722
Gurganus, T. B., 90
Guttmann, W. H., 483

H

Haage, G., 91
Haefner, A., 278
Haiman, Gy., 761
Hale, W. F., 233, (495)
Hamaker, H. C., 827
Hamann, K., 93, 105, (169), 392, 393
Hamilton, W. C., 392, 393
Hansen, R. H., 167, 514
Haq, K. E., 15
Hare, E. F., 90, 127
Harkins, W. D., 90, 166, 393, 761
Harlan, Jr., J. J., 214
Harris, Jr., J. F., 91
Hartmann, J. F., 314
Hartsuch, P. J., 722
Hata, T., 165, 167
Hausstein, R. W., 185
Healy, T. W., 53
Herd, J. M., 54
Hergenrother, P. M., 17
Hermann, J. J., 633, 647
Herring, C., 852
Hertzberg, R. W., 562
Hertzberg, W. J., 127
Hesselink, F. Th., 53
Hewitt, N. L., 278

AUTHOR INDEX

Hicks, A. E., 278
Hicks, E., 53
Hindersinn, R. R., 215
Hirano, Y., 722
Hoene, R., 105
Hoey, C. E., 466
Hofbauer, E., 198
Hofrichter, C. H., 353
Hoh, G. L. K., 185
Hohn, P., 828
Holden, G., 214
Holland, D. L., 17
Holland, L., 90
Hope, C. J., 167
Hopkins, A. J., 54
Horn, J., 93, 105
Housel, W. S., 632
Howard, G. J., 54
Hsu, E. C., 281, 288
Hudson, F. W., 827
Humphreys, K. W., 186
Hunklinger, S., 828
Hunston, D. L., 511
Huntsberger, J. R., 184, 392
Hurwitz, M. D., 449, (495)

I

Ichimi, T., 761
Ilkka, G., 165
Illinger, J. L., 217, 231, (252)
Imoto, M., 412
Irwin, G. R., 127, 168, 562, 613
Israelachvili, J. N., 828
Ito, T., 215
Iyengar, R., (357)
Iyengar, Y., 165, 185

J

Jackson, J. D., 827
Jackson, Jr., W. J., 185
Jacob, S. M., 828
Jedlinski, Z., 186
Jemain, W. A., 614
Jenkins, R. K., 186
Joesten, D. L., 233
Johanson, O. K., 288
Johnson, J. M., 186

Johnson, K. L., 15
Johnson, Jr., R. E., 90, 91, 126, 127
Jones, J. I., 198
Jones, R. L., 215, 510, 511
Jordan, H. F., 761

K

Kaas, R. L., 288
Kaden, V., 105
Kaelble, D. H., 15, 126, 167, 185, 199, 214, 288, 613, (615), (616), 663, 704, 723, 735, 761
Kalfoglou, N. K., 215
Kamat, D. V., 466
Kambour, R. P., 127, 215
Kanamaru, K., 852
Kane, P. F., 15
Kantner, G. C., 465, 466
Karasz, F. E., 231
Kardos, J. L., 288
Kay, P. M., 722
Kaye, W., 288
Keith, R. E., 215
Kelley, F. N., 515
Kelvin, Lord, 633
Kendall, K., 15, 186, 685, 704
Kendig, M., (169)
Keown, R. W., 17
Kerr, C., 428, 613
Khanna, R., (169)
Kies, J. A., 511
Killmann, E., 53
King, H., 483
Kinloch, A. J., 15, 17, 53, 165, 185, 186, (250), (251), 288, (437), 597, 613, 614, (615), (617), (618)
Kitazaki, Y., 165, 167
Kitchener, J. A., 828
Klempner, D., 215
Kling, W., 828
Klomp, J. T., 15
Kloubek, J., 15
Knibbs, R. W., 186

Knudsen, M., 514
Kobin, I., 15
Koenig, J. L., 288
Koleske, J., 249
Kolesnikov, G., 198
Koranyi, G., 393
Korolev, A. Ya, 165
Kottler, W., 827
Krafft, J. M., 514
Kraus, G., 278
Kreibich, R. E., 17
Krieger, R. B., 614
Kroker, R., 105
Kruma, R., 466
Krupp, H., 16, 827, 828
Kruse, J., 314
Kuczynski, G. C., 632
Kuenzi, E. W., 428
Kuespert, D., 14
Kuhnis, H., 198
Kuzenko, M., 53
Kwei, T. K., 53, 185, 428, 723, 852

L

Laible, R. C., 93, 105, 215
Lake, G. J., 288
LaMer, V. K., 53
Landau, L. D., 632
Langbein, D., 827, 828
Lange, H., 828
Langley, P. G., 186
Langmuir, I., 685
Lantenschlaeger, F., (616)
Larrabee, G. B., 15
Larson, L. G., 722
Latta, H., 314
Laughrey, R. A., 722
Lavelle, J. A., 734
Lavrentev, V. V., 16
Law, T. J., 705
Lawn, B. R., 514
Lawton, E. L., 16
Layne, W. S., 614
Lee, L. H., 1, 14, 16, 166, 185, 186, (356), 435, 483, (616), (617), 647, 663, (707), 711, 723, 831, 852

Lee, R. J., 493
Legge, N. R., 214
Leigh-Dugmore, C. H., 314
Leonard, R. F., 721
Levine, M., 165
Lewis, A. F., 483, 514, 563, (615), (616),
Lewis, R. W., 217, 231, 232
Lifshitz, E. M., 632, 827, 828
Lin, C. J., 185
Lingwall, R. G., 562
Lipatov, Yu. S., 53
Lipatova, T. E., 53
Livingstone, H. K., 166, 613
Litz, R. J., 17, 493
Liu, C. A., 704, 705
Locke, C. E., 186
Loeser, E. H., 166
Loffler, F., 828
Logan, T. J., 16
Lopata, S. L., 232
Lowengrub, 704
Lubowitz, H. R., 198
Ludema, K. C., 677
Lurie, R. M., 165
Lyklema, J., 53
Lyon, F., 278

M

MacDonald, N. C., 613
Mahoney, C. L., (496)
Malpass, B. W., 6, 15, 91
Manning, C. R., 90
Manson, J. A., 562, 852
Marian, J. E., 127
Mark, H., 19, 29
Marker, L., 215
Marrs, O. L., 16
Marsden, J. G., 338
Mase, G. D., 493
Mathews, J. B., 734
Matsuoka, S., 186
Maus, L., 761
May, C. A., 483
McCabe, J. M., 722
McCarvill, W. T., 16
McCrackin, F. L., 53, 54
McDonald, J. W., 185
McDonald, N. C., 16

AUTHOR INDEX

McFarlane, J. S., 632, 647
McGarry, F. G., 215, 428, 511
McLeod, L. A., 514
McNeil, M. B., 705
McSweeney, R. T., 466
Medalia, A. I., (169), (357)
Mehrota, R. C., 353
Melrose, J. C., 126, 166
Metsik, M. S., 828
Mihalik, R., 186
Mikheev, E. P., 353
Miles, M. L., 412
Minor, F. W., 647
Miron, J., 443
Mitchell, D. J., 828
Mittal, K. L., 127, 129, (170), (171), (355)
Moacannin, J., 214
Moelwyn-Hughes, E. A., 827
Molau, G. E., 214
Molvan, Jr., H. E., 185
Montgomery, D. J., 829
Montoya, O., 165, 613
Moravec, R. W., 90
Moreau, W., 721
Morrissey, B. W., 43, 54
Mostovoy, S., 428, 510, 513, 562, 613
Mottus, E. H., (252)
Muchnik, S. N., 393
Muddarris, S., 165
Mukherjee, S. K., 215
Mulherin, J. H., 514
Murayama, T., 16
Murphy, T. P., 184
Mylonas, C., 167

N

Nakano, Y., 17
Makao, K., 185
Narain, R. P., 353
Nashay, A., 215
Natarajan, R. T., 563
Nelson, R. P., 613
Neugebauer, H. E. J., 827
Neumann, A. W., 91, 127, 167
Neville, K., 435, 483
Newby, W., (251)
Newman, S. B., 127, 511, 723, 852

Newton, A., 466
Nielsen, L. E., 249
Nijboer, B. R. A., 827
Ninham, B. W., 827, 828
Nippert, C. R., 852
Nir, S., 828
Nisbet, K. D., (251), (707)
Nolland, J. S., (251), 413, (616), (618)
Nollen, K., 105
Nyce, J. L., 17

O

O'Brien, W. J., 621, 633, 635, 647
Okubo, S., 705
Olivei, A., 633
Olsen, D. A., 90
O'Rear, J. G., 429, 434, 665, 676
Orman, S., 428, 613
Orowan, E., 613
Orwoll, R. A., 126
Osteraas, A. J., 90
Otsu, T., 412
Overbeek, J. Th. G., 53, 828
Owens, D. K., 167, 761

P

Pacansky, T. J., 722
Packham, D. E., 15, 16, 91
Paddy, J. F., 632, 722
Pahoja, M. H., 428
Papahagu, L., 761
Parikh, N. M., 90
Parks, C. R., 278
Parr, F. T., (357)
Parsegian, V. A., 827, 828
Pascuzzi, B., 17
Passaglia, E., 53, 54
Patrick, E. C., 614
Patrick, R. L., 15, 167, 483, 499, 510, 562, 614, 663
Pattison, V. A., 215
Pauling, L., 828
Pav, D., 663, 723, 735
Pell, J. P., 16
Penwell, R. C., 704, 705
Perepelkin, V., 29
Peterlin, A., 29

Peters, R. Z., 16
Peterson, C. M., 186
Petke, F. D., 91, 175, 177, (250), (254)
Petrich, R. P., 288
Petrine, D., 483
Petrov, A. D., 353
Petrov, L. N., 761
Peyser, P., 54, 215, 510
Peyton, F. A., 647
Phillips, M. C., 167
Picknett, R. G., 632, 647
Pirani, M., 685
Pitaevski, L. P., 827
Plueddemann, E. P., 329, 338, (357), (358), 663, (707)
Polaski, E. L., 259
Polder, D., 827
Pomerantz, P., 90
Porter, M., 278
Post, H. W., 353
Powers, P. O., (437), 485, 493, (496)
Price, S. J., 17
Priel, Z., 54
Pritchard, W. H., 185
Progar, D. J., 187, 392
Prosser, A. P., 828
Prosser, J. L., 15

Q

Quick, J. E., 435
Quick, J. R., (496)

R

Rabenhorst, H., 827, 828
Raevskii, R. G., 165
Ranney, N. W., 338
Raraty, L. E., 165
Rasmussen, J. J., 613
Ray, B. R., 91
Reardon, J. P., 91, 429
Reilly, A. F., 353
Rein, R., 828
Reines, S. A., 434
Reinhart, T., 483
Renzow, D., 91, 127

Reynolds, G. E. J., 185
Rhee, S. K., 90, 91, 167
Richmond, P., 828
Riddiford, A. C., 167
Riew, C. K., 511
Riggs, W. M., 392
Ripling, E. J., 428, 510, 513, 562, 613
Roberts, A. D., 15
Roberts, W. J., 412
Robertson, R. E., 127
Robeson, L. M., 215
Roe, R. J., 53, 852
Rogers, J. M., 232
Roninger, F. H., 314
Rothstein, E., 54
Rounds, N. A., (495)
Rouxeville, E., 412
Rowe, E. H., 511
Royka, S. F., 827
Roylance, D. K., 232
Roylance, M. E., 231, 232
Rowland, F. W., 54
Rubin, H., 722
Rubin, R. J., 53
Ruby, I. R., 314
Ruch, R. J., 393
Ruckel, E. R., 395, 412, (437)
Rudt, H., 198
Rusch, K. C., 327
Ryan, F. W., 166, 168, 185

S

St. Clair, T. L., 187, (251), (252), 392
St. John, D. F., 829
Samuels, S. L., 215
Sandstede, G., 828
Satas, D., 186
Saunders, J. H., 483
Saxon, R., 514
Schaffert, R. M., 826
Schank, R. L., 722
Schick, M. J., 54
Schiekelman, R., 428
Schmukler, S., (357), (495)
Schnabel, W., 16, 828
Schneider, M., 105
Schneider, N. S., 231

AUTHOR INDEX

Schonhorn, H., 16, 166, 167, 168, 185, 393, 428, 514, 613, 723, 852
Schrader, M. E., 90, 614
Schram, K., 827
Schultz, J., 165, 186, 353, 514
Schure, R., (250), (251)
Schuyten, H. A., 353
Schwab, G. M., 353
Schwartz, A. M., 647
Schwartz, W. T., 215
Seger, Jr., S. G., 577
Seifer, M. I., 466
Sek, D., 186
Sell, P. J., 165, 167, 185
Sergeeva, L. M., 53
Severynse, G., 829
Sexsmith, F. H., 259, 278
Seymour, R. W., 214
Shafrin, E. G., 90, 91, 127
Sharma, S. C., (170), (253), (356)
Sharpe, L. H., 165, 166, 393, 428, 514, 577, 613, (616), (617)
Sharples, L. K., 761
Shartsis, L., 613
Shaw, V. J., 761
Shelgayeva, V., 198
Shen, M., 215
Shereshefsky, J. L., 633
Sheriff, M., 186
Shih, P. T. K., 288
Sibila, J. P., 435
Siebert, A. R., 511
Siegbahn, K., 393
Silberberg, A., 54
Silver, J. L., 721
Simpson, B. D., 16
Sites, R. D., 278
Skapski, A., 126
Skeist, I., 327, 441, 443, 483, (495)
Slessor, J. M., 327
Slysh, R., (616)
Smilga, V. P., 166, 827
Smith, A. H., 721, 722
Smith, A. W., 428
Smith, H. R., 562
Smith, J. T., 393
Smith, L. E., 53, 54
Smith, R. W., 289, 314, (356)
Smith, T. L., 215
Sneedon, I. N., 704
Snowden, K. V., 514
Soller, W., 165
Sorensen, D. P., 722
Sorkin, J. L., 722
Sparnaay, M. J., 828
Spencer, R., 763
Spinner, S., 613
Sprinkle, J. K., 165
Stark, F. O., 288
Steen, S. D., (356)
Stein, F., 828
Steinberg, H. L., 53
Steiner, G. F., 852
Steinkamp, R. A., 185
Stevens, G. H., 428
Strauss, V., 761
Strella, S., 852
Stromberg, R. R., 43, 53, 54, 184
Sugita, K., 722
Sultan, J. N., 215, 511
Swanson, F. D., 17
Sweet, J. S., (496)
Sypula, D. S., 722

T

Tabor, D., 165, 632, 647, 685, 828
Takata, A., 412
Talaey, M. V., 633
Tanaka, Y., 483
Tanford. C., 827
Tanner, W., 127
Taylor, H. F., 165
Taylor, L., 392
Temple, G., 705
Thelen, E., 90
Thies, C., 54
Thomas, A. G., 288
Thomas, G. R., 231
Thomas, J. R., 54
Tichane, R. M., 90
Timmons, C. O., 511
Timoshenko, S., 633, 647, 704
Tobolsky, A. V., 29
Tollenaar, D., 722

Toporov, Y. P., 827, 828
Toy, L. E., 16
Toyama, M., 168
Trantina, G. G., 15, 510, 511
Traskos, R. T., 721
Tscheogl, N. W., 214
Tucek, C. S., 90
Tuohey, P. F., 722
Tustanowski, S., 353
Tutas, D. J., 53
Twiss, S. B., 166

U

Ullman, R., 54
Uy, K. S., 167

V

Vadimsky, R. G., 186
Vakula, V. L., 166
Van Brederode, R. A., 185
Van Den Temple, 16
Vanderhoff, J. W., 722, 725, 734
Vandesaer, J., (253)
Van Kempen, N. G., 827
Van Oene, H., 723, 852
Van Silfhout, A., 828
VanVoorsf Vader, F., 16
Vasilienko, Ya, P., 53
Vaughan, M. F., 198
Vdovin, V. M., 353
Vermeulen, T., 127
Vertnik, L. R., 185
Vincent, B., 53
Viswanathan, N., 721
Vogel, A. I., 91
Vogel, G. E., 288
Volkova, T. S., 828
Voyutskii, S. S., 166, 514
Vullo, J., (250)
Vyverberg, R. G., 826

W

Wake, W. C., 165, 185, 278, 514, 613
Walker, K., 562
Walkup, L. E., 827
Walters, G., 828
Walters, J. P., 16
Waltons, T. R., 665, 676
Wang, T. T., 186
Warrick, E. L., 353
Watson, P. K., 827
Watson, R., 412
Weber, C. D., 16, 186
Wefers, K., 428
Weiss, G. H., 827
Weiss, L., 828
Weiss, P., 165
Weiss, R. G., 493
Wells, A. A., 127
Wendt, R. C., 167
Wenzel, R. N., 852
Wetzel, F., 493
Weyman, H. P., 412
Wheeler, J. B., 722
White, M. L., 90
White, T. E., 288
Whittaker, D., 412
Wiebull, W., 574
Wiederhorn, S. M., 515
Wight, P., 483
Wightman, J. P., 198, 365, (437), (617)
Wilcox, R. C., 614
Wilde, A. F., 232
Wilkes, G. L., 215
Wilkinson, M. C., 723, 725, 734
Williams, C. M., 412
Williams, H. L., 215
Williams, M. L., 15, 510, 515, 704
Williamson, I., 466
Willner, A. M., 428
Winterton, R. H. S., 828
Withen, T. V., 734
Wojcik, R. T., 395
Wolfe, Jr., T. R., 215
Wolfram, E., 761
Wolock, I., 127, 511
Wright, J. F., 722
Wright, W. W., 198
Wrigley, A. N., (495)
Wu, S., 165, 166, 185, 392
Wu, W. C., 562

AUTHOR INDEX

Y

Yager, B. J., 288, 614
Yannes, I., 249
Yarwood, J., 685
Yoerger, W. E., 722
Young, T., 632
Yu, C. U., 633, 647

Z

Zettlemoyer, A. C., 723, 725, 734
Zimon, A. D., 827, 828
Zisman, W. A., 15, 55, 90, 91, 126, 127, 166, 167, 185, 614, 663, 761

Subject Index

(Pp. 1-438 are found in Volume 9A, pp. 439-860 in Volume 9B.)

A

Abhesive, 715
Absorption coefficient, 793
Acrylamide, 714
Acrylic
    latex, 235
    monomers, 455
    PC laminate, 217
    self-crosslinking, 235
Adherend surface, 372
Adhesiology, 3
Adhesion, 55
    capillary, 621
    definition, 109
    elastomer-textile, 259
    free energy of, 111
    liquid bridge, 621
    metal-ceramic, 6
    metal-glass, 6
    paint-film, 6
    post-cure, 324
    specific, 247
Adhesion constant, 818
Adhesion pressure, 623
Adhesion promoter, 341, 662
Adhesion theories, 131
Adhesives
    advances of, 1
    aerospace, 9
    alloy, 477
    consumption, 3
    electronics, 10
    fusible, 464
    growth of, 443
    high solid, 12
    hot-melt, 13, 83
    laminator, 462
    microscopy of, 289
    photocured, 13
    pressure sensitive, 151
    surface energy of, 157
    synthetic, 177
    thin film, 83
Adiprene L-100, 25
Adsorption, 118
    -desorption, 104
    interdiffusion, 199
    rate of, 51
Aging stability of rubber, 409
Alpha-methylstyrene, 485
Aluminum
    -aluminum joint, 606
    γ_c, 76
    filler, 197
    oxide, 64
4-Aminophthalonitrile, 666
Anaerobic adhesives, 469
Association polymers, 715
ATR, 416
Attachment site
    density, 565
    theory of, 563
Auger spectroscopy, 4
Autophobic
    liquids, 68
    systems, 118

B

Beam
    incident, 5
    reflected, 5
Bending energy, 637

Benzophenone, 13
Betone-38, 581
Bimetallic plate, 755
Binding energy of elements, 388
Bis-2-chloro ethyl-
  vinyl phosphonate, 566
Block copolymer
  as adhesives, 199
  SB, 200
  triblock, 200
Blocked isocyanate, 481
  for lithographic plate, 718
Blow-off
  incremental, 817
Bonding failure
  of ferrite, 584
Brillouin
  doublet frequency, 631
Brittle film
  adhesion of, 687
Brominated alkanes, 71
BSDA, 189
BTDA, 188, 367
Bumper
  energy absorbing, 315
Butadiene
  methacrylic acid, 262
Butt point, 153, 568
Button test, 323

C

Cab-o-sil, 282
Calorimeter
  differential scanning, 235
Cantilever beam, 600
Capillary
  adhesion, 621
  deformable, 635
  isolated, 625
  parallel slit, 835
  V-shape, 835
Carbenium ion
  camphenic, 400
Carbon black, 295
  in toner, 844
Carbon fiber, 22
Carboxy terminated
  nitrile (CTBN), 199
Carpet backing, 450
Cascade development, 765

Cationic polymerization, 94
Centrifugal
  adhesion spectrum, 811
Charge distribution, 818
Charge transfer, 119
Charging, 764
Chelating polymer, 28
Chinon, 24
Chrome-steel plate, 726
Cinnamic ester, 714
Clay, 330, 581
  filled rubber, 334
  kaoline, 334
Cleaning, 768
Cloud point, 407
Cloud temperature, 490
Cohesion
  free energy of, 111
Cohesive energy density, 116, 720
Cohesive plateau, 565
Colloid-dichromate plate, 712
Contact adhesive, 233
Contact angle, 4, 118, 637, 650,
    739, 836
  of fracture surface, 370
  of glass, 67
  of octane-water, 385
  of Ti-6-4, 384
  reversibility, 63
  tetrabromomethane, 60
Cordelan, 24
Copper
  bronze powder, 669
  γ_c, 76
  phthalocyanine, 94, 667
Corona discharge, 765
Corrosion
  inhibiting primer, 425
Coulter counter, 808
Coupling agent, 282, 647
Crack
  closing force, 121
  detection, 677
  growth rate, 523, 607
  length, 122, 698
  strain, 688
  velocity, 607, 608
Cracking
  rate, 521
  stress corrosion, 520

SUBJECT INDEX

Craze, 123
   fibril, 125
Critical micelle
   concentration, 759
Critical surface tension, 61, 110, 130, 135
   of berylium, 77
   of boron carbide, 77
   of bulk water, 69
   of ceramics, 77
   of glasses, 77
   of metals, 73, 76
   of polymers, 137
   of polysiloxanes, 648
   of selenium, 77
   of toner, 836
Crosslinking
   degree of, 287
Cryomicrotomy, 291
CTBN, 505
   modified epoxy, 515
Cyanoacrylate, 7, 468

D

DABP, 367
Detachment
   force, 768
   of particle, 823
   work of, 285
Development, 765
Diazo oxide, 714
Diazo plate, 713
Diazo resin, 714
Dielectric film, 10
   copper-clad, 12
Diffusion
   of rubber, 6
   rate, 569
DGEBA-piperidine, 505
Diglyme, 189
Dilation deformation, 505
p-Dinitrosobenzene, 8
Dipentene, 398
Disjoining pressure, 773
Diurethane system, 268
Drakeol-10, 730
Driography, 715
Drude equation, 46
Duplex film, 58

E

Elastomer
   crosslink density, 7
   thermoplastic, 480
   to aluminum, 343
   to metal, 260
   to steel, 342
   to textile, 259
Elbonite, 261, 291, 315
Electro-adhesion, 6
Electrode cascade, 809
Electrophotographic
   image, 831
Electrostatic
   constant, 816
Ellipsometry, 6. 46
Energy
   dissipation, 122
   elastic, 122
   fracture, 123
Entanglement, 212
EPDM
   to metal bonding, 264
Epotone-403, 581
Epoxy adhesive, 473, 606
   -elastomer, 478
   -nitrile, 478, 515
   -nylon, 478, 515
   -polyamide, 476, 603
Epoxy resin, 182
   aluminum oxide, 610
   CTBN in, 199
   Epon-828, 674
   film adhesive, 470
   fluorinated, 429
   for transformer, 581
   in-situ toughened, 7
   microvoid, 7
Equation of State
   of adhesive, 568, 571
ESCA, 4
   of fracture surface, 365, 371, 415
Ethylene-propylene
   terpolymer, 8
Exposure, 765

F

Fabric
    backings, 463
    polypropylene, 464
Failure
    energy, 183
    loci of, 504, 607
Fatigue crack
    of adhesives, 554
    propagation, 523
FEP polymer, 79, 182
Ferrite
    joint of, 577
Fibers
    bicomponent, 24
    comfort, 24
    industrial, 24
    safety, 24, 25
    semiconducting, 22
Filler
    silane-treated, 329
Film
    pressure, 114, 134
    thickness, 47
Fisher equation, 624
Fixing, 768
Flash photography, 222
Flaw tolerance
    of adhesives, 513
Flexible adhesives, 475
Flexography, 718
Flocking process, 452
Fluoroanhydride
    curing agent, 429
Fluoroexpoxy
    curing of, 432
    surface tension, 433
Force
    capillary, 773
    chemical, 772
    coulombic, 775
    dipole-dipole, 770
    dipole-induced dipole, 770
    dispersion, 770
    double layer, 771
    hydrophobic, 773
    induced image, 775
    of adhesion, 806
    point charge, 778
    time dependent, 812
Fountain
    concentrate, 755
    solution, 713, 753
Fracture
    mixed mode, 502
    morphology, 531
    toughness, 520
Fracture energy, 506, 600
    of adhesives, 503, 607, 609, 611
    tensile, 211
Fracture mechanics, 600
Fracture surface energy, 420
Fractured surface, 375
Frenkel equation, 832
Friction
    of polymer, 7
Fringe order, 638
Functional
    monomers, 455

G

Geometric mean rule, 153
Germanium
    γ_c, 76
Glass
    fiber composite, 604
    -glass joint, 602
    temperature, 234, 243
Glueline, 607
Glycol
    polyethylene, 221
    polypropylene, 221
Gold, 74
    wettability, 56
Good and Girifalco's equation, 138
Graft
    anionic, 96, 99
    polymerization, 96
    radical, 96
Gravure, 712
Green strength
    open time, 234
Griffith
    crack, 754

-Irwin theory, 120
theory, 120, 737
Gruntiest theory, 124
Gum arabic, 713

H

Hamaker constant, 790
Harkins-Jordon
　correction factor, 743
Heat capacity, 110
Hexafluoroacetone, 430
High speed
　testing, 315
Hole
　desorption theory, 630
Holography, 7, 23
Hooke's Law, 690
Hot-melt, 446
　compatibility, 489
　ethylene copolymers, 486
　growth, 485
　polymer distribution, 490
　viscosity, 488
Hydrodynamic
　thickness, 50
Hydrophobic silica, 352
Hydrostatic pressure, 624

I

Impact resistance
　of acrylic/pc, 217
　of laminates, 224
Impact velocity, 222
Impedance factor, 569
Implantology, 28
Infrared spectroscopy, 371
　ATR, 50
　Bound fraction, 48
　specular reflectance, 386
　　(SRIS)
Inks
　UV-cured, 718
　water-based, 718
Insulation bonding, 482
Interaction
　matrix, 571
　parameter, 113

Interface, 108
Interfacial
　deformation, 796
　gradient, 109
　parameters, 571
　process, 107
　strain, 688, 701
　stress, 697
　tension, 142, 146
　void, 121
Ionization energy, 111
Ion-probe, 4
Iron
　γ_c, 76
　oxide, 76
Isoelectric point
　of surface, 180

J

Joint
　attachment site, 563
　durability, 597
　strength, 129, 147
　temperature effect, 183

K

Keeson force, 760
Kelvin equation, 632
Kraton-101, 201
　cavitation, 203
Kronig-Kramer
　equation, 791

L

Langmuir
　adsorption theory, 683
Lap shear strength
　aluminum, 673
　copper, 673
　titanium, 673
Laser interferometry, 630, 638
Laserite plate, 718
Latexes
　acrylic, 247
　vinyl-acrylic, 247
LEED, 4

Letterpress plate, 712
    Hylox plate, 718
    Nyloprint, 718
    Rollfex, 718
    Toplon, 718
Lifshitz constant, 822
Liquid bridge
    tensile strength, 626
Liquid surface
    tension, 743
Lithography
    mechanism of, 725
    polymers for, 711
    surface energetics, 735
    waterless, 715
London forces, 69, 760

M

Macrocarbanion, 101
Magnetic
    brush development, 768
    permeability, 805
Marine venus/mercury system, 9
Maxwell's equation, 791
McFarlane-Tabor equation, 623
Mechanical interlocking, 6, 88
Mercury, 47
    contact angle, 370
Meta-dibromobenzene, 667
Meta-methoxyphenol, 667
Metal
    ceramic bond, 6
    γ_c, 76
    glass bond, 6
    wettability, 74
1-Methylnaphthalene, 61
Microhardness, 62
Microroughness, 750
Microscopy
    optical, 292
    SEM, 312
    TEM, 305
Mineral oil, 726
Mixing
    heat of, 116
MOCA, 205
Modified acrylic, 7
Modulus
    of elasticity, 638

Molecular
    polarizability, 111
    weight, 406
Molybdenum
    γ_c, 76
Monte Carlo
    technique, 45
MRL adhesives, 515
Mylar, 155

N

Negative pressure, 623
Network
    interpenetrating, 214
Newspaper plate, 755
Nickel
    γ_c, 76
Nonwoven bonding, 461
Nylon
    -6   181, 266
    -6,6   266
    -12   181
    to SBR, 267

O

ODPA, 188
Optical properties
    of polycarbonate, 217
Orientation effect, 770

P

Pasa-Jell-107 method, 366
Peeling
    rate of, 131
    strength, 155, 263
Pendant drop, 62
Penetration, 835
    depth, 778
Peracid oxidation, 405
Phase separation, 204
Phenol formaldehyde, 715
Phenolic dispersion, 235
Photobond process, 13
Photochromism
    of polymer, 23
Photoconductor, 764
    oxidation, 800

SUBJECT INDEX

Photoelectricity, 23
Phthalocyanine resin, 665
Piezoelectricity, 23
Pinene, 397
    Alpha-, 398, 403
    Beta-, 397, 399
PMDA, 367
Poisson's ratio, 638, 688
Polar force, 599
Polyacrylonitrile, 24
Polyamic acid
    solution, 189, 364
Polyamides, 21, 94
Polybenzimidazoles, 25
Polybutadiene, 285
Poly(butyl acrylate), 161
Polycarbonate
    γ_c, 82
    PMMA laminate, 218
Polychlorotrifluoroethylene, 151
Polyethylene, 20, 147, 151
Poly(ethylene oxide), 715
Poly(ethylene terephthalate), 82
Polyimidazoquinazoline, 9
Polyimide, 9, 25
    adhesive, 187
    Ti-6-4, 370
Polyisocyanage, 275
Polymer friction, 7
Polymercaptan, 8
Polymeric/membrane, 26
Polymerization
    "dead end", 99
Polymethylene, 114
Poly(methyl methacrylate)
    adsorption of, 49
Polyoxadiazole, 25
Polyoxymethylene, 82
Poly(oxypropylene triamine), 581
Polypentenamer, 27
Polypeptide, 25
Polyphenylquinoxaline, 9
Polyphosphazene foam, 27
Polypropylene
    modified, 566
    stereoregular, 181
Polysiloxanes
    wettability, 647
Polystyrene, 151
    conformation of, 43
    lithiated, 95
    living, 96
Polysulfide adhesive, 476
Polysulfones, 26
Polytetrafluoroethylene, 58, 82, 151
Polyurethane, 204, 458, 473
    polyester block, 207, 718
Poly(vinyl alcohol), 712
    denatured, 718
    grafted, 24
Poly(vinyl carbazole), 23
Poly(vinyl pyridine), 101
Potential energy, 701
Powder adhesive, 446
Pressure sensitive
    adhesive, 151, 183, 204, 446
Pyroelectricity, 23
Pyromellitic
    dianhydride, 188

Q

Quartz
    spreading of liquids, 57
    wettability, 65

R

Radial
    block copolymer, 7
Rayleigh's principle, 701
Rayon, 266
Re-adhesion, 680
    strength, 683
Reflection spectroscopy, 6
Refractive index, 46, 638
Relative humidity, 58
    on toner adhesion, 798
Relaxation modulus, 205, 208
Resorcinol-formaldehyde, 181, 259, 265
RHEED, 4
Rheological parameter, 571
Rogovan, 24
Rubber
    to glass adhesion, 281
    to metal adhesion, 315

S

Scaling Law, 838
Scarf point, 509
Screen printing, 712
Scuming
   of ink, 726
Selenium
   film, 693
   toner adhesion, 824
SEM, 365
Sessile drop, 841
Shear strength,
   double lap, 602
   lap, 192, 205, 568
Shear stress, 691
Shift factor
   penetration, 846
   sintering, 838
   spreading, 834
Silane
   A-1120, 582
   acetoxy, 350
   diamine, 332
   epoxy, 481
   γ-glycidoxy-
    propyltrimethoxy, 604
   styryl, 333
   iso-thiuronium Cl, 332
   treated glass, 284
   vinyl dimethyl, 282
Silica, 282
Silicone
   γ_c, 82
   multiblock copolymer, 717
   RTV, 482
Silicone-elastomer
   adhesion, 339
   catalyst, 339
   radiation-cured, 347
Sintering, 832
   surface tension, 832
   viscosity, 832
Solid surface tension
   polysiloxanes, 651
Solubility parameter, 130, 648, 659
Specialty adhesives, 467, 471
Specular reflectance, 365

Spiroacetal amine, 604
Spreading, 834
   kinetics, 834
   on borosilicate, 57
   on glass, 57
   on quartz, 57
   on sapphire, 57
   shift factor, 834
   velocity, 835
Spreading and receding, 729
Spreading coefficient, 57, 69, 130, 145, 161, 739, 754
Static fatigue, 609
Strain energy
   rate release, 515
   tensile, 698
Stress
   intensity factor, 600
Structural adhesives, 413
   fracture of, 501
Styrene-butadiene
   rubber, 451
Styrene-n-butyl
   methacrylate copolymer, 807
Superconducting polymers, 23
Surface
   entropy, 110
   excess, 134
   force, 740
   reaction, 93
   roughness, 779, 797
   tension, 133, 625
Surface free energy, 114, 136, 179
   of adhesives, 157, 159
   of aluminum oxide, 599
   of silica, 599
Surfactant, 7

T

Tack, 184, 396
   contact, 235
Tackifier, 239, 397
   hot-melt, 486
Tantalum
   γ_c, 76
TATUS, 717
Tear strength, 287
   threshold, 287

SUBJECT INDEX

Temperature
  effect on γ_c, 82
Tensile ligament, 565
Tertiary-butyl phenol, 246
Terpene resins, 395
Tetrafluoroethylene-
  perfluoro (propyl
  vinyl ether), 83
Textile adhesives, 449
Thermal fixing, 831
Thermo-mechanical analysis
  (TMA), 189, 208
Tinting
  of ink, 726, 733
TIPSI, 717
Tire cord
  polyester, 308
Titanium, 189
Toner
  agglomeration, 801
  density, 787
  monolayer, 784
  multilayer, 788
  particle, 6, 763
Toner adhesion, 763
  air blow, 804
  centrifuge, 802
  impulse, 803
  mechanical, 804
  spring balance, 802
  vibration, 803
Torsional
  braid analysis, 189, 229
  test, 7
Toughness
  arrest, 525
  initiation, 525
Transfer, 768
  efficiency, 819, 821
Transformer
  assembly, 582
  core, 578
Transition temperature
  SCC, 542
Triboelectric series, 765
Triboelectricity, 6
Triboelectrification, 782
Trimethoxysilane
  acetoxypropyl, 649
  γ-chloropropyl, 660

cumyl, 649
cyanoethyl, 649
γ-(epoxycyclohexyl)ethyl, 660
γ-glycidoxypropyl, 660
γ-mercaptopropyl, 660
γ-methacryloxypropyl, 660
methyl, 649
phenyl, 649
propyl, 649
trifluoropropyl, 649
trimethoxysilylpropyl, 660
vinyl, 660
Tungsten
  γ_c, 76

U

Ulmann condensation, 667
Ultracryotomy, 221
Ultrasonic
  coefficient, 680
  transmissibility, 677
Ultraviolet light
  cured inks, 718
Universal tester, 319
Urethane adhesives, 473
  moisture-cured, 480
  silane-modified, 8

V

Vacuole formation, 308
Van der Waal forces, 770
  constant, 794
  retardation, 771
Versamide-140, 674
Vinyl pyridine
  styrene-butadiene, 265
Viscoelastic
  energy dissipation, 183
Viscoelasticity, 844
Vistalon-404, 282
Volatile
  condensate material (VCM), 10

W

Wagner-Meerwein isomerization of
  pinenes, 400

Water
    γ_c, 73
Weak boundary layer, 131, 563, 573
Weak fluid boundary layer, 720
Weibull distribution function, 567
Weissenberg rheogoniometer, 836
Wettability
    envelop, 180, 739
    of fracture surface, 391
    of glass, 66
    of metals, 74
    of PE, 149
    of polymers, 82
    of polysiloxanes, 647
Wetting pressure, 142
WLF superposition, 833
Wood adhesives, 8
Work function, 771
Work of adhesion, 73, 119, 130, 133, 142, 179, 598, 651, 737

X

Xerography, 764
X-ray diffraction, 6

Y

Yield value, 845
Young's equation, 133
Young-Dupre equation, 119, 134, 621, 637
Young's modulus, 688

Z

Zirconium
    γ_c, 76

MIX
Papier aus verantwortungsvollen Quellen
Paper from responsible sources
FSC® C105338

If you have any concerns about our products,
you can contact us on
ProductSafety@springernature.com

In case Publisher is established outside the EU,
the EU authorized representative is:
**Springer Nature Customer Service Center GmbH
Europaplatz 3, 69115 Heidelberg, Germany**

Printed by Libri Plureos GmbH
in Hamburg, Germany